Flux
Control
in Biological
Systems

Physiological Ecology
A Series of Monographs, Texts, and Treatises

Series Editor
Harold A. Mooney
Stanford University, Stanford, California

Editorial Board
Fakhri Bazzaz   F. Stuart Chapin   James R. Ehleringer
Robert W. Pearcy   Martyn M. Caldwell   E.-D. Schulze

List continues at the end of this volume

Flux Control in Biological Systems

From Enzymes to Populations and Ecosystems

Edited by

E.-D. Schulze

Lehrstuhl für Pflanzenökologie
Universität Bayreuth
Bayreuth, Germany

Academic Press, Inc.
A Division of Harcourt Brace & Company
San Diego   New York   Boston
London   Sydney   Tokyo   Toronto

Front cover photograph: Casparian bands in the endodermis and exodermis of a maize root section stained with the fluorescent dye, berberine sulfate, according to the procedure of Brundrett *et al.* (1989) (courtesy of C. A. Peterson, University of Waterloo, Ontario, Canada). For details see Chapter 8, Figure 10B.

This book is printed on acid-free paper. ∞

Academic Press, Inc.
525 B Street, Suite 1900, San Diego, California 92101-4495

United Kingdom Edition published by
Academic Press Limited
24–28 Oval Road, London NW1 7DX

Library of Congress Cataloging-in Publication Data

Flux control in biological systems : from enzymes to populations and
    ecosystems / edited by E. -D. Schulze.
        p.   cm. —(Physiological ecology)
    Includes bibliographical references and index.
    ISBN 0-12-633070-0
    1. Biological control systems.   2. Ecophysiology.   3. Plant
physiology.   I. Schulze, E. -D. (Ernst-Detlef), Date   II. Series.
    QH508.F58   1994
581 . 1'88--dc20   93-4172
                                                                                    CIP

PRINTED IN THE UNITED STATES OF AMERICA
93 94 95 96 97 98   QW   9 8 7 6 5 4 3 2 1

Contents

3. Controlling the Effects of Excessive Light Energy Fluxes: Dissipative Mechanisms, Repair Processes, and Long-Term Acclimation   37

C. Schäfer

Part II
Flux Control at the Organismic Level

4. Plant Growth, Storage, and Resource Allocation: From Flux Control in a Metabolic Chain to the Whole-Plant Level   57

M. Stitt and E.-D. Schulze

Part III
Flux Control at the Soil–Organism Interface

9. Patterns and Regulation of Organic Matter Transformation in Soils: Litter Decomposition and Humification   303
W. Zech and I. Kögel-Knabner

Part IV
Flux Control at the Population and Ecosystem Level

Part V
Flux Control in Biological Systems: A Comparative View

Contributors

Numbers in parentheses indicate the pages on which the authors' contributions begin.

E. Beck (3, 57, 471), Lehrstuhl für Pflanzenphysiologie, Universität Bayreuth, 95440 Bayreuth, Germany

R. Horn (335) Institut für Pflanzenernährung und Bodenkunde, Christian-Albrechts-Universität, 23 Kiel, Germany

U. Jensen (447), Lehrstuhl für Pflanzenökologie und Systematik, Universität Bayreuth, 95440 Bayreuth, Germany

I. Kögel-Knabner (303) Lehrstuhl für Bodenkunde und Bodenökologie, Fakultät für Geowissenschaften NA6/134, Ruhr-Universität Bochum, 463 Bochum, Germany

E. Komor (153), Lehrstuhl für Pflanzenphysiologie, Universität Bayreuth, 95440 Bayreuth, Germany

C. Schäfer (37), Lehrstuhl für Pflanzenphysiologie, Universität Bayreuth, 95440 Bayreuth, Germany

R. Scheibe (3), Lehrstuhl Pflanzenphysiologie, Universität Osnabrück, 45 Osnabrück, Germany

E.-D. Schulze (57, 203, 421, 471), Lehrstuhl für Pflanzenökologic, Universität Bayreuth, 95440 Bayreuth, Germany

E. Steudle (237, 471), Lehrstuhl für Pflanzenökologie, Universität Bayreuth, 95440 Bayreuth, Germany

M. Stitt (13, 57, 471), Botanisches Institut, Universität Heidelberg, 69 Heidelberg, Germany

W. Zech (303), Lehrstuhl für Bodenkunde, Universität Bayreuth, 95440 Bayreuth, Germany

H. Zwölfer (365, 421, 447, 471), Lehrstuhl für Tierökologie, Universität Bayreuth, 95440 Bayreuth, Germany

Preface

Many factors have influenced our understanding of global biomass production. First, the International Biological Program (IBP, 1966–1976) assessed biomass in ecosystems representative of all the climactic regions of the globe, and estimated turnover of biomass. Later, when forest damage by acid rain became apparent, ecologists began to appreciate that the flux of material was at least as important as pool sizes. The fluxes of water, nutrients, and carbon explained the long-term effects of anthropogenic depositions of air pollutants on ecosystems. However, we were still unable to model these systems because we lacked an understanding of the control of fluxes. At present, this lack of understanding affects efforts to use modeling in many areas of research, from agriculture to ecosystem studies. In particular, it restricts us to predicting interactions of terrestrial ecosystems with the boundary of the atmosphere in terms of the International Geosphere Biosphere Program.

In 1980, the Deutsche Forschungsgemeinschaft decided to establish a "Sonderforschungsbereich" (Collaborative Research Center) at the University of Bayreuth. Our aim has been to define general principles of flux control by comparing patterns and mechanisms of regulation at various levels of organization, from enzymes to populations and ecosystems. We think that knowledge of such principles will enable us to better predict human influences on ecosystems. By depicting general patterns and principles of controls, we hope that we may be able to predict more quickly and more precisely how ecosystems will respond in a world of global change.

In this volume, botanists, microbiologists, soil scientists, and zoologists investigate patterns and mechanisms of matter transfer regulation in biological systems of different complexity. Based mainly on the experimental results of the long-term research at Bayreuth, this volume contains significant contributions made by numerous scientists from all parts of the world. We are very grateful to the Deutsche Forschungsgemeinschaft for its support of this lengthy research. I should also like to recognize Steve Halgren for editorial help.

E.-D. SCHULZE

I

Flux Control at the Cellular Level

1

The Malate Valve: Flux Control at the Enzymatic Level

R. Scheibe and E. Beck

I. Introduction

The photosynthetic machinery in the thylakoids of autotrophic organisms converts light energy into the biochemical energy equivalent ATP and the reducing power equivalent NADPH. Linear electron transport and photophosphorylation are strictly coupled reactions. The two primary products of the light reactions, ATP and NADPH, are required to drive CO_2, NO_3^-, and SO_4^{2-} reduction as well as a great variety of biosynthetic processes. Despite the large capacity for the generation of ATP and NADPH, there are two major problems connected with the transfer of photon flux into biochemical fluxes: (i) The rate of production must fit the demand and (ii) the turnover of the ATP and NADPH pools must be controlled separately to compensate for the coupled production and to attain flexibility with respect to changing requirements by the various biosynthetic activities in the chloroplast.

II. The ATP to NADPH Balance

Let us consider conditions where there is unlimited photon flux into the photosynthetic system: A balanced ratio of the ATP and NADPH production rates should be dictated by the need to achieve an adequate rate of production of the one metabolite which is in higher demand, and a concomitant disposal of the excess of the other (see also Stitt, this volume, Chapter 2). At light-saturated CO_2 fixation, the limiting compo-

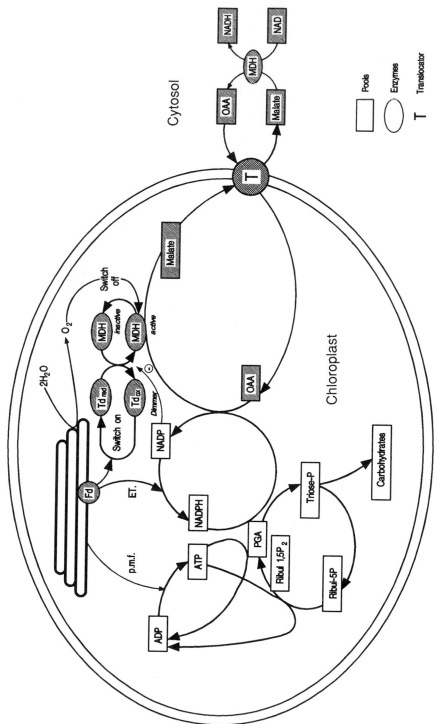

Figure 1 Indirect transport of reducing equivalents across the chloroplast envelope.

nent appears to be ATP. Since ATP production by linear electron flow is limited by the availability of NADP, cyclic photophosphorylation in combination with the Mehler reaction (pseudocyclic photophosphorylation) may support ATP production when the NADPH pool is fully reduced (Steiger and Beck, 1981). The hydrogen peroxide formed is finally detoxified by recycling the electrons from NADPH using the "soluble electron transport chain" composed of the ascorbate/glutathione redox system (Groden and Beck, 1979; Halliwell, 1981). However, it seems likely that this system is only resorted to in the absence of any other electron acceptor, i.e., in cases of emergency when the system has already reached a highly reduced ("overreduced") state. This is evident from the fact that addition of oxaloacetate to isolated chloroplasts as an alternative acceptor immediately decreases the rate of H_2O_2 production (Steiger and Beck, 1981).

The malate valve system, namely the conversion of excess NADPH into malate by the chloroplast enzyme NADP-malate dehydrogenase can apparently serve to unload the chloroplast from excess reducing equivalents via export into the cytosol (Heber, 1974). The chloroplast dicarboxylate translocator localized in the inner envelope membrane can exchange malate generated in the stroma for oxaloacetate from the cytosol with high efficiency (Ebbighausen *et al.*, 1987). This reaction leads to reoxidation of NADPH inside the chloroplast and export of reducing equivalents in the form of malate into the cytosol (Fig. 1). Thus, by involving the cytosol, regulation of the ATP-to-NADPH ratio in the chloroplast can be achieved (Fig. 2). Further use of the exported reducing equivalents either in photorespiration (reduction of hydroxy-pyruvate in the peroxi-

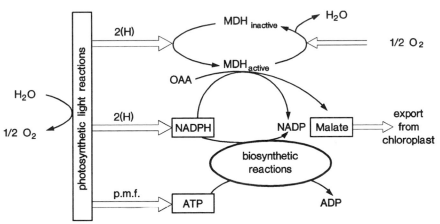

Figure 2 Poising of the ATP/NADPH ratio by a self-controlled export system for reducing equivalents ("malate valve").

somes; Ebbighausen *et al.*, 1987) or in generating energy through uptake by the mitochondria is then possible. The latter is suggested from the observation that specific inhibition of the respiratory chain by oligomycin decreases photosynthetic CO_2 fixation in isolated protoplasts (Krömer *et al.*, 1988).

The full capacity of the system is rather high: 100 to 200 μmol of reducing equivalents per milligram chlorophyll per hour can be converted into transportable malate in C_3 plants. It follows that the potential export of photosynthetic reducing power from the chloroplast via the malate valve must be strictly controlled, otherwise this export would deplete stromal NADPH and inhibit the reductive assimilatory steps.

III. Redox Control of the Malate Valve

In the following the components and mechanisms which mediate the flux through the "malate valve" are described. In contrast to the NAD (and NADP)-dependent malate dehydrogenases of microorganisms, animals, and even the extrachloroplastic isoenzymes of plants, the strictly NADP-dependent plastid enzyme is under redox control. Reduction of a special regulatory disulfide bridge converts the inactive protein into the catalytically competent enzyme (Scheibe, 1987). Electrons required for the reductive activation are sequestered from the photosynthetic electron transport chain via the ferredoxin/thioredoxin system (Fig. 3) (Buchanan, 1984). As with some other key enzymes of chloroplastic carbohydrate metabolism (in addition to the NADP-malate dehydrogenase, the Calvin-cycle enzymes fructose-1,6-bisphosphatase, sedoheptulose-1,7-bis-phosphatase, and phosphoribulosekinase and the coupling factor CF_1 are converted into catalytically competent forms in the light) the reductive light modulation acts as an on/off switch (Anderson, 1985). In contrast to NADP-malate dehydrogenase and the other light-activated enzymes, the chloroplast glucose-6-phosphate dehydrogenase is converted by photosynthetic reduction into a form without significant affinity for its sub-

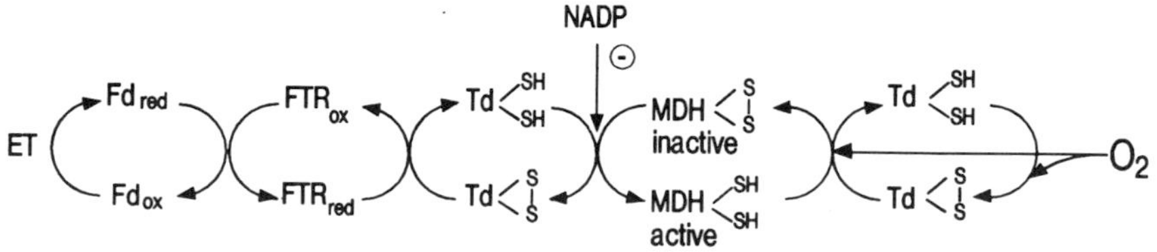

Figure 3   Redox modulation of chloroplast enzymes mediated by the ferredoxin/thioredoxin system.

strate glucose-6-phosphate (Scheibe *et al.*, 1989). Thus functioning of the oxidative pentose phosphate cycle is prevented at the first step, as long as photosynthesis is operating (see also Schäfer, this volume, Chapter 3).

IV. A Futile Cycle Provides the Machinery for Flux Regulation

During photosynthesis, oxygen evolution and the consequent continuous reoxidation of the light-generated proteinaceous thiols will lead to a futile cycle of reduction and oxidation of these enzyme forms (see also Komor, this volume, Chapter 6). The futile cycle, however, creates the basis for flux control: At a constant rate of reoxidation the portion of reduced active enzyme is determined by the rate of disulfide reduction by photosynthetically energized electrons (Fig. 4). The rate of electron flow to the target enzyme must be under metabolite control. A comparable principle has been described for a protein phosphorylation/dephosphorylation system (Stadtman and Chock, 1977). A protein kinase phosphorylates its target enzyme, e. g., mammalian glycogen phosphorylase or glycogen synthase, while at the same time a protein phosphatase regenerates the

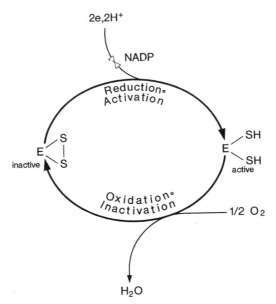

Figure 4 "Futile cycle" of NADP-MDH (E) which allows adaptation of the enzyme activity to the actual requirement of the chloroplast for export of reducing equivalents.

unphosphorylated form. Since both interconvertible forms exhibit different properties, the relative rates of the two converting steps determine the amount of active enzyme. Certain effectors can influence these rates and thus a sensitive regulatory system or "cascade" is constituted (Stadtman and Chock, 1977; see also Stitt, this volume, Chapter 2).

Returning to the malate valve system, NADP strongly inhibits the reductive interconversion of inactive-to-active NADP-malate dehydrogenase (Scheibe and Jacquot, 1983), thus acting as a negative effector of the crucial converter step. A high chloroplastic demand for NADPH, as signaled by its rapid reoxidation, will cause the enzyme to be largely in its oxidized form, and thus export of reducing equivalents is prevented. If, on the other hand, the level of NADPH increases and NADP decreases due to lack of ATP, the NADP-malate dehydrogenase is activated and overreduction of the electron transport chain is prevented because the active enzyme now catalyzes the reduction of oxaloacetate and export of malate. Such a transitory opening of the valve can be seen in leaves upon a rapid dark-to-light transition. Transient phases of overreduction of the electron transport chain are indicated by a decrease in the rate of oxygen evolution and a change of chlorophyll fluorescence which, however, is readily overcome by the activation of NADP-malate dehydrogenase and the concomitant release of reduction equivalents through malate (Scheibe and Stitt, 1988). The fact that, even under steady-state conditions, significant rates of malate production can be observed suggests the use of the malate valve also for draining reducing equivalents into the cytoplasm, even when overreduction does not occur (Ebbighausen et al., 1987).

V. Extra Sequences in the Polypeptide Are Responsible for Redox Control

A comparison of amino acid sequences shows that the chloroplast malate dehydrogenase is very similar to all other malate dehydrogenases of bacteria, animals, and plants which are nonregulatory (Fickenscher et al., 1987). The polypeptide chain of the chloroplast enzyme is, however, longer by 54 amino acids. This is due to extensions of 17 amino acids at the C-terminus and of 37 amino acids at the N-terminus of the enzyme. Consequently the subunit molecular mass of the chloroplast enzyme is about 41 kDa, as compared to 35 kDa found in nonregulatory counterparts. The N-terminal extension carries the two cysteine residues which provide the basis for the redox modification of the enzyme. The active, reduced form is the dithiol form while in the oxidized enzyme a disulfide bond bridges both sulfur atoms (Scheibe et al., 1991). It is important to note that oxidation of cysteines is the crucial point allowing control of the flux by this enzyme: First, it allows complete inactivation (switching off) of the enzyme in the dark, thus preventing any export of reducing equivalents. Second, it provides the basis for the interconvertible enzyme cascade that allows the continuous readjustment of the activity to the actual metabolic redox-state (NADPH/NADP ratio) in the chloroplast. Simultaneously, other light-modulated enzymes located at control steps in the Calvin cycle can be modulated independently, since their redox modulations are controlled by other signal metabolites (Scheibe, 1990). An individual control of each redox-modulated enzyme is a prerequisite for the effective regulation of photosynthesis under a variety of conditions.

VI. Energy Expenditure for Redox Control

What are the energy costs of this control mechanism? It is obvious that a continuous redox cycle mediating the interconversion of active and inactive enzyme molecules will require a certain amount of photosynthetically energized electrons to drive the cycle. Conceivably, this type of energy is adapted to the unique situation in the chloroplast. In comparable cycles that are driven by protein phosphorylation/dephosphorylation the energy originates from ATP hydrolysis (Goldbeter and Koshland, 1987). The alternative system of protein modification found in the chloroplast which consumes reducing equivalents instead of ATP is obviously well adapted to the higher availability of electrons as compared to ATP in this particular biological system.

However, there is another aspect of the cost of flux control. As stated above, the chloroplast NADP-malate dehydrogenase is characterized by specific sequences that carry the redox-addressable thiols. In contrast to the other thiol-regulated enzymes, whose regulatory sequence is an insert in an otherwise broadly homologous sequence to the nonplastid enzyme, the regulatory structures of the NADP-malate dehydrogenase are terminal additions and therefore easily accessible at the protein level. Digestion strategies using amino- and carboxypeptidases were applied in order to generate truncated forms of the native enzyme. Proteolytic removal of seven to eight amino acids of the C-terminus of the oxidized NADP-malate dehydrogenase was accompanied by a considerable increase of catalytic activity. Dithiothreitol converted the truncated oxidized enzyme into a form that was even more active than the intact enzyme (Fickenscher and Scheibe, 1988). A novel enzyme species that could no longer be inactivated completely upon oxidation, but was substantially more active than the chloroplast enzyme in its reduced form, has thus been created. Similarly, the use of aminopeptidase K gave rise to a novel active enzyme that had lost its regulatory N-terminal sequence (Ocheretina *et al.*, 1993). However, even this form was still responsive to a change in the redox potential. Taken together, these facts indicate that either part of the extra-sequences is not exclusively involved in the regulation of the enzyme. Since the nonregulatory counterparts of the malate dehydrogenase exhibit mostly a higher specific activity than the chloroplast enzyme, it can be put forward that a strategy termed "adaptability at the expense of activity" is being pursued.

VII. Conclusions

Biological systems exhibit the intrinsic capability of self-regulation which may respond to exogenous factors or environmental conditions. The malate valve represents such a self-regulating system which adjusts the production of photosynthetic primary metabolites to the actual rates of their consumption; in other words, it represents a system which controls fluxes. Clearly, such a system renders that degree of metabolic flexibility which is a prerequisite for a sessile organism to survive in an environment subjected to continuous changes.

Adjustment of photosynthetic fluxes to changing requirements poses the more general question of how such a biological system recognizes the necessity of adapting its metabolic fluxes. In complex systems, such as whole plants and animals, hormones are usually involved as signals which, by complex chain reactions, create a message that is perceptible by the addressed cells. For short-term flux control, as in photosynthesis, such signals have never been demonstrated. In these cases metabolic imbalances can provide an effective principle of flux control. Obviously the malate valve is implied because the photosynthetic production of ATP and NADPH is not balanced with respect to their utilization. By periodically operating the malate valve, the inhibitory effect of this imbalance is relieved. Consequently, the imbalance itself provides the basis for the control system. Therefore, in order to allow regulation of metabolic fluxes, imbalances must be maintained. Perfect homeostasis, on the other hand, characterizes a stable system of biochemical reactions and rather is indicative of inflexibility. The well-known steady state of photosynthetic reactions, however, must be considered as the integration of oscillating imbalances of the actual rates, caused by the fact that some degree of overreduction of the NADP pool is a prerequisite for the operation of the malate valve. Such minute oscillations of oxygen evolution have been detected by differential analysis of the signal of a sensitive oxygen electrode (Walker *et al.*, 1983).

References

Anderson, L. E. (1985). Light/dark modulation of enzyme activity in plants. *Adv. Bot. Res.* **12**, 1–46.

Buchanan, B. B. (1984). The ferredoxin/thioredoxin system: A key element in the regulatory function of light in photosynthesis. *BioScience* **34**, 378–383.

Ebbighausen, H., Hatch, M. D., Lilley, R. McC., Krömer, S., Stitt, M., and Heldt, H. W. (1987). On the function of malate-oxaloacetate shuttles in a plant cell. *In* "Plant Mitochondria: Structural, Functional and Physiological Aspects" (A. L. Moore and R. B. Beechey, eds.), pp. 171–180. Plenum, New York.

Fickenscher, K., and Scheibe, R. (1988). Limited proteolysis of inactive tetrameric chloroplast NADP-malate dehydrogenase produces active dimers. *Arch. Biochem. Biophys.* **260**, 771–779.

Fickenscher, K., Scheibe, R., and Marcus, F. (1987). Amino acid sequence similarity between malate dehydrogenases (NAD) and pea chloroplast malate dehydrogenase (NADP). *Eur. J. Biochem.* **168**, 653–658.

Goldbeter, A., and Koshland, D. E., Jr. (1987). Energy expenditure in the control of biochemical systems by covalent modification. *J. Biol. Chem.* **262**, 4460–4471.

Groden, D., and Beck, E. (1979). H_2O_2 destruction by ascorbate-dependent systems from chloroplasts. *Biochim. Biophys. Acta* **546**, 426–435.

Halliwell, B. (1981). "Chloroplast Metabolism" pp. 179–205. Oxford Univ. Press (Clarendon), Oxford.

Heber, U. (1974). Metabolite exchange between chloroplasts and cytoplasm. *Annu. Rev. Plant Physiol.* **25**, 393–421.

Krömer, S., Stitt, M., and Heldt, H.-W. (1988). Mitochondrial oxidative phosphorylation participating in photosynthesis metabolism of a leaf cell. *FEBS Lett.* **226**, 352–356.

Ocheretina, O., Harnecker, J., Rother, T., Schmid, R., and Scheibe, R. (1993). Effects of N-terminal truncations upon chloroplast NADP-malate dehydrogenases from pea and spinach. Biochim. Biophys. Acta *1163*, 10–16.

Scheibe, R. (1987). $NADP^+$-malate dehydrogenase in C_3 plants: Regulation and role of a light-activated enzyme. *Physiol. Plant.* **71**, 393–400.

Scheibe, R. (1990). Light/dark modulation: Regulation of chloroplast metabolism in a new light. *Bot. Acta* **103**, 327–334.

Scheibe, R., and Jacquot, J.-P. (1983). NADP regulates the light activation of NADP-dependent malate dehydrogenase. *Planta* **157**, 548–553.

Scheibe, R., and Stitt, M. (1988). Comparison of NADP-malate dehydrogenase activation, Q_A reduction and O_2 evolution in spinach leaves. *Plant Physiol. Biochem. (Paris)* **26**, 473–481.

Scheibe, R., Geissler, A., and Fickenscher, K. (1989). Chloroplast glucose-6-phosphate dehydrogenase: K_m shift upon light modulation and reduction. *Arch. Biochem. Biophys.* **274**, 290–297.

Scheibe, R., Kampfenkel, K., Wessels, R., and Tripier, D. (1991). *Biochim. Biophys. Acta* **1076**, 1–8.

Stadtman, E. R., and Chock, P. B. (1977). Superiority of interconvertible enzyme cascades in metabolic regulation: Analysis of monocyclic systems. *Proc. Natl. Acad. Sci. U.S.A.* **74**, 2761–2765.

Steiger, H.-M., and Beck, E. (1981). Formation of hydrogen peroxide and oxygen dependence of photosynthetic CO_2 assimilation by intact chloroplasts. *Plant Cell Physiol.* **22**, 561–576.

Walker, D. A., Sivak, M. N., Prinsley, R. T., and Cheesbrough, J. K. (1983). Simultaneous measurement of oscillations in oxygen evolution and chlorophyll a fluorescence in leaf pieces. *Plant Physiol.* **73**, 542–549.

2

Flux Control at the Level of the Pathway: Studies with Mutants and Transgenic Plants Having a Decreased Activity of Enzymes Involved in Photosynthesis Partitioning

M. Stitt

I. Introduction

Biochemical interconversions occur via a series of sequentially ordered small changes, each catalyzed by a different enzyme. Together, these reactions constitute a metabolic pathway (e.g., glycolysis). The aim of the following contribution is to compare different methodologies for studying the control of flux in metabolic pathways, to illustrate their application using photosynthetic carbon metabolism as an example, and to attempt to formulate some general conclusions about the control of metabolic flux. The reader is referred to other recent reviews for a more detailed discussion of the regulation mechanisms themselves (Stitt *et al.,* 1987a,b; Stitt and Quick, 1989; Stitt, 1991; see also Scheibe and Beck, this volume, Chapter 1; Schäfer, this volume, Chapter 3). A simplified illustration of the relevant pathways is given in Fig. 1.

Flux Control in Biological Systems

13

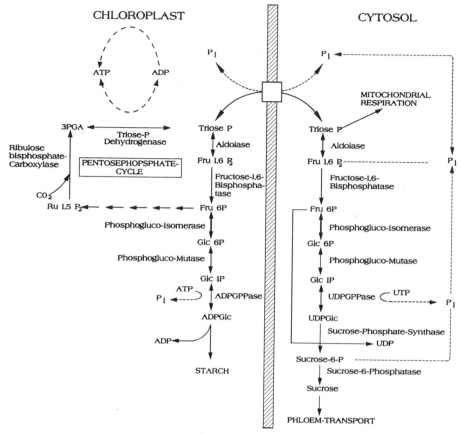

Figure 1 Simplified presentation of the pathways of photosynthetic carbon assimilation.

II. The Traditional Approach to the Study of Regulation

The earliest and most widely used approach to study the control of metabolism involves the concept of a "pacemaker" or "limiting" enzyme, which regulates the flux through the whole pathway (see Scheibe and Beck, this volume, Chapter 1). A series of criteria was developed to allow such "regulatory" reactions to be identified (Rolleston, 1972; Newsholme and Start, 1973).

a. The limiting enzyme should be present at activities which are not greatly above ("in excess of") the required flux through the pathway.

b. The enzyme should possess regulatory properties, e.g., allosteric sites, phosphorylation sites, or product inhibition.
c. The enzyme should catalyze a nonequilibrium ("irreversible") reaction. When a reaction lays close to its thermodynamic equilibrium (i.e., is freely reversible *in vivo*) the enzyme is considered to be in "excess," and a small decrease should not have a significant effect on flux through the pathway.
d. There should be characteristic changes of the substrate concentration, reciprocal to the change of flux (by analogy, cars accumulate at the point in the street where the flow of traffic has been stopped by a red light or slowed down by road repair or a badly parked car).

These criteria are of varying value. For example, criteria a was not very useful because the *in vivo* activity of an enzyme (especially a highly regulated enzyme) has little relation to its maximal activity *in vitro*.

III. Metabolic Regulations Illustrated by Photosynthetic Sucrose Synthesis

Application of these criteria led to a series of conclusions, which I illustrate using photosynthetic sucrose synthesis as an example.

First, a metabolic pathway often contains *more than one regulatory enzyme*. The pathway of sucrose synthesis contains several nonequilibrium reactions, including those catalyzed by the cytosolic fructose-1,6-bisphosphatase (Fru1,6Pase), sucrose-phosphate synthase (SPS), sucrose-6-phosphatase, and, probably, the hydrolysis of pyrophosphate (Stitt *et al.*, 1987b). At least two of these enzymes have regulatory kinetic properties (Fru1,6Pase and SPS), and their substrate concentrations charge in a reciprocal direction to flux under some experimental conditions (Stitt *et al.*, 1980, 1983; Gerhardt *et al.*, 1987).

Second, *the activity of a regulatory enzyme can be regulated by several different effectors.* The activity of the cytosolic Fru1,6Pase may be regulated by the signal metabolite fructose-2,6-bisphosphate (Fru2,6bisP), substrate concentration, AMP, pH, Mg^{2+}, phosphate (P_i) (Stitt *et al.*, 1985, 1987b), and, possibly, Ca^{2+} (Brauer *et al.*, 1990). SPS can be regulated by protein phosphorylation (Huber *et al.*, 1989; Siegl *et al.*, 1990), allosteric regulation by glucose-6-phosphate and P_i (Doehlert and Huber, 1984), and, in some species, sucrose (Stitt *et al*, 1987b). The various effectors are highly interactive in their effect on the individual enzymes (see, e.g., Stitt *et al.*, 1985).

Third, regulation mechanisms often allow *amplification*. For example, SPS is regulated via the Glc6P/P_i quotient. Since Glc6P and P_i often change reciprocally *in vivo*, changes of the quotient will amplify changes in the individual metabolites. Amplification can also be achieved via minicascades. For example, fructose-6-phosphate (Fru6P) activates Fru6P,2-kinase (the enzyme which synthesises Fru2,6bisP) and inhibits Fru2,6Pase (the enzyme which degrades Fru2,6bisP) (Stitt *et al.*, 1970, 1987b). A relatively small increase of Fru6P can thereby lead to a much larger increase of the regulator metabolite Fru2,6bisP, allowing sensitive feedback regulation of the cytosolic Fru1,6Pase (Stitt, 1989; Neuhaus *et al.*, 1989).

Fourth, regulation often involves *balances*. For example, the feedback loop involving Fru2,6bisP which was discussed in the previous paragraph is counteracted by a feedforward loop. When the rate of photosynthesis increases, the Fru2,6bisP concentration is driven downward, because a rising glycerate-3-phosphate (PGA)/P_i ratio also inhibits Fru6P,2-kinase (Stitt *et al.*, 1970, 1987b). The Fru2,6bisP concentration therefore reflects the balance between the supply of fixed carbon for sucrose synthesis and the demand for further sucrose synthesis. A similar balance appears to regulate the phosphorylation state of SPS (Stitt *et al.*, 1988) but the details of the mechanism have not yet been characterized (for ATP/NADPH balance, see Chapter 1).

Fifth, the various enzymes in a metabolic pathway are highly coordinated. For example, Fru1,6Pase and SPS both are inhibited by P_i directly (Stitt *et al.*, 1985; Doehlert and Huber, 1984) and indirectly (via changes of Fru2,6bisP and phosphorylation state; Stitt *et al.*, 1987b, 1988; J. L. A. Huber, unpublished). There is also a robust negative correlation between inactivation (phosphorylation) of SPS and an inhibition of the cytosolic Fru1,6Pase by increased Fru2,6bisP (references in Brauer and Stitt, 1990).

IV. Regulation and Control

Why is this all so complicated?

In considering the regulation of metabolic flux, we need to distinguish between two different reasons why enzymes are regulated (see also Chapter 1). On the one hand, regulation is obviously needed to allow the flux through a pathway to be altered in response to "external" factors (these can be changes in other pathways or signals coming from outside the cell or organism). In addition, it is also necessary to maintain a functional state within the pathway. Since every enzyme in a linear pathway catalyzes the same reaction rate at steady-state pathway flux, it is obvious that the activity of the various enzymes must be coordinated. Moreover, it is also important that the concentration of shared substrates and cofactors are poised in a range where stable operation of all the enzymes in the pathway is possible. In the simplest case, a linear path of three enzymes, the middle enzyme will have its own "maximal" activity when its substrate concentration is high and product negligible; however, its substrate is the product of the first enzyme, likewise its product is the substrate of the third enzyme. It is therefore obvious that the *pathway* flux will be less than the theoretical capacity of each individual enzyme (see also Scheibe and Beck, this volume, Chapter 1). This requirement will become even more important in branched or circular pathways and with feedback loops.

The reductionist approach, which has been described so far, allows us to identify potential mechanisms for regulation (i.e., for influencing the activity of an enzyme). However, it does not unambiguously identify their significance: does a particular mechanism (a) operate to allow an enzyme to respond to charges occurring elsewhere in the pathway, or is it (b) operating to alter the activity of *other* enzymes and, thus, *control* the flux through the pathway? Put another way, how can we decide which enzyme is the slave and which is the master?

Logically, the simplest way to answer this question would be to decrease the amount of selected enzyme by a small amount (e.g., 20%), without altering the amount of any other enzymes in the pathway, and to monitor the resulting change in flux through the pathway. If the flux does not change, the enzyme is evidently not contributing to control of pathway flux. If the flux changes by 20%, then the enzyme is controlling the flux (is limiting). If the flux changes by an intermediate value, control is shared with other enzymes. This exercise has been formalized by Kacser and Burns (1973; see also Kacser and Porteous, 1987) as the Flux Control Coefficient,

$$C_{E_i} = \frac{dJ/J}{dE/E}, \tag{1}$$

where dJ/J is the fractional change of pathway flux which results from the fractional change dE/E of the amount of the enzyme in question. This approach has the advantage that it is quantitative and it can be applied without requiring assumptions about the structure of the system or the nature of regulation. The disadvantage (which was immediately pounced on by critics) is that it is difficult to realize experimentally.

V. Measurement of Flux Control Coefficients Using Mutants and Genetically Manipulated Plants

One way to decrease the activity of an enzyme is to use specific inhibitors. However the "specificity" of such compounds is often a matter of debate. It is also effectively impossible to decide what the concentration is at the site of action, i.e., how much enzyme has been "titrated out." For these reasons, the best and most direct way to measure a flux control coefficient is to decrease the amount of an enzyme by using genetics and then to measure the resulting change in flux through the pathway. The flux control coefficient is yielded by the slope $(dJ/J)/(dE/E)$ of the J vs E plot in the region corresponding to the wild type.

This approach requires a series of plants with a stepwise reduction in the amount of the enzyme in question. One source for such material is classical mutagenesis programs. However, there are practical limitations to the use of classical mutants. First, this approach is only possible when a simple screen exists to generate reduced-activity mutants for that particular enzyme (e.g., iodine staining for starchless leaves). Second, it is not applicable for enzymes where a null mutation would be lethal. Third, mutagenous programs usually yield null mutants, or mutants with very low activity, and these are not suitable (see below) for carrying out control analysis. At best, crossing the null mutant with the wild type leads to an intermediate amount of enzyme in the heterozygote, provided that the mutation is in the structural gene itself (Kruckeberg *et al.*, 1989; Neuhaus and Stitt, 1990).

In the past 2–3 years a new source of material has become feasible, by using antisense-DNA technology (Van der Krol *et al.*, 1988). The gene, or a portion of the gene containing the ribosome binding and the initiation sequences, with a strong promotor ligated at the wrong end of the gene, is introduced into the plant genome by conventional transformation techniques (e.g., *Agrobacterium*). The promotor directs synthesis of an mRNA which is read from the wrong strand of DNA. This "antisense" RNA is complementary to the normal "sense" mRNA and is thought to hybridize with it, inhibiting or preventing translation (the precise mechanism may be more complex and is still controversial). Because the quantitative expression of DNA introduced by *Agrobacterium* varies greatly (depending on the number of copies introduced and on position effects, see Willmitzer, 1988), a series of plants can be produced with varying expression of antisense and, thus, with a progressive decrease in the expression of the target protein (Rodermel *et al.*, 1988; Quick *et al.*, 1991a), e.g., Ribulose-1,5-bisphosphate-carboxylase oxygenase (Rubisco; Fig. 2).

Figure 2 *Nicotiana Tabacum* exhibiting decreased Rubisco activity at a photon flux density of 750 μmol m^{-2}s^{-1}.

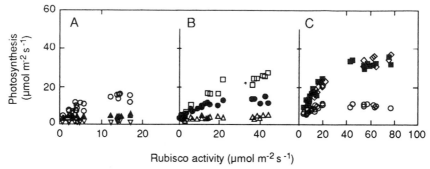

Rubisco activity (μmol m⁻² s⁻¹)

Figure 3 Relationship between Rubisco content and the rates of photosynthesis in tobacco grown at three different irradiances. All plants were grown at 28°C, with a high N supply (5 mM NH$_4$NO$_3$). Each point corresponds to a different plant. Plants were grown at 100 (A), 300 (B), or 1000 (C) μmol·m^{-2}·s^{-1} irradiance. Photosynthesis was measured at 50 ($\bigtriangledown$) 100 ($\triangle$,$\blacktriangle$) 300 ($\bigcirc$,$\bullet$) 1000 ($\square$,$\blacksquare$) or 2000 ($\diamond$) μmol·m^{-2}·s^{-1} irradiance in the measuring cuvette. The closed symbols ($\blacktriangle$,$\bullet$,$\blacksquare$) represent the rate of photosynthesis in the irradiance corresponding to that at which the plants were growing. The results are from Lauerer *et al.* (1993).

solely (limited) by one enzyme, there will be a direct linear relation between the enzyme amount and the pathway flux, and the slope of J vs E will equal unity in a normalized plot. If the enzyme exerts no control, a decrease in the amount of enzyme in the range corresponding to the wild-type amount will have no effect on pathway flux. Partial or shared control can be readily identified and quantified, because the relevant enzymes would in this case have control plots in which the slope in the range corresponding to the wild-type enzyme content exhibits a value which is somewhere between 0 and 1. The relative contribution of the enzymes can be gauged by comparing their respective control coefficients. Incidentally, it should be noted that when the amount of one enzyme is substantially lowered, relative to the remainder of the pathway, its control coefficient will tend to rise (as we see in the following examples). For this reason, I have emphasized that the control coefficient must be estimated from plant material in which the enzyme amount is varied around a point which is similar to that found in the wild type. In the following plots, the data have not been mathematically normalized, but the axis scale has been selected so that the figure is nevertheless comparable to a control plot (in later examples I present normalized data).

We first measured the rate of photosynthesis under the conditions in which the plants were growing (shown in Figs. 3A–3C and 4A–4C as solid symbols) to reveal whether Rubisco limits photosynthesis in the wild

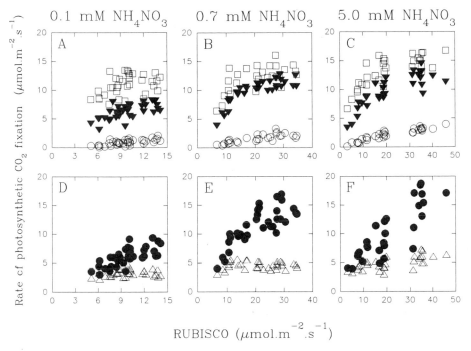

Figure 4 Relationship between Rubisco content and the rate of tobacco photosynthesis in tobacco grown at three different nitrogen concentrations. All the plants were grown at 330 μmol·m^{-2}·s^{-1} irradiance and 20°C. Each point corresponds to a different plant. The plants were grown at 5 mM (A,D), 0.7 mM (B,E), or 0.1 mM (C,F) NH$_4$NO$_3$. Photosynthesis was measured in ambient CO$_2$ and the growth irradiance ($\blacktriangledown$, A–C), or at growth irradiance with enhanced (1000 μbar, $\square$) and decreased (100 μbar, $\circ$) CO$_2$ (Figs A–C), or at ambient CO$_2$ with increased (1000 μmol·m^{-2},$\bullet$) and decreased (100 μmol, m^{-2}·s^{-1},$\triangle$) irradiance (Figs. D–F). The results are from Quick *et al.* (1993).

type under growth conditions. Remember, the plants were grown at three different irradiances, or at three different NH$_4$NO$_3$ supplies. It is evident that Rubisco is never totally limiting under growth conditions. Indeed, under most growth conditions it exerts remarkably little control. In general, Rubisco exerted more control when plants were grown in higher light or in limiting nitrogen.

The rate of photosynthesis was also measured at irradiances lying above and below the growth irradiance and at CO$_2$ concentrations lying above and below ambient. It is evident that the flux control coefficient of Rubisco increases when the irradiance is increased and when the CO$_2$ concentration is decreased.

These results emphasize (a) that the control of photosynthetic rate is *shared* between Rubisco and other proteins and (b) that the contribution made by Rubisco *depends on the conditions*—both short-term and long-term. This contradicts the idea that a fixed limiting enzyme controls flux through pathways.

Comparison of these results with the mechanistic models which have previously been used to interpret gas-exchange measurements in terms of regulation by limiting factors (e.g., Farquhar and von Caemmerer, 1982) confirmed some of their general predictions (Stitt *et al.*, 1991). However, the assumption of a single limiting step which is implicit in these models is clearly oversimplistic. It also became evident that some of the assumptions made in interpreting gas-exchange responses were not always valid; for example, Farquhar and von Caemmerer (1982) assumed that the light-saturated rate of photosynthesis in ambient CO_2 is always limited by Rubisco. Comparison of the rate of photosynthesis at 1000 and 2000 $\mu mol \cdot m^{-2} \cdot s^{-1}$ irradiance (see Fig. 4C) shows that this need not always be the case.

It is known that the allocation of nitrogen and other resources changes, depending on the environmental conditions in which the plants are growing (see, e.g., Evans, 1988, 1989). It has often been assumed that this acclimation represents an optimization of resource allocation. The transgenic plants with decreased expression of Rubisco allow us to test whether this is the case. More nitrogen was invested in Rubisco than was needed to avoid a strict limitation of photosynthesis by this protein (i.e., a control coefficient approaching unity under growth conditions), provided they were well fertilized with NH_4NO_3. However, if they were grown on low NH_4NO_3 they contained just enough Rubisco to avoid a one-sided limitation of photosynthesis, i.e., they did not overinvest N in Rubisco. This shows that nitrogen allocation to Rubisco is being optimized with respect to nitrogen availability (see also Chapter 4).

A different picture emerges during acclimation to low light. When tobacco is grown at lower light intensities, although allocation of nitrogen to Rubisco was decreased the change was relatively small and Rubisco is present in a large excess. In this case the allocation of nitrogen changes, but it would be incorrect to say the distribution has been "optimized." Indeed, the results in Figs. 3C, 4E, and 4F indicated that when photosynthesis is measured at low irradiance, leaves with a genetically engineered reduction of Rubisco content may have higher rates of photosynthesis than those of the wild type, presumably because the N which is not "wasted" in Rubisco can be invested in other more useful (at least under these conditions) proteins. In agreement, the leaves contained more chlorophyll.

VII. Flux Control Coefficients of Four Enzymes in a Linear Pathway: Photosynthetic Starch Synthesis

Enzymes are linked in sequences, or pathways. Reduced-activity mutants or antisense transformants for other Calvin-cycle enzymes are not yet available, so we could not extend our analysis of this particular pathway. However, it is relatively easy to screen for starch-deficient leaves, and a series of classical mutants are already available with decreased activities of several enzymes required for starch synthesis. These include decreased-activity mutants for the plastid phosphoglucoisomerase (pPGI) *Clarkia xantiana* (Jones *et al.*, 1986b), the plastid phosphoglucomutase (pPGM), and ADP glucose pyrophosphorylase (ADPGlc-PPase) in *Arabidopsis thaliana* (Caspar *et al.*, 1989; Lin *et al.*, 1988) and decreased starch-branching enzyme in pea (Bhattacharyya *et al.*, 1990). These represent four of the five enzymes involved in the pathway of starch synthesis, providing material to investigate how the different enzymes of a linear pathway contribute to the control of flux.

The rate of sucrose and starch synthesis was measured in leaf discs by supplying $^{14}CO_2$ for 20–30 min in the presence of low ($125\ \mu mol.m^{-2}.s^{-1}$) or high ($1000\ \mu mol.m^{-2}\ s^{-1}$) irradiance and is summarized in a normalized control plot (Fig. 5). Starch synthesis is shown as closed symbols (for discussion of the effect on the rate of sucrose synthesis, see below). When fluxes were measured in low irradiance (Figs. 5A–5D), only ADPGlc-PPase exerted significant control over the rate of starch synthesis. When fluxes were measured in high irradiance (Figs. 5E–5H), some control was also exerted by pPGI and pPGM. These results reemphasize the previous conclusions (a) that control can be shared between several enzymes in a pathway and (b) that the contribution of an enzyme to control is flexible and depends on the conditions (Kruckeberg *et al.*, 1989; Neuhaus *et al.*, 1989; Smith *et al.*, 1990; Neuhaus and Stitt, 1990).

It is interesting to compare the results of these experiments with the conclusions reached by the more traditional approach to metabolic regulation outlined at the beginning of this chapter. Using these criteria, ADPGlc-PPase has been identified as a regulatory enzyme because it catalyzes an irreversible reaction *in vivo* (Bassham and Krause, 1969), because it is regulated at an allosteric site by glycerate-3-phosphate and P_i (Preiss, 1982), and because there is evidence from comparisons of fluxes and metabolites that this mechanism operates *in vivo* (Heldt *et al.*, 1977; Neuhaus and Stitt, 1990). pPGI and pPGM both catalyze readily reversible reaction (for example, both operate in the glycolytic direction during phosphorolytic starch mobilization; Stitt and ap Rees, 1980; Stitt

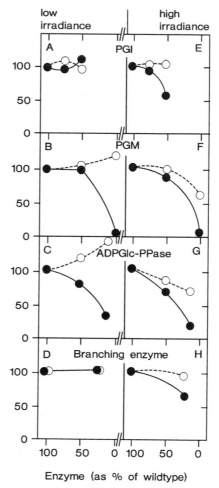

Figure 5 Effect of a decreased dosage of four different enzymes in the pathway of starch synthesis on the rate of starch and sucrose synthesis in leaves. The rate of starch synthesis was measured in low irradiance (Figs. A–D) and in high irradiance (Figs. E–H). The rate of starch synthesis (●) and sucrose synthesis (○) are normalized on the flux in the wild type (= 100%). (A,E) Decreased plastid PGI in *Clarkia xantiana* (data from Kruckeberg *et al.*, 1989). (B,F) Decreased plastid PGM in *Arabidopsis* (data from Neuhaus and Stitt, 1990). (C,G) Decreased ADPGlc-PPase in *Arabidopsis* (data from Neuhaus and Stitt, 1990). (D,H) Decreased starch-branching enzyme in pea (data from Smith *et al.*, 1990).

and Heldt, 1981). It is therefore rather surprising to find both of these enzymes exerting control in conditions of high flux. This shows that they are not present in large excess (see below for further discussion).

VIII. Control of Partitioning to Sucrose and Starch: An Example of Control at a Branch Point in Metabolism

Metabolic pathways do not exist in isolation *in vivo*. Instead, they join and branch. The control of flux between alternative pathways at a metabolic branch point is of crucial importance, because it ultimately determines which products are formed and in what proportion they accrue. For example, although a portion of the photosynthate is retained in the chloroplast and stored temporarily as starch, most of the photosynthate is transferred as triose-phosphates to the cytosol, where it is converted to sucrose and exported via the phloem to the rest of the plant. The plant material described in the last section, plus an additional series of mutants with a decreased amount of the cytosolic phosphoglucoisomerase in *C. xantiana* (Jones *et al.*, 1986a), allowed us to investigate the control of metabolic fluxes at this branch point between sucrose and starch. Photosynthate partitioning provides a particularly interesting model system, because the chemical reactions leading to sucrose and starch are essentially identical, except for the final steps. Despite this chemical similarity, the pathways are distinct because they occur in different cellular compartments.

The effect of a decreased rate of starch synthesis on the flux to sucrose is summarized in Fig. 5. (Starch synthesis is shown as filled symbols and sucrose synthesis as open symbols.) The result is quite complicated. In low irradiance, there is a compensating stimulation of sucrose synthesis, and the rate of photosynthesis (not shown) is unaltered. In high irradiance, sucrose synthesis remained unaltered or was inhibited and the rate of photosynthesis (not shown) decreased.

A different picture emerged when sucrose synthesis was inhibited, using the series of plants with decreased amounts of cytosolic PGI (Fig. 6). First, in contrast to the results with plastid PGI (Figs. 5A and 5E), decreased cytosolic PGI had more effect on the fluxes in low irradiance than on those in high irradiance. Second, starch synthesis was stimulated in low irradiance and in high irradiance. This allowed partitioning to be altered without decreasing the rate of photosynthesis.

These results can be compared with current ideas about the regulation of partitioning (reviewed in Stitt *et al.*, 1987b; Stitt and Quick, 1989; Stitt, 1991). Biochemical studies have provided evidence that partitioning is controlled from the cytosol and that the chloroplast responds (see Fig. 1). When sucrose accumulates in the leaf, feedback regulation mechanisms including phosphorylation of SPS and an increase of Fru2,6bisP come into operation and sucrose synthesis is inhibited. As a result the glycerate-3-phosphate/P_i ratio increases, ADPGlc-PPase is activated, and starch

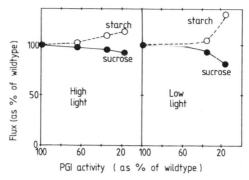

Figure 6 The relationship between dosage of cytosolic PGI and flux to sucrose and starch. Plots are shown for decreased PGI cytosol of *Clarkia xanthiana*. The fluxes to sucrose and starch were measured at 1000 and 125 μmol·m^{-2}s^{-1} irradiance and saturating CO_2. Sucrose synthesis is shown as closed symbols ($\bullet$) and starch synthesis as open symbols ($\bigcirc$) (data from Kruckeberg *et al.*, 1989).

synthesis is stimulated. The results described above with the mutants are fully consistent with the idea that the regulatory network in the leaf allows the chloroplast to respond consistently to changes in the cytosol, but not vice versa.

Relating these results to the control of fluxes in biological systems, we can see that even this comparatively simple system with just one branchpoint has already generated a complex pattern of control, which cannot easily be predicted from the properties of the individual enzymes. Also, an important simplifying theoretical constraint on the behavior of the flux control coefficients has been lost. In a linear pattern, the flux control coefficient sums to unity (the Summation Theory, Kacser and Burns, 1973). In a branched pattern, the flux control coefficients of both branches must also sum to unity but, since enzymes in one branch often have a negative flux control coefficient for the fluxes in the other branches (e.g., decreased cytosol PGI leads to increased starch synthesis), the flux control coefficient of the individual enzymes can adopt a far larger range of values.

IX. What Determines the Flux Control Coefficient of an Enzyme?

The question now arises, what factors determine the flux control coefficient of an enzyme? It is evidently not simply the extent to which the reaction is removed from equilibrium per se, nor does it depend on the enzyme possessing regulatory properties, because in some conditions we observe significant (even though small) flux control coefficients for enzymes like PGM and PGI, which satisfy neither of these criteria. To answer this question, Kacser and Burns (1973) introduced the concept of the elasticity coefficient

$$\varepsilon \frac{E_i}{S_i} = \frac{dV/V}{dS/S},$$ (2)

where dV/V is the fractional change in enzyme activity which is produced by dS/S, a fractional change in the concentration of one of its ligands (e.g., substrates, products, inhibitors, activators) while keeping all the other ligands constant. Equation (2) defines the abbreviation for the elasticity coefficient of an enzyme E_i for a particular ligand, S_i.

In vivo, several ligands may change simultaneously and, in this case, the net change in the activity of the enzyme is given as the sum of each individual elasticity coefficient

$$\frac{dV}{V} = \varepsilon \frac{E_1}{S_1} * \frac{dS_1}{S_1} + \varepsilon \frac{E_2}{S_2} * \frac{dS_2}{S_2} + \cdots + \varepsilon \frac{E_n}{S_n} * \frac{dS_n}{S_n}.$$ (3)

The elasticity coefficient represents a local property, and the flux control coefficient is a system property. Nevertheless, these two coefficients are linked. When a component in the system is changed (e.g., an alteration in the amount or activity of an enzyme) the concentrations of ligands which interact with this enzyme (e.g., substrates, products, regulators) will change. These changes of the metabolites (which constitute variables in the system) will then affect the activity of further enzymes. The flux control coefficients of the individual enzymes emerge from their elasticity coefficients. This can be illustrated by taking a simple pathway,

$$\begin{array}{c} \underline{\hspace{2cm} \text{feedback} \hspace{2cm}} \\ \downarrow \quad\quad \text{inhibition} \quad\quad | \\ \ominus \\ S_1 \;\rightarrow\; S_2 \;\rightarrow\; S_3 \;\rightarrow\; S_4\,, \\ \;\; E_1 \quad\quad E_2 \quad\quad E_3 \end{array}$$ (4)

and increasing the amount of the first enzyme (dE_1/E_1). This will generate changes in the concentration of its ligands $(dS_1/S_1, dS_2/S_2, dS_4/S_4)$ which, in turn, will lower the activity of E_1. They will also affect the activity of the other enzymes, E_2 and E_3.

These interactions are described by the elasticity coefficients for each enzyme and metabolite and will determine how much of the initial change (dE_1/E_1) is transmitted as a change of flux (dJ/J) through the whole pathway. For example, if $\varepsilon \frac{E_1}{S_1}$ and $\varepsilon \frac{E_1}{S_4}$ are large and negative (i.e., E_1 is very sensitive to product inhibition and feedback inhibition), and $\varepsilon\frac{E_2}{S_2}$ and $\varepsilon\frac{E_3}{S_3}$ are small and positive (i.e., E_2 and E_3 are only weakly stimulated by rising substrate concentrations), then the resulting increase in the concentrations of S_2, S_3, and S_4 will strongly inhibit E_1, without significantly stimulating the following two enzymes, and the flux control coefficient of E_1 would be very low.

The relationship between flux control coefficients and the elasticity coefficients is summarized in the connectivity theorem (Kacser and Burns, 1973) which states that the flux control coefficient of two adjacent enzymes is inversely related to the ratio of their elasticity coefficients for a shared substrate,

$$\frac{C_{E_1}}{C_{E_2}} = -\frac{\varepsilon\frac{E_2}{S_x}}{\varepsilon\frac{E_1}{S_x}}, \tag{5}$$

where C_{E_1} and C_{E_2} are the flux control coefficients of two "adjacent" enzymes and $\varepsilon\frac{E_1}{S_x}$ and $\varepsilon\frac{E_2}{S_x}$ are their elasticity coefficients for a shared metabolite, S_x.

X. Measurement of Elasticity Coefficients

If elasticity coefficients could be measured, it would be possible to understand, and indeed predict, values for flux control coefficients. However, elasticity coefficients are difficult to measure experimentally. Three approaches are possible. I only describe the general principle; for details the reader is referred to the original articles which are cited below.

First, the elasticity coefficients can be estimated by inspecting the kinetics of the isolated enzyme (Stitt, 1989). However, the value of each elasticity coefficient varies, depending on the conditions. For example, the elasticity coefficient for a substrate of an enzyme with Michaelis-Menten kinetics will vary from 1 to 0 when substrate concentration is increased in the absence of products or regulators. An almost infinite range of values is possible, if other ligands are also present.

Second, the elasticity coefficients can be estimated from the theoretical equilibrium constant (k_{eq}) and the *in vivo* product/substrate ratio using equations developed by Groen *et al.* (1982). This approach is only applicable to relatively simple enzymes with Michaelis-Menten kinetics (i.e., it is unfortunately inapplicable to most of the "interesting" enzymes for regulation which have cooperative kinetics). Table I provides an example of this approach, where the elasticity coefficients of cytosolic PGI for its substrate $\left(\varepsilon_{Fru6P}^{cPGI}\right)$ are estimated from the measured Glc6P/Fru6P ratio in *C. xantiana* mutants with decreased cytosolic PGI (Kruckeberg *et al.*, 1989). It is evident that the elasticity coefficient decreases when the amount of enzyme decreases and the reaction is displaced slightly further away from its theoretical k_{eq} of Glc6P/Fru6P = 3.3.

The elasticity coefficients in the wild type are much greater than unity (the maximum expected for a Michaelic-Menten enzyme in the absence of product). This can be explained because, *in vivo*, the net flux (V) is the difference between the forward (V^{+1}) and reverse (V^{-1}) flux (Rolleston, 1972). A relatively small change of V^{+1} caused by a small change of the substrate concentration can therefore have a large effect on the net flux, V. For this reason, enzymes which catalyze near-equilibrium reactions have quite high elasticity coefficients and, hence, tend to have low flux control coefficients (but see below for further discussion).

Third, elasticity coefficients can be measured *in vivo* using the dual-modulation method (Kacser and Burns, 1979a). The principle is as follows. If we take the simplest case of an enzyme with two effectors, S_1 and S_2, the change in activity, dV/V after perturbing the cell and allowing a new steady state to be reached can be expressed by

$$\frac{dV}{V} = \frac{dJ}{J} = \varepsilon_{S1}\frac{d_{S_1}}{S_1} + \varepsilon_{S2}\frac{d_{S_2}}{S_2}, \tag{6}$$

where dJ/J is the fractional change of the steady-state flux, dS_1/S_1 and dS_2/S_2 are the fractional changes in the steady-state lebels of the metabo-

Table I Estimation of the Elasticity Coefficient of Cytosolic PGI for Fru6P and Glc6P from the Relationship between the Measured Product/Substrate Ratio and the Theoretical Equilibrium Constant (=3.3)

PGI (% wild type)	Limiting irradiance			Saturating irradiance		
	Glc6P/Fru6P	ε^{cPGI} Fru6P	ε^{cPGI} Glc6P	Glc6P/Fru6P	ε^{cPGI} Fru6P	ε^{cPGI} Glc6P
100	2.87	+7.6	−6.9	2.61	+4.6	−4.9
64	2.92	+8.4	−8.4	2.43	+3.2	−3.3
32	2.61	+5.2	−4.9	2.23	+2.3	−2.5
18	2.19	+2.6	−2.1	1.54	+1.2	−1.5

Results are from Kruckeberg *et al.* (1989). The elasticity coefficient for Glc6P is negative because increased product will decrease the net rate of conversion of Fru6P to Glc6P.

Table II Elasticity Coefficients of Spinach Leaf Cytosolic Fru1,6Pase for Triose-phosphate and Fru2,6bisP

	Cytosolic Fru1,6Pase			
	$\varepsilon^{\text{Fru1,6Pase}}_{\text{Triose P}}$	$\varepsilon^{\text{Fru1,6Pase}}_{\text{Fru2,6bisP}}$	α	$\alpha \cdot \varepsilon^{\text{Fru1,6Pase}}_{\text{Fru2,6bisP}}$
Low irradiance	$+0.18$	-0.61	4.7	-2.9
High irradiance	$+0.68$	-0.27	12.1	-3.3

Values were obtained via the dual-modulation method (see Neuhaus *et al.*, 1990). The term α refers to the amplification in the Fru2,6bisP regulator cycle (Stitt, 1989; Neuhaus *et al.*, 1989, 1990). The combined term $\alpha \, \varepsilon^{\text{Fru1,6Pase}}_{\text{Fru2,6bisP}}$ represents a nominal elasticity coefficient of the cytosolic Fru1,6Pase for Fru6P (see text and Fig. 7).

lites, and ε_{S_1} and ε_{S_2} are the elasticity coefficients of the enzyme for the metabolites S_1 and S_2. Empirical values can be obtained for dJ/J, dS_1/S_1 and dS_2/S_2 by measuring the steady-state fluxes and metabolite levels before and after metabolism is peturbed. The elasticity coefficients, ε_{S_1} and ε_{S_2}, represent unknowns. If two independent perturbations are carried out, two simultaneous equations can be written and solved for ε_{S_1} and ε_{S_2}. This approach has the disadvantage that it makes simplifying assumptions about the enzyme (i.e., that it has a small number of effective ligands). The approach is also subject to error, because it requires relatively small changes of metabolites and fluxes, and the data will be subject to experimental error.

Table II summarizes apparent elasticity coefficients estimated using the dual modulation method for the response of the cytosolic Fru1,6Pase to triose-phosphate and its inhibitor, Fru2,6bisP (Neuhaus *et al.*, 1990). Triose-phosphates are in equilibrium with Fru1,6bisP, the true substrate, but are technically easier to measure (Gerhardt *et al.*, 1987). The term $\varepsilon^{\text{Fru1,6Pase}}_{\text{triose-P}}$ therefore represents the response of the Fru1,6Pase to substrate availability. The term $\varepsilon^{\text{Fru1,6Pase}}_{\text{Fru2,6bisP}}$ represents the sensitivity to inhibition by Fru2,6bisP. It is apparent that Fru1,6Pase becomes more sensitive to its substrate concentration and less sensitive to the inhibitor *in vivo* under conditions of high light.

XI. Comparison of Flux Control Coefficients and Elasticity Coefficients for PGI and the Cytosolic Fru1,6Pase

It must be stressed that these estimates of elasticity coefficients are only approximations. Nevertheless, they can be used to illustrate how the

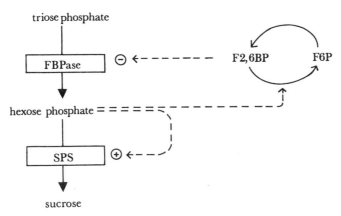

triose phosphate

hexose phosphate

sucrose

Figure 7 Scheme of the interaction between the cytosolic PGI and the cytosolic Fru1,6Pase. From Stitt (1989). The metabolic fluxes of carbon are shown as solid lines (———) and regulator of interactions are shown as dashed lines (– – –).

flux control coefficients emerge from the elasticity coefficients of the individual enzymes, using the interaction between the cytosolic PGI and cytosolic Fru1,6Pase as an example (Fig. 7).

The shared metabolite (S_x) for these enzymes is Fru6P. An interesting complication arises because the cytosolic Fru1,6Pase catalyzes an irreversible reaction and, under *in vivo* conditions direct product inhibition by Fru6P can be ignored (Stitt *et al.*, 1985). Instead, Fru6P feeds back via a loop involving the regulator metabolite Fru2,6bisP. Rising Fru6P stimulates Fru6P,2-kinase and inhibits fructose-2,6-bisphosphatase, leading to an increase in the Fru2,6bisP concentration and, thus, inhibition of the cytosolic Fru1,6Pase. The effect of an increase of Fru6P on Fru1,6Pase activity is therefore described by the term $\alpha\, \varepsilon\, \dfrac{\text{Fru1,6Pase}}{\text{Fru2,6bisP}}$ where α is a term describing the amplification in the Fru2,6bisP regulator cycle (Stitt, 1989; Neuhaus *et al.*, 1989). The *in vivo* value of α can be determined empirically by measuring the relative change of Fru6P and Fru2,6bisP. Measured values for α lay between 2 in *C. xantiana* (Neuhaus *et al.*, 1989) and 4–12 in spinach (Neuhaus *et al.*, 1990). These are in agreement with the kinds of value expected from the properties of Fru6P,2-kinase and Fructose-2,6-bisphosphatase (Stitt, 1989).

The value of $\alpha\, \varepsilon\, \dfrac{\text{Fru1,6Pase}}{\text{Fru2,6bisP}}$ in *Clarkia* is therefore about -0.9, whereas $\varepsilon\, \dfrac{\text{PGI}}{\text{Fru6P}}$ in the wild type is between 4.6–7.6 (see Table I). Since

$$\frac{C_{\text{PGI}}}{C_{\text{Fru1,6Pase}}} = -\frac{\alpha\,\varepsilon_{\text{Fru2,6bisP}}^{\text{Fru1,6Pase}}}{\varepsilon_{\text{Fru6P}}^{\text{PGI}}}, \tag{7}$$

the flux control coefficient of PGI will be negligible compared to that of the cytosolic Fru1,6Pase. In contrast, mutants with a fivefold decrease of PGI have an estimated value for $\varepsilon_{\text{Fru6P}}^{\text{PGI}}$ between 1.2–2.6, predicting that PGI will now have a flux control coefficient which is 35–75% that of the cytosolic Fru1,6Pase. This is in agreement with the observation that PGI begins to exert partial control over sucrose synthesis in these mutants. Control is shared between PGI and Fru1,6Pase, even though PGI catalyzes a near equilibrium reaction and Fru1,6Pase catalyzes a reaction which is far removed from equilibrium. This is understandable because the Fru1,6Pase responds quite sensitively to an increase of Fru6P, due to the amplification occurring in the Fru2,6bisP regulator cycle.

Inspection of the estimated elasticity coefficients for the cytosolic Fru1,6Pase also explains the rather unexpected decrease of C_{PGI} in high irradiance, compared to low irradiance. This is done by comparing the values of $\alpha\,\varepsilon_{\text{Fru2,6bisP}}^{\text{Fru1,6Pase}}$ and $\varepsilon_{\text{Triose-P}}^{\text{Fru1,6Pase}}$ in high and low irradiance (Table II). In low light, $\alpha\,\varepsilon_{\text{Fru2,6bisP}}^{\text{Fru1,6Pase}}$ is much larger than $\varepsilon_{\text{Triose-P}}^{\text{Fru1,6Pase}}$, i.e., feedback control will operate effectively and will not be overriden by increasing concentrations of substrates. In high irradiance, $\alpha\,\varepsilon_{\text{Fru2,6bisP}}^{\text{Fru1,6Pase}}$ decreases relative to $\varepsilon_{\text{Triose-P}}^{\text{Fru1,6Pase}}$, i.e., feedback inhibition will be overriden by a small increase of triose-phosphates. The cytosolic PGI will exert less control in high irradiance, because under these conditions an increase of Fru6P has a smaller effect on flux through the pathway.

XII. Conclusions

On the basis of the results obtained in these studies, the following general conclusions are proposed.

1. The two approaches discussed in the first part of the chapter are complementary. The traditional approach is a useful way to identify

and characterize *potential mechanisms* for regulation. By using genetically manipulated plants to decrease the amount of chosen enzymes, it is possible to identify which enzymes actually *control* pathway flux.

2. Regulation mechanisms allow a high degree of coordination of the individual enzymes of a pathway. For example, it is evident that several enzymes in the pathway of sucrose synthesis can be regulated, that each enzyme can be regulated in several ways, and that there will be a close interaction between the various enzymes in the pathway. This coordination is essential if the flux through the pathway is to be altered without this entailing large changes in the concentrations of the various intermediates (which would represent a potential disturbance in the operation of the system). It will also be important to allow regulation to respond flexibly to different demands and under different conditions.

3. Regulation often involves amplification, but also often involves balances which dampen the response. This, and the coordination between the individual enzymes, may be important because it allows flexibility of response to be reconciled with stability of operation.

4. The control of flux can be shared between several enzymes in a pathway, and their individual contributions to control varies, depending on the short-term and the long-term conditions.

5. The traditional separation of pathways into "irreversible and regulatory" and "reversible and nonregulatory" reactions is oversimplistic. Enzymes which catalyze nonregulatory reversible reactions are not necessarily present in large excess. Indeed, in some circumstances, they can contribute to the control of pathway flux. Control in metabolic pathways is not a dictatorship, nor is it an Athenian democracy; rather, it is a halfway house in which (to quote Orwell) "all pigs are equal but some are more equal than others."

6. These results have interesting implications for adaptation and evolution. For example, if control is shared between several enzymes, there will be limits to the extent to which a particular flux or the resulting phenotype can be modified by altering the expression of any one single protein. The finding that many proteins can be reduced by half without a major effect on pathway flux also provides a simple explanation for the phenomenon of dominance (Kacser and Burns, 1979b). Interesting problems are also raised with respect to the mechanisms of natural selection. It appears that some enzymes (e.g., ADPGlc-PPase) have a higher flux control coefficient than others. These are also often the enzymes with regulatory properties. Perhaps it is "advantageous" to concentrate control at enzymes which have complex regulatory patterns and can therefore respond to changing metabolic, physiological, or environmental conditions. In this case, natural selection would operate to prevent

overexpression or increased catalytic rates of these proteins (or, conversly, to prevent the other enzymes from decreasing so far that they exert major control over pathway flux).

Acknowledgments

The data in this chapter have been collected by a number of co-workers and guest scientists to whom I express my sincere gratitude. These co-workers are M. Brauer, Dr. J. E. Dancer, Dr. H. E. Neuhaus, Dr. W. P. Quick, and G. Siegl. The following scientific guests also contributed to this work: Dr. A. L. Kruckeberg and Dr. A. M. Smith.

References

Bassham, J. A., and Krause, G. H. (1969). Free energy changes and metabolic regulation in steady state photosynthetic metabolism. *Biochim. Biophys. Acta* **189**, 207–221.

Bhattacharyya, M. K., Smith, A. M., Ellis, T. N. H., Hedley, C., and Martin, C. (1990). The wrinkled-seeded character of peas deseribed by Mendel is caused by a transposon-like insertion in a gene encoding starch-branching enzyme. *Cell (Cambridge, Mass.)* **60**, 115–122.

Brauer, M., and Stitt, M. (1990). Vanadate inhibits fructose-2,6-bisphosphatase and leads to an inhibition of sucrose synthesis in barley leaves. *Physiol. Plant.* **78**, 568–573.

Brauer, M., Sanders, D., and Stitt, M. (1990). Regulation of photosynthetic sucrose synthesis: A role for calcium? *Planta* **182**, 236–243.

Caspar, T. M., Lin, T.-P., Monroe, J., Bernhard, W., Spilatro, S., Preiss, J., and Somerville, C. R. (1989). Altered regulation of -amylase activity in mutants of Arabidopsis with lesions in starch metabolism. *Proc. Natl. Acad. Sci. U.S.A.* **86**, 5830–5833.

Doehlert, D. C., and Huber, S. C. (1984). Phosphate inhibition of spinach leaf sucrose-phosphate synthase as affected by glucose-6-phosphate and phosphoglucose isomerase. *Plant Physiol.* **76**, 250–253.

Evans, J. R. (1988). The relationship between electron transport components and photosynthetic capacity in pea leaves grown at different irradiances. *Aust. J. Plant Physiol.* **14**, 157–170.

Evans, J. R. (1989). Photosynthesis and nitrogen relationship in leaves of C-3 plants. *Oecologia* **78**, 9–19.

Farquhar, G. D., and von Caemmerer, S. (1982). Modelling of photosynthetic response to environmental conditions. *Enzycl. Plant Physiol.* **12B**, 549–587.

Gerhardt, R., Stitt, M., and Heldt, H. W. (1987). Subcellular metabolite levels in spinach leaves. Regulation of sucrose synthesis during diurnal alterations in photosynthesis. *Plant Physiol.* **83**, 399–407.

Groen, A. K., Van der Meer, R., Westerhoff, H. V., Wanders, R. T. A., Akerboom, T. P. M., and Tager, J. M. (1982). Control of metabolic fluxes. *In* "Control of Metabolic Compartmentation" (H. Sies, ed.), pp. 9–37. Academic Press, London.

Gutteridge, S. (1990). Limitations of the primary events of CO_2 fixation in photosynthetic organisms; the structure and mechanisms of Rubisco. *Biochim. Biophys. Acta* **1015**, 1–14.

Heldt, H. W., Chon, C. J., Maronde, D., Herold, A., Stankovic, Z. S., Walker, D. A., Kraminer, A., Kirk, M. R., and Heber, U. (1977). Role of orthophosphate and other factors in the regulation of starch formation in leaves and isolated chloroplasts. *Plant Physiol.* **59**, 1146–1155.

Huber, J. L. A., Huber, S. C., and Nielsen, F. (1989). Protein phosphorylation as a mechanism for regulation of spinach leaf sucrose-phosphate synthase activity. *Arch. Biochem. Biophys.* **270**, 681–690.

Jones, T. W. A., Pichersky, E., and Gottlieb, L. D. (1968a). Enzyme activity in EMS-induced null mutations of duplicated genes encoding phosphoglucose isomerase in *Clarkia. Genetics* **113**, 101–114.

Jones, T. W. A., Pichersky, E., and Gottlieb, L. D. (1986b). Reduced enzyme activity and starch level in an induced mutant of chloroplast phosphoglucose isomerase. *Plant Physiol.* **81**, 367–371.

Kacser, H., and Burns, J. A. (1973). The control of flux. *Symp. Soc. Exp. Biol.* **27**, 65–107.

Kacser, H., and Burns, J. A. (1979a). Molecular democracy: Who shares the control? *Biochem. Soc. Trans.* **7**, 1149–1161.

Kacser, H., and Burns, J. A. (1979b). The molecular basis of dominance. *Genetics* **97**, 639–666.

Kacser, H., and Porteous, J. W. (1987). Control of metabolism; what do we have to measure. *Trends Biochem. Sci.* **12**, 5–14.

Kruckeberg, A. L., Neuhaus, H. E., Feil, R., Gottlieb, L. D., and Stitt, M. (1989). Decreased activity mutants of phosphoglucose isomerase in the cytosol and chloroplast of *Clarkia xantiana. Biochem. J.* **261**, 457–467.

Lauerer, M., Saftic, D., Quick, W. P., Labate, C., Fichtner, K., Schulze, E. D., Rodermel S. R., Bogorad, L., and Stitt, M. (1993). Decreased ribulose-1,5-bisphosphate carboxylase-oxygenase in tobacco transformed with antisense *rbc*S. VI. Effect on photosynthesis in plants grown at different irradiance. *Planta* **190**, 332–345.

Lin, T.-P., Caspar, T., Somerville, C. R., and Preiss, J. (1988). A starch deficient mutant of *Arabidopsis thaliana* with low ADP-glucose pyrophosphorylase activity lacks one of the two subunits of the enzyme. *Plant Physiol.* **88**, 1175–1181.

Neuhaus, H. E., and Stitt, M. (1990). Control analysis of photosynthate partitioning. Impact of reduced activity of ADP-glucose pyrophosphorylase or plastid phosphoglucomutase on the fluxes to starch sucrose in *Arabidopsis thaliana* (1.) Heynh. *Planta* **182**, 445–454.

Neuhaus, H. E., Kruckeberg, A. L., Feil, R., and Stitt, M. (1989). Reduced activity mutants of phosphoglucose isomerase in the cytosol and chloroplast of *Clarkia xantiana*. II. Studies of the mechanisms which regulate phosphosynthate partitioning. *Planta* **178**, 110–122.

Neuhaus, H. E., Quick, W. P., Siegl, G., and Stitt, M. (1990). Control of photosynthetic sucrose synthesis: Analysis of the interaction between feedforward and feedback regulation of sucrose synthesis. *Planta* **181**, 583–592.

Newsholme, E. A., and Start, C. (1973). "Regulation in Metabolism." Wiley, New York.

Preiss, J. (1982). Regulation of the biosynthesis and degradation of starch. *Annu. Rev. Plant. Physiol.* **33**, 431–454.

Quick, W. P., Schurr, U., Scheibe, R., Schulze, E.-D., Rodermel, S. R., Bogorad, L., and Stitt, M. (1991a). Decreased ribulose-1, 5-bisphosphate carboxylase oxygenase in transgenic tobacco transformed with "antisense" rbcS. I. Impact on photosynthesis in ambient growth conditions. *Planta* **183**, 542–554.

Quick, W. P., Schurr, U., Fichtner, K., Schulze, E.-D., Rodermel, S. R., Bogorad, L., and Stitt, M. (1991b). The impact of decreased *Rubisco* on photosynthesis, growth, allocation and storage in tobacco plants which have been transformed with antisense rbcS. *Plant J.* **1** 51–58.

Quick, W. P., Fichtner, K., Schulze, E.-D., Wendler, R., Leegood, R. C., Mooney, H., Rodermel, S. R., Bogorad, L., and Stitt, M. (1992). Decreased ribulose-1,5-bisphosphate carboxylase-oxygenase in transgenic tobacco transformed with antisense *rbc* S. IV. Impact on photosynthesis in conditions of altered nitrogen supply. *Planta* **188**, 522–531.

Rodermel, S. R., Abbott, M. S., and Bogorad, L. (1988). Nuclear-organelle interactions: Nuclear antisense gene inhibits ribulose bisphosphate carboylase enzyme in transformed tobacco plants. *Cell (Cambridge, Mass.)* **55**, 673–681.

Rolleston, F. S. (1972). A theoretical background to the use of measured intermediates in the study of the control of metabolism. *Curr. Top. Cell Regul.* **5**, 47–75.

Sharkey, T. D. (1989). Evaluating the role of Rubisco activation in photosynthesis in C-3 plants. *Philos. Trans. R. Soc. London, Ser. B* **323**, 435–448.

Siegl, G., Mackintosh, C., and Stitt, M. (1990). Sucrose phosphate synthase is dephosphorylated by protein phosphatase 2A in spinach leaves. *FEBS Lett.* **270**, 198–202.

Smith, A. M., Neuhaus, H. E., and Stitt, M. (1990). The impact of decreased activity of starch-branching enzyme on photosynthetic starch synthesis in leaves of wrinkled-seeded peas. *Planta* **181**, 310–315.

Stitt, M. (1989). Control analysis of photosynthetic sucrose synthesis: Assignment of elasticity coefficients and flux control coefficients to the cytosolic fructose-1, 6-bisphosphatase and sucrose phosphate synthase. *Philos. Trans. R. Soc London, Ser. B* **323**, 435–443.

Stitt, M. (1991). Rising CO_2 levels and their potential significance for carbon flow in photosynthetic cells. *Plant, Cell Environ.* **14**, 741–762.

Stitt, M., and ap Rees, T. (1980). Carbohydrate breakdown by chloroplasts of *Pisum sativum*. *Biochim. Biophys. Acta* **627**, 131–143.

Stitt, M., and Heldt, H. W. (1981). Physiological rates of starch breakdown in isolated intact spinach chloroplasts. *Plant Physiol.* **68**, 755–761.

Stitt, M., and Quick, W. P. (1989). Photosynthetic carbon partitioning its regulation and possibilities for manipulation. *Physiol. Plant.* **77**, 633–641.

Stitt, M., Wirtz, W., and Heldt, H. W. (1980). Metabolite levels in the chloroplast and extrachloroplast compartments of spinach protoplasts. *Biochim. Biophys. Acta* **637**, 348–359.

Stitt, M., Wirtz, W., and Heldt, H. W. (1983). Regulation of sucrose synthesis by cytoplasmic fructose-1, 6-bisphosphatase and sucrose phosphate synthase during photosynthesis in varying light. *Plant Physiol.* **72**, 767–774.

Stitt, M., Herzog, B., and Heldt, H. W. (1985). Control of photosynthetic sucrose synthesis by fructose-1, 6-bisphosphatase. V. Modulation of spinach leaf cytosoloc fructose-1, 6-bisphosphatase *in vitro*. *Plant Physiol.* **79**, 590–598.

Stitt, M., Gerhardt, R., Wilke, I., and Heldt, H. W. (1987a). The contribution of fructose-2, 6-bisphosphate to the regulation of sucrose synthesis during photosynthesis. *Physiol. Plant.* **69**, 377–386.

Stitt, M., Huber, S. C., and Kerr, P. (1987b). Control of photosynthetic sucrose synthesis. *In* "The Biochemistry of Plants" (M. D. Hatch and N. K. Boardman, eds.), Vol. 10, pp. 327–409. Academic Press, Orlando, FL.

Stitt, M., Quick, W. P., Schurr, V., Schulze, E.-D., Rodermel, S. R., and Bogorad, L. (1991). Decreased ribulose-1,5-bisphosphate carboxylase in transgenic tobacco transformed with "antisense" rbcS II flux control coefficients for photosynthesis in varying light, CO_2 and air humidity. *Planta* **183**, 555–566.

Stitt, M., Wilke, I., Feil, R., and Heldt, H. W. (1988). Coarse control of sucrose phosphate synthase in leaves: alterations of the kinetic properties in response to the rate of photosynthesis and the accumulation of sucrose. *Planta* **174**, 217–230.

Van der Krol, A. R., Mol, J. N. M., and Stuitje, A. R. (1988). Antisense genes in plants; an overview. *Gene* **72**, 45–50.

Willmitzer, L. (1988). The use of transgenic plants to study gene expression. *Trends Genet.* **4**, 13–18.

Woodrow, I. E., and Berry, J. A. (1988). Enzymatic regulation of photosynthetic CO_2 fixation in C_3 plants. *Annu. Rev. Plant Physiol. Plant Mol. Biol.* **39**, 533–594.

3

Controlling the Effects of Excessive Light Energy Fluxes: Dissipative Mechanisms, Repair Processes, and Long-Term Acclimation

C. Schäfer

I. Introduction

The conversion of light energy to chemical energy by oxygenic photosynthesis occurs in cyanobacteria, algae, and higher plants. The production of organic matter in this process represents the main energy input for the development of food chains and ecosystems. In algae and higher plants the photosynthetic reaction sequences are located in the chloroplast. Light energy is absorbed and converted to chemical energy which is used to reduce CO_2 and to synthesize carbon skeletons for growth and metabolism.

The conversion of light energy into chemical energy takes place at special pigment protein complexes in the thylakoid membranes within the chloroplasts. *In vitro*, photosynthetic pigments have a high susceptibility to photodestruction, but *in vivo* photodestruction can only be detected under extreme conditions (Miller and Carpentier, 1991). The following reasons for lack of damage by excess light are conceivable: (i) The primary photosynthetic reactions "unload" the excited pigments; (ii) Protective mechanisms exist which help to avoid harmful effects of photochemical side reactions; and/or (iii) repair mechanisms are operative in order to compensate for any damage that occurs.

Frequently, excess light energy does not result in pigment loss, but in a reduction of the efficiency of photosynthetic energy conversion which is only slowly reversible. This type of damage is called photoinhibition (Powles, 1984).

The risk of an overexcitation of the photosynthetic apparatus does not primarily depend on the light intensity but on the balance between the absorption of light energy and its consumption in photosynthetic reactions (Björkman and Demmig-Adams, 1993). Ribulose-1,5-bisphosphate-carboxylase-oxygenase (Rubisco) is the central enzyme of CO_2 fixation, and photosynthetic capacity (the light-saturated photosynthetic rate) correlates over a broad range with the activity of this enzyme. Quantitatively Rubisco is also the predominant enzyme of green tissues and a considerable fraction of cellular nitrogen is tied up in this protein (Björkman, 1981). Hence an increase in the capacity for photosynthetic CO_2 fixation would put a considerable strain on the cellular nitrogen budget if additional Rubisco is needed. These rather high costs of photosynthetic machinery would suggest that the amount of photosynthetic proteins not only depends on the available light energy but also on the nitrogen budget and the growth requirements of the organism.

The first part of this chapter deals with long-term acclimatory processes which occur when the balance between the absorption of light energy and its photosynthetic utilization is affected. The second part considers more closely a protective mechanism (the xanthophyll cycle) and the repair processes which proceed during high-light stress.

II. Experimental Approach

In higher plants studies on the molecular structure and function of the photosynthetic apparatus are complicated by the fact that the leaf structure is not homogeneous with respect to the cell type and the light environments of the cells (internal shading). Furthermore it is difficult to apply effectors, and their cellular concentrations cannot be manipulated in a well-defined way. The advantage of using suspension-cultured cells in studies of the photosynthetic apparatus of higher plants lies in the constant mixing of the cells which assures a homogeneous treatment with light and with effectors. Photoautotrophically growing cell cultures have been obtained from several higher plant species (Yamada, 1985). Culture cells grow as small aggregates in mineral medium (Fig. 1). Most of the cell lines need some vitamins and an elevated CO_2 concentration (2%). Several studies indicate that the basic features of the photosynthetic apparatus in cultured cells are comparable to those of intact plants (Carrier *et al.*, 1989). We used a photoautotrophic cell line of *Chenopodium rubrum* L. (Fig. 1; Hüsemann and Barz, 1977) in our experiments.

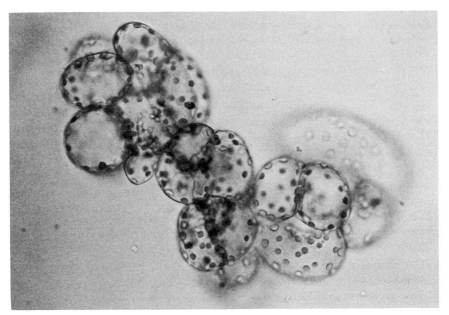

Figure 1   Photoautotrophic culture cells of *Chenopodium rubrum* L. × 1350.

The approach was based on the scheme which is shown in Fig. 2. We expected that the organization of the photosynthetic apparatus is determined not only by the input of photosynthetic energy but also by the potential to synthesize photosynthetic proteins and by the cellular requirements for carbon skeletons. Disturbances of each of these factors should have a specific effect on the amount of excessive light.

The availability of photosynthetic energy was manipulated by changing the light intensity. The potential to synthesize photosynthetic machinery was reduced by nitrogen limitation and the demand for photosynthetic products was reduced by feeding glucose to the cells. The effects of these manipulations on the maximum photosynthetic rate, the functioning of photosystem 2 (PS2), the occurrence of photoinhibition, and the total photosynthetic capacity of the cell are discussed in the following.

III. Changes in Photosynthetic Capacity

A. General Observations

In cultured *C. rubrum* cells all manipulations which disturbed the balance between energy absorption and photosynthetic energy consumption (increase in PFD, nitrogen deficiency) or which reduced the necessity for photosynthetic energy conversion (glucose feeding) resulted in a reduc-

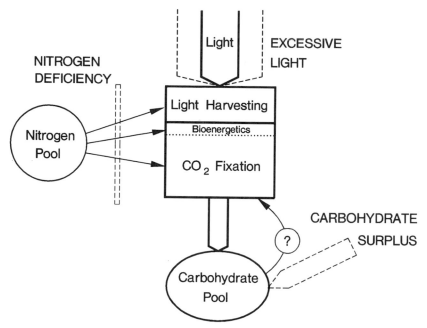

Figure 2   Schematic diagram showing the photosynthetic reaction sequence and possible points of impact for excessive light, for a surplus of carbohydrates, and for nitrogen deficiency (broken lines). The central square is subdivided into the different steps of photosynthesis: light harvesting, bioenergetics (electron transport and ATP synthesis), and photosynthetic CO_2 fixation. The area of each rectangle is proportional to the allocation of nitrogen to the respective components. The necessity for nitrogen to synthesize these components and a possible feedback effect of carbohydrate accumulation are indicated by thin arrows.

tion of the light-saturated photosynthetic rate per cell weight (photosynthetic capacity, Table I). This indicates that a down-regulation of the photosynthetic apparatus was possible when the energy input surpassed the demands. An up-regulation (increase in photosynthetic capacity on a weight basis), e.g., when the availability of light energy was increased, could not be observed under these experimental conditions. Comparable results were described for the unicellular alga *Dunaliella tertiolecta* (Sukenik *et al.*, 1990), whereas in intact plants considerable increases in photosynthetic capacity were observed after transfer from low- to high-light conditions (Björkman, 1981).

The components of the photosynthetic apparatus can be subdivided into an energy-absorbing fraction (light-harvesting complexes with their pigments, mainly chlorophylls) and an energy-converting fraction (proteins of electron transport, ATP synthesis, and CO_2 fixation). At light saturation, photosynthesis is limited by energy conversion and not by

Table I   Changes in Photosynthetic Capacity and Efficiency following Manipulations of the Balance between the Availability of and the Requirements for Light Energy

| | Photosynthetic capacity | | | |
Manipulation	Weight^{-1}	Chlorophyll^{-1}	Fv/Fm	Reference
Moderate high light				
350 μmol m^{-2} s^{-1}, 6 days treatment (A)	70.4%	136.8%	80.6%	Schäfer and Heim (1992)
Strong high light				
1000 μmol m^{-2} s^{-1}, 6 days treatment (B)	68.5%	95.2%	78.2%	Schäfer and Schmidt (1991)
Nitrogen deficiency				
60 μmol m^{-2} s^{-1}, 6 days treatment (C)	54.7%	66.3%	97.5%	Schäfer and Heim (1992)
Glucose feeding				
60 μmol m^{-2} s^{-1}, 4 days treatment (D)	34.6%	96.6%	95.0%	Schäfer *et al.* (1992)

The incident light energy was increased from low light (60 or 150 μmol m^{-2} s^{-1}) to moderate high light (A) or to strong high light (B). Nitrogen deficiency was induced by subculturing the cells in nitrogen-free medium (C) and a surplus of carbohydrates was introduced by feeding glucose to the cells (D). Photosynthetic capacity was determined as oxygen evolution at light and CO_2 saturation and calculated on weight basis (dry weight basis in A–C, fresh weight basis in D) and chlorophyll basis. The dark values of variable over maximum fluorescence (Fv/Fm,) were used as estimates for photochemical efficiency of PS2. All data are calculated as percentages of untreated controls.

light absorption. Changes in pigment content should not have an effect on light-saturated photosynthetic rate (Björkman, 1981). Hence, changes in the photosynthetic capacity on a chlorophyll basis would indicate that the ratio of light-harvesting capacity to electron transport and CO_2 fixation capacity was affected. Table I shows that this occurs under specific conditions. In moderate high light, photosynthetic rate on a chlorophyll basis increased, indicating that the loss in chlorophyll content per cell dominated. Under nitrogen deficiency this parameter decreased, indicating that the components which determine the oxygen evolution capacity decreased predominantly. These reactions of the cells may help to reduce the stress intensity. In the first case (light stress) energy absorption is reduced and in the second case (nitrogen deficiency) where the risk of light stress is small due to the low-light regime, photosynthetic proteins could be used as nitrogen sources to slow down the development of nitrogen deficiency. A reduction in Rubisco content during periods of nitrogen deficiency is also typical for intact plants (e.g., Medina, 1971; Ferreira and Davies, 1987; Sage *et al.*, 1987; compare also Stitt and Schulze, this volume, Chapter 4).

A third physiological parameter which was measured for all three manipulations was the photochemical efficiency of PS2. PS2 is known to be particularly sensitive to photoinhibition (Critchley, 1981) and the ratio of variable over maximum fluorescence yield (Fv/Fm) is considered as a suitable parameter to estimate its photochemical efficiency (Demmig and Björkman, 1987). Table I shows that the photochemical efficiency only decreased during light stress but was unaffected by the chlorophyll losses which accompanied nitrogen deficiency and the transition to photomixotrophic growth. Changes in Fv/Fm are discussed in more detail below.

In summary the data from Table I suggest that a balanced reduction in light-energy-absorbing and -converting components of the photosynthetic apparatus was only observed after glucose feeding. Under nitrogen deficiency light-harvesting pigment–protein complexes were reduced to a smaller extent than the proteins of photosynthetic carbon fixation. High-light stress was accompanied by photoinhibition of PS2. Although all manipulations resulted in a reduction of photosynthetic performance the points of impact of acclimation and/or damage were different.

B. Molecular Basis of Photosynthetic Capacity Changes

Measurements of Rubisco activity after glucose feeding showed that the maximum activity of this enzyme decreased proportionally to the chlorophyll content when glucose was added to the medium of photoautotrophic cells (Fig. 3; Schäfer *et al.*, 1992). This maximum activity, which is measured after full activation, was shown to be proportional to the total amount of the enzyme (Quick *et al.*, 1991). Rubisco is considered one of the components which could potentially limit the maximum photosynthetic capacity. Therefore a proportional reduction in Rubisco and in chlorophyll content is in agreement with the observed reduction in photosynthetic capacity on a weight basis as well as the observed constancy of photosynthetic capacity on a chlorophyll basis. Further studies showed that with the reduced amounts of Rubisco the contribution of photosynthesis to growth was still significant (Schäfer *et al.*, 1992). The reduction in chlorophyll content upon glucose feeding seems to be a general phenomenon. It has also been detected in several other cell cultures (Edelman and Hanson, 1971; Pamplin and Chapman, 1975; Dalton and Street, 1977) and in algae (Shihira-Ishikawa and Hase, 1965). Studies by Krapp *et al.* (1991) showed that also in intact spinach leaves glucose feeding results in reductions of chlorophyll and Rubisco contents. It is possible to demonstrate that these changes in Rubisco activity are due to changes at the mRNA level (compare Stitt and Schulze, this volume, Chapter 4). It could be demonstrated that glucose feeding caused a rapid reduction (time scale: hours) of the mRNA which codes for the small subunit of Rubisco (SSU-mRNA) (Fig. 4). This change occurred before reductions

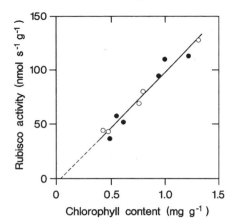

Figure 3 Correlation of chlorophyll content and maximum (fully activated) Rubisco activity in *C. rubrum* cells after the addition of glucose (75 mM) to the medium. Chlorophyll content and maximum Rubisco activity per fresh weight were determined over 0–5 days (○) and 0–8 days (●). They decreased during this period. Data are from two experiments (○, ●) and each sample was from a separate culture (after Schäfer *et al.*, 1992).

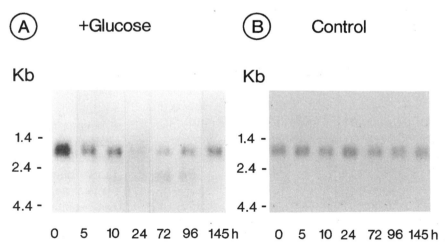

Figure 4 Time course for the relative content of mRNA which codes for the small subunit of Rubisco (SSU-mRNA) in *C. rubrum* cell cultures after the addition of glucose (50 mM at time 0; A) and in controls without glucose addition (B). Northern blots were performed using rbcS-cDNA for detection of SSU-mRNA. The same amount of RNA was applied in each blot. Hence the band intensity gives an estimate for the relative content of SSU-mRNA. The location of markers for RNA length (kb, *10^3 basis) is also indicated (Hofmann, 1991).

in photosynthetic capacity were detectable and the effect was reversible whenever the applied glucose was metabolized.

Further studies of this regulation mechanism showed that the reduction in SSU-mRNA was related to the accumulation of carbohydrates but was not directly dependent on the saccharide concentration in the cell. It also occurred in the dark and was independent of photosynthetic performance and osmotic effects (Hofmann, 1991). The latter could be tested by adding nonmetabolizable solutes to the medium (e.g., sorbitol), which did not have a significant effect on SSU-mRNA level. Possibly the change in mRNA level was triggered by a metabolic signal which is affected by the general carbon balance of the cell. Photosynthetic input of assimilates then would be only one of several triggers. The components of this signal chain still need to be elucidated.

It is possible that some of the effects of nitrogen deficiency were caused by the same reaction sequence as those of glucose feeding, because nitrogen deficiency also resulted in elevated intercellular carbohydrate levels (Schäfer and Heim, 1992). Another trigger might be responsible for the high-light effects as high light alone did not result in a pronounced increase in carbohydrate levels (Schäfer and Heim, 1992). Furthermore a reduction in photosynthetic efficiency was only observed in high light. Finally it must be considered that hormonal signals may affect the organization of the photosynthetic apparatus also (Parthier, 1979; Longo *et al.*, 1990; Ohya and Suzuki, 1990; compare also Beck, this volume, Chapter 5).

IV. Changes in Photosynthetic Efficiency

A. General Aspects

High light was the only growth condition which caused a considerable reduction in the photosynthetic efficiency of PS2 (up to 50%). Reductions in photosynthetic efficiency of the cells might be caused by the net loss of PS2 reaction centers and/or a reduction in the efficiency of remaining centers. Concerning the latter mechanism many results point at the significance of zeaxanthin and antheraxanthin formation from violaxanthin during periods of light stress (xanthophyll cycle; Demmig-Adams, 1990; Demmig-Adams and Adams, 1992). It was argued that zeaxanthin could somehow absorb excessive light energy and dissipate it as heat (Demmig-Adams, 1990). Many correlative data support this view but the molecular basis of this process remains unclear. Furthermore there is not yet a definite answer concerning the exact localization of zeaxanthin in the thylakoid membrane and its association with specific pigment–protein complexes.

We were able to demonstrate that the xanthophyll cycle operates in photoautotrophically cultured cells (Schäfer and Schmidt, 1991). The contents of zeaxanthin and antheraxanthin increased and the violaxanthin content decreased during periods of light stress (Fig. 5). The maximum levels of zeaxanthin and antheraxanthin which could be obtained were quite low, however, and the potential to convert violaxanthin to zeaxanthin was not increased during long-term acclimation to high-light intensities. It is conceivable that the cells lost part of their potential for zeaxanthin accumulation during the long-term culture in low light. It may be that in *C. rubrum* cells there are further mechanisms besides zeaxanthin accumulation which improve the high-light resistance. When looking for such mechanisms it is necessary to focus on the turnover of the D1 protein and its significance for the repair of photoinhibitory damage.

The D1 protein has a central position in the PS2 reaction center (Fig. 6; Andersson and Styring, 1991; Trebst, 1993). Together with the D2 protein it binds P680, the chlorophyll involved in charge separation. It contains the primary electron donor to PS2 (a tyrosine residue in position 161) and it binds the secondary electron acceptor (Q_B, a plastoquinone molecule). It has been known for some time that turnover of the D1 protein can be considerably higher than that for any other thylakoid protein (Mattoo *et al.*, 1984) although it is only present in small amounts (about 0.5% of total cellular protein, calculated from Evans and Seemann, 1989). D1 protein turnover is detectable under low-light conditions and

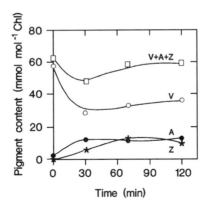

Figure 5   Changes in xanthophyll cycle pigment content after transfer of *C. rubrum* cell cultures from low light (70 μmol m^{-2} s^{-1}) to photoinhibitory light (800 μmol m^{-2} s^{-1}). Before and during the photoinhibitory treatment cell samples were taken from the culture vessels and pigments (violaxanthin, V; antheraxanthin, A; and zeaxanthin, Z) were quantified by high-pressure liquid chromatography after pigment extraction. Data are means of two cultures (redrawn from Schäfer and Schmidt, 1991).

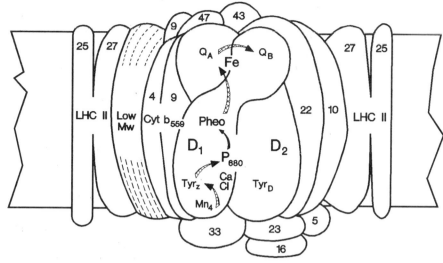

Figure 6 A current model of PS2. The PS2 subunits are given with their names or molecular masses (in kilodaltons). The central D1/D2 heterodimer is shown with the bound redox components. The charge separation between P680 and pheophytin (Pheo) and the secondary electron transfers are symbolized by filled and hatched arrows, respectively. Tyr_Z, Tyr_D, Q_A, Q_B, and the acceptor-side iron are located at their approximate positions in the reaction center (after Andersson and Styring, 1991).

it increases several fold if cells are exposed to photoinhibitory light intensities. According to a contemporary model photoinhibitory light causes an inactivation of the D1 protein which rapidly becomes irreversible. The repair mechanism includes (i) the degradation of inactivated D1 protein, (ii) its resynthesis, and (iii) the reassembling of functional PS2 reaction centers (Ohad *et al.*, 1990). The D1 protein is synthesized in the chloroplast and its synthesis can be blocked by inhibitors of chloroplastic protein synthesis (chloramphenicol, streptomycin, etc.).

B. D1 Protein Turnover in Photoautotrophic Cell Cultures

We measured D1 protein turnover in pulse/chase experiments with $[^{35}S]$methionine. We could confirm the observation that D1 protein turnover is light-intensity dependent and increases with increasing photon flux density (Fig. 7). From the fraction of $[^{35}S]$methionine which was incorporated into D1 protein we estimated that D1 protein synthesis could reach about 10% of total protein synthesis of the cell in high-light intensities (V. Schmid and C. Schäfer, unpublished). This estimate corresponds to the data of other studies (Raven and Samuelsson, 1986). Hence the energetic costs of D1 protein turnover are quite considerable although its share of total cellular protein is only small.

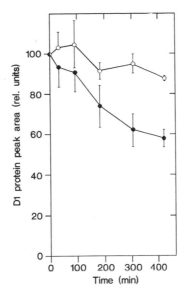

Figure 7 Comparison of D1 protein degradation at 120 μmol m^{-2} s^{-1} (○) and at 900 μmol m^{-2} s^{-1} (●). A [^{35}S]methionine pulse (40 min pulse length) was applied for prelabeling. Then the cells were washed, excess unlabeled methionine was added to prevent further label incorporation, and the cultures were returned to the respective light intensities. The relative D1 peak area was determined from autoradiographs after electrophoresis of whole-cell homogenates. The relative D1 peak areas at the beginning of chase experiments were set to 100%. Data are means±SD of three experiments (after Schmid and Schäfer, 1993).

It may be assumed that the D1 protein level reflects the balance between D1 protein degradation and resynthesis and that photoinhibition occurs when the rate of inactivation and degradation surpasses its resynthesis (Kyle and Ohad, 1986). Quantification of D1 protein content by immunoblotting and atrazine-binding site determination (binding of [^{14}C]atrazine to the Q_B binding site of D1 protein; Paterson and Arntzen, 1982) indicated a net loss of D1 protein during exposure of cells to photoinhibitory light (Schäfer *et al.*, 1993). It amounted to about 30% after 5 h of photoinhibition and it was largely reversible (Table II). The light intensities in this experiment were in a physiological range and *C. rubrum* cells survived in this light regime for prolonged periods of time (Table I).

Photoinhibition resulted in an increase of the minimum chlorophyll fluorescence yield (Fo, Table II), which is interpreted as a reduction of the rate constant for photochemical energy conversion (Demmig and Björkman, 1987). Photoinhibitory loss of D1 protein in high light was also accompanied by a reduction of the maximum chlorophyll fluorescence yield (Fm), which is referred to as nonphotochemical quenching (Schreiber *et al.*, 1986). However, we could not observe such a decrease

Table II Effect of Photoinhibition in Strong High Light (800 μmol m^{-2} s^{-1}, 5 h) and Recovery in Low Light (60 μmol m^{-2} s^{-1}, 5 h, Following the Photoinhibitory Treatment) on Fluoresence Parameters, the Number of Atrazine-Binding Sites, and the D1 Protein Immunoblot Signal in *C. rubrum* Cells

	Percent changes	
	Inhibition (5 h, 800 μmol m^{-2} s^{-1})	Recovery (5 h, 60 μmol m^{-2} s^{-1})
Fv/Fm	$-$ 24.7 $\pm$ 11.7 (7)	$+$ 28.9 $\pm$ 22.6 (7)
Fo	$+$ 15.8 $\pm$ 10.3 (7)	$-$ 3.9 $\pm$ 9.8 (7)
Fm	$-$ 33.9 $\pm$ 9.3 (7)	$+$ 46.3 $\pm$ 20.5 (7)
Atrazine binding	$-$ 35.9 $\pm$ 20.1 (4)	$+$ 40.8 $\pm$ 12.1 (4)
Immunoblot D1 protein	$-$ 24.6 $\pm$ 8.2 (6)	$+$ 28.7 $\pm$ 31.7 (6)

Data were calculated as percentages of the values which were obtained at the beginning of the inhibition and the recovery period, respectively. Means $\pm$ SD (n) are shown (after Schäfer *et al.*, 1993).

in Fm when D1 protein loss was induced in moderate light intensities by blockage of chloroplastic protein synthesis (Schäfer *et al.*, 1993). Hence nonphotochemical quenching could be an additional high-light effect and not a consequence of D1 protein loss. It could rather indicate the accumulation of zeaxanthin (Demmig-Adams, 1990). The fate of residual PS2 complexes after D1 protein degradation is still an open question. Hundal *et al.* (1990) suggested from *in vitro* studies that the residual PS2 complex dissociates into subunits and that some of these migrate from the appressed to the nonappressed thylakoid regions. According to *in vitro* studies by Barbato *et al.* (1992) photoinhibition results in partial breakdown of the D1 protein and the loss of CP 43 from the PS2 core complex. They conclude that the residual PS2 core complex—including the D1 protein breakdown products—moves without further dissociation to the nonappressed thylakoid regions were the D1 breakdown products are replaced by new D1 protein.

Inactivation and breakdown of PS2 reaction centers in *C. rubrum* resulted in reductions of the efficiency of photosynthetic oxygen evolution. However, the light-saturated rate of photosynthetic oxygen evolution was only barely affected by these changes (Schäfer and Schmidt, 1991). Hence the partial loss of functional PS2 reaction centers could be compensated by the residual centers if sufficient light energy was available and the energy conversion in PS2 did not limit photosynthetic electron transport. This observation confirms the present model of photosynthetic electron transport which includes the coupling of many PS2 and cytochrome b$_6$f complexes via the plastoquinone pool (Cramer *et al.*, 1991).

Besides the short-term effects on D1 protein content, long-term changes occur in the capacity for D1 protein turnover. Assuming that D1 protein inactivation is an unavoidable flaw of photosynthetic performance one may hypothesize that prolonged light stress might induce an increase of the maximum capacity of D1 protein turnover. Indications for the occurrence of such a mechanism have been obtained for cyanobacteria (Samuelsson *et al.*, 1987). This can be studied by comparing the effect of chloramphenicol on the reduction in photochemical efficiency during light stress in cells which had been precultured in photoinhibitory light (PIL) for about a week (PIL cells) with controls which had been precultured in moderate light intensities. Chloramphenicol enhanced the reduction in Fv/Fm during light stress and this effect was more pronounced in PIL cells then in controls (Schmid and Schäfer, 1992). Considering the observation that D1 protein is the main product of chloroplastic protein synthesis during light stress, this result indicates that the absence of D1 protein synthesis had more harmful effects in PIL cells than in controls. Therefore, a likely explanation for this observation would be that the D1 protein synthesis rate was higher in PIL cells. A direct quantification of this effect from pulse experiments was not yet possible, because total label uptake and incorporation was also affected by culture conditions.

Chase experiments with both cell types indicated that the half-life for the D1 protein was reduced by preculture in photoinhibitory light (Schmid and Schäfer, 1992). Hence both components of D1 protein turnover, the degradation of inactivated proteins and their resynthesis, were enhanced in PIL cells (Schmid and Schäfer, 1992). It is possible that a partial limitation of the D1 protein repair process which occurs at the degradation step of inactivated D1 protein is reduced in the long-term acclimated cells.

It may be summarized that characteristic features of D1 protein turnover were detectable in photoautotrophically cultured cells of *C. rubrum*. The turnover rate was not only dependent on the actual light intensity but also on the preculture conditions of the cells. D1 protein synthesis was a considerable fraction of total protein synthesis during periods of light stress. A net loss of D1 protein was observed under these conditions, showing that degradation of PS2 surpassed its resynthesis.

V. Effects of Multiple Stress

In the above studies we considered only the physiological reactions to manipulations of single growth parameters (high light, nitrogen deficiency, and carbohydrate content). Under natural conditions parallel

changes of various enviornmental factors occur which may cause synergistic effects.

Subjecting *C. rubrum* cells to changes in nitrogen supply and light
intensity showed that effects of light stress and nitrogen deficiency on
chlorophyll content and photosynthetic capacity were additive when both
parameters were manipulated together (Schäfer and Heim, 1992). However, the development of photoinhibition was considerably enhanced
under conditions of multiple stress (Fig. 8). The additional effect of
nitrogen deficiency increased with increasing light intensity but it was
not detected at 1000 μmol m^{-2} s^{-1}. Probably under these conditions of
extreme light stress the additional effect of nitrogen deficiency was negligible.

A synergistic effect of light stress and nitrogen deficiency was also
observed in other studies (Ferrar and Osmond, 1986; Henley *et al.*, 1991).
It could indicate that the rate of PS2 inactivation was increased in
nitrogen-deficient cells and/or that the rate of D1 protein resynthesis was
affected by nitrogen deficiency due to a general suppression of protein
synthesis. In nitrogen-deficient cells the changes in atrazine-binding sites
during 5 h of photoinhibition (800 μmol m^{-2} s^{-1}) and 5 h recovery (60μmol
m^{-2} s^{-1}) were $-$ 31.3 and $+$ 37.0%, respectively (Schäfer *et al.*, 1993).
These numbers are comparable to those obtained in cells well supplied

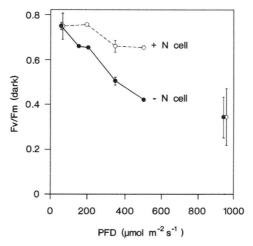

Figure 8 Enhancement of the photoinhibitory effect of high light by nitrogen deficiency. Cultures of nitrogen-repleted and nitrogen-deficient *C. rubrum* cells were exposed
to the indicated light intensities for at least five consecutive light periods. Then the Fv/
Fm ratios of darkened cells were determined as estimates for the photochemical efficiency
of PS2. Data from single experiments or means ± SD of three to five experiments are
shown (after Schäfer and Heim, 1992).

with nitrogen (Table II). A limitation of D1 protein synthesis by restricted nitrogen availability could not be detected. Probably the recycling of amino acids during D1 protein turnover reduces the nitrogen demand of this process.

In the short-term experiments over several hours the reduction in Fv/Fm was slightly higher in nitrogen-deficient as compared to nitrogen-repleted cells (Schäfer and Heim, 1992; Schäfer *et al.*, 1993). These small differences could add up during prolonged culture in high light, thus leading to the observed synergistic effects of nitrogen deficiency.

VI. Conclusions

In photoautotrophic culture cells the potentials for light energy absorption and CO_2 fixation and the balance between these components of photosynthetic performance are influenced in a specific way by light and nutrition. The following regulatory processes could be distinguished (Fig. 9):

- Excessive accumulation of carbohydrates results in a balanced downward regulation of light absorption and CO_2 fixation. At least in the case of Rubisco-SSU this change is probably caused by a reduction in the respective mRNA level (Fig. 9A).
- Nitrogen deficiency leads to a predominant reduction in the CO_2-fixing enzymes and this may represent a mobilization of nitrogen for other synthetic processes. These changes increase the susceptibility to photoinhibition (Fig. 9B).

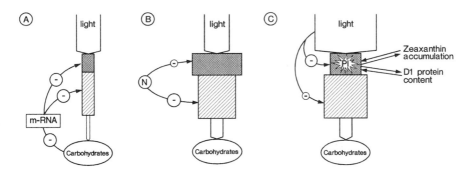

Figure 9 Summarizing sheet for the effects of carbohydrate surplus (A), nitrogen deficiency (B), and excessive light (C) on the photosynthetic apparatus of *C. rubrum* cells. The contents of light-harvesting proteins (cross-hatched) and Calvin-cycle proteins (hatched) after treatment are indicated by the sizes of the respective areas and thin arrows show impacts of regulation. The occurrence of photoinhibition (PI) is also indicated.

- Excessive light results in a predominant loss of light-absorbing pigments and in photoinhibition. The zeaxanthin cycle and the D1 protein repair cycle probably help to reduce the extent of photoinhibition (Fig. 9C).

Acknowledgments

The data discussed in this chapter were obtained in collaboration with R. Heim, B. Hoffmann, V. Schmid, E. Schmidt, and G. Vogg and I thank them for their enthusiasm and many valuable discussions. I further thank H. Simper for excellent technical assistance and Dr. F. Koenig (Frankfurt) for critical reading of the manuscript.

References

Andersson, B., and Styring, S. (1991). Photosystem II: Molecular organization, function and acclimation. *Curr. Top. Bioenerg.* **16,** 1-81.

Barbato, R., Friso, G., Rigoni, F., Dalla Vecchia, F., and Giacometti, G.M. (1992). Structural changes and lateral redistribution of photosystem II during donor side photoinhibition of thylakoids. *J. Cell Biol.* **119,** 325-337.

Björkman, O. (1981). Responses to different quantum flux densities. *Encycl. Plant Physiol. New Ser.* **12A,** 57-107.

Björkman, O., and Demmig-Adams, B. (1993). Regulation of photosynthetic light energy capture, conversion and dissipation in leaves of higher plants. *Ecol. Stud.* **100** (in press).

Carrier, P., Chagvardieff P., and Tapie, P. (1989). Comparison of the oxygen exchange between photosynthetic cell suspensions and detached leaves of *Euphorbia characias* L. *Plant Physiol.* **91,** 1075-1079.

Cramer, W. A., Furbacher, P. N., and Szczepaniak, A. (1991). Electron transport between photosystem II and photosystem I. *Curr. Top. Bioenerget.* **16,** 179-222.

Critchley, C. (1981). Studies on the mechanism of photoinhibition in higher plants. I. Effects of high light intensity on chloroplast activities in cucumber adapted to low light. *Plant Physiol.* **67,** 1161-1165.

Dalton, C. C., and Street, H. E. (1977). The influence of applied carbohydrates on the growth and greening of cultured spinach (*Spinacia oleracea* L.) cells. *Plant Sci. Lett.* **10,** 157-164.

Demmig, B., and Björkman, O. (1987). Comparison of the effect of excessive light on chlorophyll fluorescence (77K) and photon yield of O_2 evolution in leaves of higher plants. *Planta* **171,** 171-184.

Demmig-Adams, B. (1990). Carotenoids and photoprotection in plants: A role for the xanthophyll zeaxanthin. *Biochim. Biophys. Acta* **1020,** 1-24.

Demmig-Adams, B., and Adams, W. W., III (1992). Photoprotection and other responses of plants to high light stress. *Annu. Rev. Plant Physiol.* **43,** 599-626.

Edelman, J., and Hanson, A. D. (1971). Sucrose suppression of chlorophyll synthesis in carrot callus cultures. *Planta* **98,** 150-156.

Evans, J. R., and Seemann, J. R. (1989). The allocation of protein nitrogen in the photosynthetic apparatus; costs, consequences, and control. *In* "Photosynthesis" (W. R. Briggs, ed.), pp. 183–205. Alan R. Liss, New York.

Ferrar, P. J., and Osmond, C. B. (1986). Nitrogen supply as a factor influencing photoinhibition and photosynthetic acclimation after transfer of shade-grown *Solanum dulcamara* to bright light. *Planta* **168,** 563-570.

Ferreira, R. B., and Davies, D. D. (1987). Protein degradation in *Lemna* with particular reference to ribulose bisphosphate carboxylase. II. The effect of nutrient starvation. *Plant Physiol.* **83,** 878-883.

Henley, W. J., Levavasseur, G., Franklin, L. A., Osmond, C. B., and Ramus, J. (1991). Photoacclimation and photoinhibition in *Ulva rotundata* as influenced by nitrogen availability. *Planta* **184,** 235-243.

Hofmann, B. (1991). Die Regulation der Photosynthese durch Kohlenhydrate: Untersuchungen mit photoautotrophen *Chenopodium rubrum* Kulturzellen. Diplomarbeit, University of Bayreuth.

Hundal, T., Virgin, I., Styring, S., and Andersson, B. (1990). Changes in the organization of photosystem II following light induced D1-protein degradation. *Biochim. Biophys. Acta* **1017,** 235-241.

Hüsemann, W., and Barz, W. (1977). Photoautotrophic growth and photosynthesis in cell suspension cultures of *Chenopodium rubrum. Physiol. Plant.* **40,** 77-81.

Krapp, A., Quick, W. P., and Stitt, M. (1991). There is a dramatic loss of Rubisco, other Calvin cycle enzymes, and chlorophyll, when glucose is supplied to mature spinach leaves via the transpiration stream. *Planta* **186,** 58-69.

Kyle D. J., and Ohad, I. (1986). The mechanism of photoinhibition in higher plants and green algae. *Encycl. Plant Physiol. New Ser.* **19,** 468-475.

Longo, G. P. M., Bracale, M., Rossi, G., and Longo, C. P. (1990). Benzyladenine induces the appearance of *LHCP-m-RNA* and of the relevant protein in dark-grown excised watermelon cotyledons. *Plant Mol. Biol.* **14,** 569-573.

Mattoo, A. K., Hoffmann-Falk, H., Marder, J. B., and Edelman, M. (1984). Regulation of protein metabolism coupling of photosynthetic electron transport to in vivo degradation of the rapidly metabolized 32-kilodalton protein of the chloroplast membranes. *Proc. Natl. Acad. Sci. U.S.A.* **81,** 1380-1384.

Medina, E. (1971). Effect of nitrogen supply and light intensity during growth on the photosynthetic capacity and carboxydismutase activity of leaves of *Atriplex patula* ssp. *hastata. Year Book—Carnegie Inst. Washington* **70,** 551-559.

Miller, N., and Carpentier, R. (1991). Energy dissipation and photoprotection mechanisms during chlorophyll photobleaching in thylakoid membranes. *Photochem. Photobiol.* **54,** 465-472.

Ohad, I., Adir, N., Koike, H., Kyle, D. J., and Inoue, Y. (1990). Mechanism of photoinhibition in vivo. *J. Biol. Chem.* **265,** 1972-1979.

Ohya, T., and Suzuki, H. (1990). Benzyladenine- and light-stimulated plastid protein synthesis in excised cucumber cotyledons. *Plant Physiol. Biochem (Paris)* **28,** 27-35.

Pamplin, E. J., and Chapman, J. M. (1975). Sucrose suppression of chlorophyll synthesis in tissue culture: Changes in the activity of the enzymes of the chlorophyll biosynthetic pathway. *J. Exp. Bot.* **26,** 212-220.

Parthier, B. (1979). The role of phytohormones (cytokinins) in chloroplast development. *Biochem. Physiol. Pflanz.* **174,** 173-214.

Paterson, D. R., and Arntzen, C. J. (1982). Detection of altered inhibition of photosystem II reactions in herbicide-resistant plants. *In* "Methods in chloroplast molecular biology" (M. Edelmann, R. B. Hallick, and H. H. Chua, eds.), pp. 109–118. Elsevier Biomedical Press, Amsterdam.

Powles, S. B. (1984). Photoinhibition of photosynthesis induced by visible light. *Annu. Rev. Plant Physiol.* **35,** 15-34.

Quick, W. P., Schurr, U., Scheibe, R., Schulze, E.-D., Rodermel, S. R., Bogorad, L., and Stitt, M. (1991). Decreased ribulose-1,5-bisphosphate carboxylase-oxygenase in transgenic tobacco transformed with "antisense" rbcS. I. Impact on photosynthesis in ambient growth conditions. *Planta* **183,** 512-554.

Raven J. A., and Samuelsson, G. (1986). Repair of photoinhibitory damage in *Anacystis nidulans* 625 (*Synechococcus* 6301): Relation to catalytic capacity for, and energy supply to, protein synthesis, and implications for μ_{max} and the efficiency of light-limited growth. *New Phytol.* **103,** 625-643.

Sage, R. F., Pearcy, R. W., and Seemann, J. R. (1987). The nitrogen use efficiency of C_3 and C_4 plants. III. Leaf nitrogen effects on the activity of carboxylating enzymes in *Chenopodium album* (L.) and *Amaranthus retroflexus* (L.). *Plant Physiol.* **85,** 355-359.

Samuelsson, G., Lönneborg, A., Gustafsson, P., and Öquist, G. (1987). The susceptibility of photosynthesis to photoinhibition and the capacity of recovery in high and low light grown cyanobacteria, *Anacystis nidulans. Plant Physiol.* **83,** 438-441.

Schäfer, C., and Schmidt, E. (1991). Light acclimation potential and xanthophyll cycle pigments in photoautotrophic suspension cells of *Chenopodium rubrum. Physiol. Plant.* **82,** 440-448.

Schäfer, C., and Heim, R. (1992). Nitrogen deficiency exacerbates the effects of light stress in photoautotrophic suspension cultured cells of *Chenopodium rubrum. Photosynthetica* **27** (in press).

Schäfer, C., Simper, H., and Hofmann, B. (1992). Glucose feeding results in coordinated changes of chlorophyll content, ribulose-1,5-bisphosphate carboxylase-oxygenase activity and photosynthetic potential in photoautrophic suspension cultured cells of *Chenopodium rubrum. Plant, Cell Environ.* **15,** 343-350.

Schäfer, C., Vogg, G., and Schmid, V. (1993). Evidence for D1 protein loss during photoinhibition of *Chenopodium rubrum* L. culture cells. *Planta* **189,** 433-439.

Schmid, V., and Schäfer, C. (1992). Analysis of D1 protein turnover in photoautotrophic suspension cultured cells of *Chenopodium rubrum*. I. Effects of light intensity and growth light regime. *Photosynthetica* **27,** 119-128.

Schreiber, U., Schliwa, U., and Bilger, W. (1986). Continuous recording of photochemical and non-photochemical chlorophyll fluorescence quenching with a new type of modulation fluorometer. *Photosynth. Res.* **10,** 51-62.

Shihira-Ishikawa, I., and Hase, E. (1965). Effects of glucose on the process of chloroplast development in *Chlorella protothecoides. Plant Cell Physiol.* **6,** 101-110.

Sukenik, A., Bennett, J., Mortain-Bertrand, A., and Falkowski, P. G. (1990). Adaption of the photosynthetic apparatus to irradiance in *Dunaliella tertiolecta. Plant Physiol.* **92,** 891-898.

Trebst, A. (1993). Dynamics in photosystem II structure and function. *Ecol. Stud.* **100** (in press).

Yamada, Y. (1985). Photosynthetic potential of plant cell cultures. *In* "Advances in Biochemical Engineering/Biotechnology" (A. Fiechter, ed.), pp. 110-176. Springer-Verlag, Berlin.

II

Flux Control at the
Organismic Level

4

Plant Growth, Storage, and Resource Allocation: From Flux Control in a Metabolic Chain to the Whole-Plant Level

M. Stitt and E.-D. Schulze

I. Introduction

This chapter addresses the assimilation and storage of carbon (C) and nitrogen (N), in order to identify sites at which whole plant growth is regulated. In this context the key questions are (i) what determines whether increased assimilation of C or N actually leads to increased growth, or merely to accumulation of stored forms of these substances; (ii) what determines where growth occurs (root, shoot, or other organs); and (iii) when storage occurs, is it competitive with growth or does it just involve a "surplus" which is not needed or cannot be used for growth at the time of formation. Storage can occur for different reasons, such as accumulation, reserve formation, and defense (reviewed by Chapin *et al.*, 1990). We need to be able to distinguish these functions, but how? Further, assimilation, allocation, and storage will interact in a complex and highly interactive manner.

We first give a general outline of the interactions between assimilation, storage, and growth and then describe how we have used two different approaches to investigate how these interactions may be regulated. In one approach, we have exploited genetically manipulated plants. Here, a selected and defined aspect can be altered, and the direct and indirect effects (ramifications) through the whole system can then be followed. This approach has the advantage that we know what has been altered (provided we use isogenic lines) and any change will be a direct or indirect consequence of the initial perturbation, so it will often be possible to assign causality. The disadvantage is that we can only apply this approach for a limited number of species (usually only annuals) and may get a one-sided or nonrepresentative answer. Therefore in a second and complementary approach we have carried out comparative studies contrasting species and plant life forms. This approach has the disadvantage that it relies heavily on correlative interpretations and, in a complex and highly interactive system, it may not always be possible to distinguish among direct, indirect, and trivial relationships. However, it has the advantage that a wider range of species can be studied.

II. What Is Assimilation and How Is It Related to Growth?

This chapter concentrates on CO_2 and nitrogen, while Schäfer (this volume, Chapter 3) and Schulze (this volume, Chapter 7) deal with light and water, respectively. Nitrogen and CO_2 must be assimilated, i.e., they are acquired in one chemical form and must be changed (reduced) into another form before they can be invested for growth. Figure 1 summarizes the interaction between acquisition and investment for four of the major inputs for plant growth, namely, CO_2, nitrogen, water, and light.

In the short-term (see Fig. 1), acquisition of CO_2 involves a coupled loss of water through the stomata. Also, since CO_2 fixation involves the use of preexisting structures to absorb and process light energy, and to fix and reduce CO_2, in the long-term, it will require investment of fixed carbon, nitrogen, and water to produce new leaf area, supporting structures, and biochemical machinery.

Assimilated carbon can be exported as sucrose to the growing sink tissues or stored as starch in the leaf. The extent to which the products of photosynthesis are exported, or retained as starch in the mature leaf, will affect the relationship between assimilation and growth. In addition, some of the fixed carbon is needed to provide the carbon skeleton for the assimilation of inorganic nitrogen. The distribution of fixed carbon between carbohydrate and amino acid formation therefore represents a second very important branchpoint at which fluxes will need to be regulated to maintain a correct balance for growth.

The sucrose which is exported from the leaf can be used for production of new leaves, supporting stems, or roots. Investment in production of leaf area will allow more light and CO_2 to be absorbed and will therefore allow more photosynthesis and growth. This is, however, only true if the nitrogen and water supply to the plant is adequate. For this reason, it is also essential that some of the photosynthate is invested in root growth

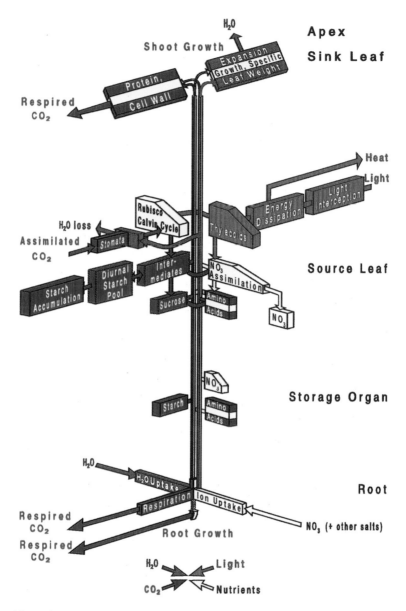

Figure 1 General outline of the interactions between assimilation of C and N and the utilization of water and irradiance for storage and growth.

(reviewed by Schulze and Chapin, 1987). On the one hand, photosynthesis and further growth will require an adequate supply of nitrogen and water; on the other hand, overinvestment in roots will, of course, decrease growth because fewer leaves are produced and because the increased respiratory load of the nonphotosynthetic organs will depress whole-plant photosynthesis. Regulation of allocation of photosynthate to form new leaves or roots can therefore be expected to play an important role in regulating the rate of plant growth. Vegetative growth will also usually require the investment of photosynthate in supporting structures, e.g., the stem, because otherwise the leaf area will be produced in the wrong place (i.e., in shade, where little light can be absorbed; Hirose and Werger, 1987). Here again, regulation is necessary.

An analogous series of interactions occurs during the assimilation and investment of N. Following uptake in the roots as NO_3^- or NH_4^+, the N must be reduced and/or converted to amino acids before it can be used for growth. The partitioning of N between storage as inorganic N or assimilation into organic compounds constitutes a potential important site for regulation. On the one hand it will be important that enough N is assimilated; on the other hand this requires use of light energy and also requires an input of C building blocks, which could compete with the formation of storage and structural carbohydrate.

Once the N has been assimilated, it can be invested into growth. As already discussed, this can occur in various structures. In the case of N, investment in root growth will promote further acquisition of N. Investment in leaves will decrease N acquisition, but in the long-term also this is essential for the use of N because it will guarantee a future supply of light energy and C skeletons to allow the plant to assimilate and use more N (Schulze and Chapin, 1987).

Within a particular organ, the allocation of N to different structural components needs to be regulated. This can be illustrated by considering the investment of N in photosynthetic apparatus. On the one hand, the proteins for the so-called "dark reactions" are required for the reactions directly involved in CO_2 fixation; on the other hand, thylakoid proteins are required for light harvesting, electron transport, and photophosphorylation, to provide energy for the fixation reactions. The regulation of allocation between these two groups of proteins is further exacerbated because, e.g., changes in the growth irradiance will alter the amount of light absorbed and processed by a given unit of thylakoid protein (i.e., a different balance would be needed for "optimal" use of N in low and high irradiance).

How will the assimilation of C and N and their use in leaf, stem, and root formation interact? Carbon assimilation in leaves, the acquisition of water and nutrients in the roots, and the assimilation of nitrogen will all be directly linked because they all feed into an open transport system (xylem and phloem) from which allocation to new growth takes place (Komor, this volume, chapter 6). This open transport system also serves as the starting point for formation of support structure, as well as reserves, defense substances, and the development of storage structures (see next section). Since plants are open systems, we may regard the growth of leaves and roots as independent events which, nevertheless, compete with each other for the available resources. Equally, new growth of leaves and roots feeds into a pool of existing leaf or root biomass and will provide more resources for the transport system. With respect to nutrients and amino acids not all resources are used, and there may be a permanent flux of substance recycling via phloem and xylem. In this sense, the xylem and phloem are the basis for a strong futile cycle of nutrients (Fig. 1)

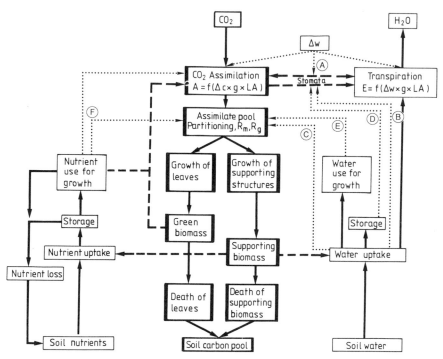

Figure 2 Schematic presentation of the internal carbon, water, and nutrient fluxes and their partitioning into different organs (thick lines). Interactions within the carbon fluxes arising from partitioning into foilage are shown as thick dashed lines, and the interactions between the carbon, water, and nutrient fluxes as the following dotted lines: (A) effects of external weather, (B,D) effects of water and (F) nutrient status on CO_2 assimilation, and effects of (C,E) water and (F) nutrient status on assimilate partitioning. Effects of water and nutrient status on aging are not shown (after Schulze and Chapin III, 1987).

from which resources are acquired (Komor, this volume, chapter 6). In the case of sucrose, there is usually negligible recycling via the xylem.

It becomes clear that a number of control mechanisms regulate the functioning of the whole plant (Fig. 2). Formation of new leaves has a strong positive feedback on carbon acquisition, and formation of roots feeds back positively on water and nutrient acquisition. The formation of permanent biomass (stems, woody structures) will feed back on resource acquisition, positively by exposing new leaves to light and roots to unexploited soil layers, but also negatively by causing self-shading to existing biomass. Growing leaves or roots will also exert direct feedback (via changing sink size) on the allocation scheme.

Complex interactions of this sort can hardly rely purely on passive control (Beck, this volume, chapter 5), because this would be slow and unstable. For example overinvestment in leaf growth will impair leaf function (e.g., by increasing water loss beyond the capacity of uptake) and this, in turn, will decrease photosynthetic rate, but will not (of itself) lead to a compensating increased allocation to the root. It will therefore also be necessary to regulate growth rates and allocation by pool concentrations, e.g., by direct modulation of development or function by sucrose or N-metabolites. The system may need to be additionally regulated by signal compounds (such as phytohormones) which are transported from the root to the shoot and vice versa (Beck, this volume, chapter 5; Schulze, this volume, chapter 7). In addition, accumulated products could serve as an intermediate buffer to balance acquisition and demand, i.e., accumulation sequesters those substances which may overload the futile cycle (e.g., nitrate) or which may cause an interruption of the plant internal flow system (osmotic gradient in the phloem).

III. What Is Storage and How Is It Related to Growth?

Storage is a major plant function, along with resource acquisition, transport, growth, defense, and reproduction. In this chapter we follow and extend the concept of the review by Chapin *et al.* (1990) who defined storage broadly as resources that build up in the plant and can be mobilized in the future to support biosynthesis for growth or other plant functions. We may distinguish, as a starting point, between three kinds of storage, namely accumulation, reserve formation, and recycling from growth or defense compounds (Fig. 3):

• Accumulation is the increase in a compound that does not directly compete with growth at the time at which it is acquired or formed. Accumulated compounds may be remobilized to support growth at a later stage or may be lost as litter.

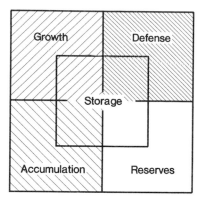

Figure 3 Storage includes reserves and components of defense, growth, and accumulation (after Chapin *et al.*, 1990).

- Reserve formation involves the metabolically regulated synthesis and compartmentation of compounds, using resources which would otherwise directly promote growth. Thus, reserve formation competes with growth and defense for resources at a given point in time. The reserves may later support growth or may be lost as litter; it may also include the formation of structure for storage (Steinlein *et al.*, 1993).
- Recycling is the reutilization of compounds whose immediate physiological function contributes to growth or defense, but which can subsequently be broken down to support future growth, otherwise these substances would be lose as litter.

Describing the relations between storage, growth, reserves, and defense in a more general flow diagram, it becomes clear (Fig. 4) that accumulation occurs when acquisition and material from recycling (flux 1 + 11) exceeds the requirement for growth, defense, and reserve formation (flux 2> flux 4, therefore flux 3 = flux 2 − 4). In a physiological context, accumulation is not purely a passive process, since it requires biosynthesis of new compounds. These pathways are activated only when the demand for growth, defense, and reserves decreases below the supply. Accumulated substances may be recycled and support growth at a later stage (flux 8), but plants may not be able to remobilize all accumulated substances at a later stage (such as starch in old rhizomes) for various physiologial reasons, and these may be lost as litter. Growth, defense, and reserve formation compete for the same resource (flux 5 + flux 6 + flux 7 = flux 4). Similar to accumulated substances, part of the reserves and defense compounds may be recycled and support growth at a later stage (flux 9

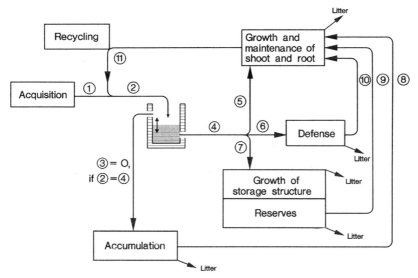

Figure 4 Interrelationships among pools (boxes) and fluxes (numbered arrows) associated with storage, growth, and litter formation. See text for explanation (after Chapin *et al.*, 1990).

and 10), while the rest is lost as litter. Recycling will also involve breakdown of components of growth to form a pool or recycled materials (flux 11) that either support growth again or enter into the compartment of accumulation.

In an ecological context all forms of storage should be evaluated in terms of alternative patterns of allocation. In this sense, growth is the formation of those components of biomass that themselves directly promote further acquisition and transport of resources, with the exception of growing structures for storage. Growth will include structure, biochemical machinery, and intermediate metabolic and transport pools. It excludes compounds whose major function is reserve formation, defense, or accumulation (starch or tannins). In this strict sense, growth is different from biomass accumulation. It will not always be possible (technically) to separate these fractions of biomass but, as we see later, the distinction may be important for correct interpretation of growth data. A relative growth rate could be calculated on a total biomass basis (including storage and potential litter) or on a structural biomass basis (minus reserves and accumulated compounds). In this case defense is regarded as a necessary side-product to support and maintain growth. In the latter case it would be interesting to try to quantify the importance of defense for growth, but we do not consider this aspect further in this chapter.

Chapin *et al.* (1990) have already pointed out some major problems with these definitions:

- The same compound may be formed in more than one compartment, e.g., assimilatory starch in leaves and reserve formation of starch in tubers may interact with growth in quite different ways.
- A given compound may serve more than one function. One interesting compound in this respect appears to be ribulose-1,5-bisphosphate-carboxylase-oxygenase (Rubisco), the essential photosynthetic protein. We will have to decide what level of Rubisco supports directly the acquisition of carbon and thus supports growth. Excess Rubisco might be regarded as nitrogen storage and may also have additional important functions, e.g., in avoiding photoinhibition under extreme conditions, such as light flecks, and thus protecting the whole production machinery (in this sense it is a process of defense; Schäfer, this volume, chapter 3); or in altering water use efficiency because the same net rate of carboxylation can be maintained at a higher CO_2, thus allowing stomatal conductance to be decreased with an ensuing reduction in water loss. In an ecological context the opportunity cost, i.e., the gain which may have been achieved, if the product was used in a different allocation scheme, is decreased, if the compound serves more than one process. We will see later in the chapter that this is the case if we store N in the form of Rubisco.
- The time scale will influence the interpretation. We distinguished accumulation from reserve formation, in that reserves compete with growth at the time the store is produced. However, logically, all stores can potentially promote growth in the long-term. For example, assimilatory starch accumulates during the daytime, but supports growth in a day/night cycle. Starch in a tuber accumulates in late summer but will support growth in the next spring, and thus contribute to growth over a 2-year cycle. Eventually all accessible storage compounds may be remobilized in a reproductive phase. Distinguishing between different storage compounds does not contain an evaluation with respect to an evolutionary selective value of one or the other. In fact recycling and utilization of storage compounds may be extremely important in an ever-changing competitive world, providing the opportunity cost of storage was low. This is the case for carbohydrate and organic nitrogen but is not the case, e.g., for water.
- Concentrations of storage compounds do not describe the pool size of these substances. The total pool size will also depend on the total mass of the organ. The pool size describes the potential store to support future growth, while the concentration of the pool may be im-

portant for and provide insight into the processes that control its formation.

• Formation of storage structures, such as tubers, may precede the actual storage process and at that time compete with growth of leaves or roots. The opportunity cost therefore arises during the production of this structure, and this process should be regarded in a similar way as reserve formation, although the structure will be lost as litter.

IV. Analysis of Whole-Plant Growth Using Mutants and Transgenic Plants

We have used two different kinds of genetic change to investigate the regulation of whole-plant growth. First, reduced-activity mutants and plants transformed with "antisense" - DNA can be used to specifically inhibit one pathway and investigate how this modifies storage, allocation, and growth in the whole plant. We describe how plants with decreased amounts of Rubisco or decreased amounts of starch-synthesizing enzymes have been used to investigate the significance of photosynthetic rate and starch storage in the leaf for whole-plant growth. This represents an extension of the experimental approach outlined by Stitt (this volume, chapter 2), which described how genetics can be used to analyze the distribution of control between the enzymes within a pathway. Second, it is possible to introduce a completely new (heterologous) enzyme into the plant. This enzyme may catalyze a process which normally does not occur, or it may perturb a preexisting process. We illustrate the use of heterologous gene expression by describing experiments with plants which have been transformed with yeast invertase, targeted to the cell wall to inhibit apoplastic phloem loading. These plants have allowed us to investigate whether accumulation of carbohydrate leads to inhibition of photosynthesis and what mechanisms are involved.

There is an important difference between experiments using isogenic plants with single gene changes and experiments in which inhibitors are used. First, the genetically manipulated plant has grown up from seed and the plant has had time to adjust its economy. The long-term effect and interactions are therefore observed, in contrast to inhibitors which reveal the immediate response including short-term dynamic effects. Second, provided the genetics has been carried out carefully, use of transgenic plants avoids problems associated with inhibitor specificity (or lack of it).

A. Analysis of the Growth of Transgenic Tobacco Plants with Decreased Expression of *rbcS*

Plants transformed with antisense *rbcS* have decreased expression of Rubisco. In chapter 2 (this volume) we described how these plants can be used to investigate the contribution of Rubisco to the control of photosynthetic rate. We now discuss how these plants allow us to (i) investigate the significance of Rubisco for nitrogen and water utilization in the plant, and (ii) analyze the interactions between photosynthetic rate, storage of carbohydrate and nitrate, biomass allocation, and whole-plant growth.

1. Rubisco and Nitrogen Use   Rubisco represents up to 40% of leaf protein (Woodrow and Berry, 1987). Accordingly, it might be expected that allocation of N to Rubisco will play an important role in the overall N utilization of a plant. Nevertheless wild-type tobacco leaves contain more Rubisco than they need, strictly speaking, to carry out photosynthesis under growth conditions. Often, 30 to 40% of the Rubisco could be removed before the rate of photosynthesis was affected (Figs. 5A, 5D, and 5G). From the point of view of photosynthesis this represents a "waste" and it will decrease the nitrogen-use efficiency of the plant (carbon gain per nitrogen).

However, at the whole-plant level further aspects need to be considered. First, this "excess" investment of Rubisco was observed in plants which had grown at a high-nitrogen supply anyway (Figs. 5A and 5B). Under these conditions nitrogen is stored in the plant anyway and allocation of N to Rubisco may provide a better opportunity cost than would accrue from accumulating NO_3 or a nonfunctional storage protein. For example, the excess Rubisco allows (i) the rate of photosynthesis to be increased in response to short-term increases in irradiance (see Stitt, this volume, chapter 2, for experimental data, and Schäfer, this volume, chapter 3), and (ii) the water-use efficiency to be increased (see below). Second, when tobacco is grown on limiting nitrogen this additional storage function is reduced or abolished, with the result that the amount of Rubisco in the wild type is now just high enough to balance the investment in other components of the photosynthetic apparatus (Fig. 5G), i.e., storage of N in Rubisco is decreased when N is limiting.

Allocation of N to Rubisco is clearly regulated. On the one hand, overinvestment is minimized when N is a scarce resource. On the other hand, the balance between Rubisco and other components of the photosynthetic machinery is adjusted in response to other changes in growth conditions, e.g., irradiance. Rubisco exerts a large limitation on the rate of photosynthesis if low-light-grown plants are suddenly transferred into high light, but it only exerts marginal control if the plants have grown at high light throughout their life. In high light a slightly larger propor-

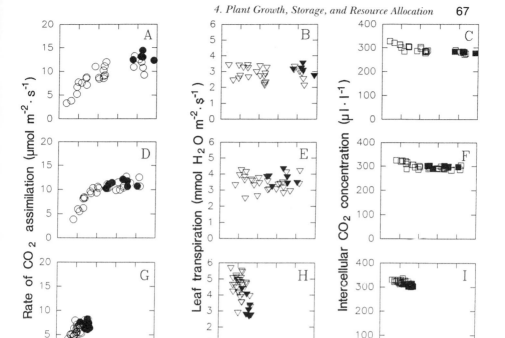

Rubisco activity (μmol m$^{-2}\cdot$s^{-1})

Figure 5 The effect of changing Rubisco activity on the rate of CO_2 assimilation in tobacco plants at three levels of N supply. Three groups of tobacco plants, each comprising several wild-type individuals (closed symbols) and a mixture of different *rbcS* anti-sense transformants (open symbols) were grown at high (5 mM NH$_4$MO$_3$, A–C), medium (0.7 mM NH$_4$NO$_3$, D–F), or low (0.1 NH$_4$NO$_3$, G–I) nitrogen supply at 20°C and 330 μmol m^{-2}s^{-1} irradiance for 6–8 weeks. The rate of photosynthesis and transpiration were then measured at ambient CO_3 and growth irradiance. (A,D,G) Photosynthesis; (B,E,H) transpiration rate; (C,F,I) estimated internal CO_2 concentration. Each point represents a single plant. The results are from Quick *et al.* (1993), Fichtner *et al.* (1993), and Fichtner (1991).

tion of the leaf N is allocated to Rubisco, and less to, e.g., thylakoid protein (see Schäfer, this volume, chapter 3; Stitt, this volume, chapter 2).

Although allocation of N to Rubisco is regulated, we should not be tempted to equate this with "optimization." For example, when wild-type plants grow in low light their overall performance may be impaired because of overinvestment of N into Rubisco. This became apparent when the rate of photosynthesis was measured in low irradiance; the rate of photosynthesis in the wild type was sometimes actually lower than that in antisense plants with three to fourfold less Rubisco. The leaves of antisense plants contain more chlorophyll and an increased chlorophyll b/chlorophyll a ratio, indicating that N has been reallocated toward thylakoid proteins, including the light-harvesting complex (Quick *et al.,* 1992). They also contain more thylakoid protein (Lauerer *et al.,* 1993).

2. Rubisco and Water-Use Efficiency   The uptake of CO_2 and loss of water are coupled processes which occur via the stomata. We might expect the relative rate of water loss to be affected by manipulating the enzyme which is immediately responsible for the fixation of the CO_2. For example, a decrease in Rubisco might be partially compensated for by increasing the internal CO_2 concentration in the leaf (C_i). This in turn, would decrease the rate of CO_2 diffusion into the leaf, because the diffusion gradient from outside is reduced. The rate of water loss, however, would not decrease; indeed, if stomatal conductances were increased to allow more CO_2 uptake and an even higher steady-state C_i, the transpiration rate would rise. Does this happen, and what are the consequences?

Gas-exchange measurements showed that the instantaneous water-use efficiency is indeed lower in plants with decreased amounts of Rubisco (i.e., less CO_2 is fixed per unit of water transpired: Fig. 6A; Quick *et al.,* 1991a; K. Fichtner, M. Lauerer, E.-D. Schulze, and M. Stitt, unpublished). These short-term measurements were confirmed by analysis of the ^{13}C/^{12}C ratio in the plant biomass; this ratio is decreased in antisense plants, showing that long-term water-use efficiency has been decreased (Fig. 6B; K. Fichtner, E.-D. Schulze, and M. Stitt, unpublished). Our gas-exchange measurements also showed that the decrease in water-use efficiency occurs because (a) as a consequence of the decreased Rubisco the plants indeed have a higher mesophyll internal CO_2 concentration (c_i; Figs. 5C,

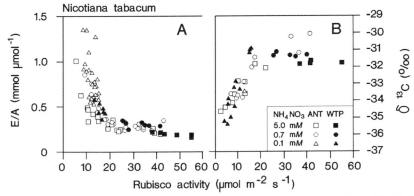

Figure 6   (A) Water-use efficiency (*E/A*) as related to Rubisco activity. (B) Carbon isotope ratio (δ-^{13}C) as related to Rubisco. The data are from the same set of plants presented in Fig. 8, WTP, wild type; ANT, antisense plants.

5F, and 5I) and (b) in some cases the stomatal conductance is increased in the antisense plants (Fig. 5H).

The poor water-use efficiency in antisense plants has consequences for plant water relations. On the one hand, we might expect the low-Rubisco plants to be more susceptible to water stress during periods of dryness. This still has to be tested. On the other hand, the high rate of transpiration means that increased amounts of salts will be carried up into the shoot. Elemental analysis has provided direct evidence that there is indeed a two- to threefold increase in salts in the antisense plants (K.Fichtner, E.-D Schulze, and M. Stitt, unpublished). As we will see, this has unexpected consequences for plant growth, because it leads to a decreased specific leaf weight and, thus, an increased leaf area.

3. Rubisco and Accumulation of Carbohydrate and Nitrate   Wild-type tobacco plants contain considerable pools of carbohydrate in their leaves. Only part of this stored carbohydrate turns over diurnally, indicating that the wild type is "sink" limited and is not able to utilize all of the carbohydrate which is being produced by photosynthesis (Quick *et al.*, 1991b). This accumulation is especially large when the plants are grown on low N (Fichtner *et al.*, 1993) or at a suboptimal temperature (Lauerer 1992; Lauerer *et al.*, 1993). In contrast, wild-type leaves contain only small amounts of NO_3, irrespective of whether the plants are growing with a low or a high N supply (Fichtner *et al.*, 1993).

As expected, antisense plants with less Rubisco and lower rates of photosynthesis contain less carbohydrate in their leaves (Figs. 7A–7C; Quick *et al.*, 1991b). Two aspects need to be distinguished in interpreting the functional significance of this decrease. The volume of the diurnal turnover was actually maintained in plants with two- to threefold lower rates of photosynthesis (Fig. 7A). This indicates that formation of assimilatory starch for subsequent remobilization during the night can be adjusted independent of the photosynthetic rate. It is presumably regulated in response to the requirement for growth during the night but more studies are needed to find out how the plant measures its requirement. The decreased starch content in leaves of antisense plants was primarily due to a large decrease in the amount of starch left at the end of the night. The carbohydrate left in the leaf at the end of the night might be seen as an accumulated pool, which is surplus to growth. Thus, the antisense plants adjust to a lower rate of photosynthesis by selectively cutting back on that portion of starch which represents wasteful accumulation of carbohydrate. It is also interesting that large pools of hexoses, which are found in the wild type at the end of the day, are absent in the antisense plants (Fig. 7C). These hexoses may act as a signal for excess carbohydrate (see below).

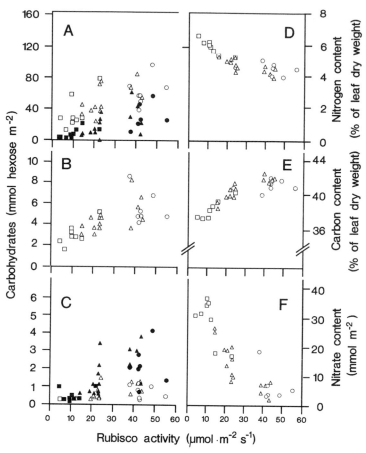

Figure 7 Impact of decreased Rubisco on leaf carbohydrate and nitrogen (A) starch content measured at the beginning (closed symbols) and end (open symbols) of the photoperiod. (B) Sucrose measured at the end of the photoperiod. (C) Glucose (closed symbols) and fructose (open symbols) as measured at the end of the photoperiod. (D) % Nitrogen, (E) % carbon, and (F) nitrate content at the left (after Quick *et al.*, 1991a). The plants were grown at 5 m*M* NH$_4$NO$_3$ at 330 μmol m^{-2}s^{-1} irradiance.

Decreased rates of photosynthesis and lower carbohydrate levels, in turn, act to modify N storage. Although amino acids rise slightly in plants with intermediate amounts of Rubisco (Fig. 7D), plants with low Rubisco contain rather low amino acid pools. Instead they accumulate strikingly large amounts NO$_3^-$, rising to 7% of the dry weight (Fig. 7F; Fichtner *et al.*, 1993; Quick *et al.*, 1991b). Nitrate reductase activity measured by the *in vitro* test decreases (K. Fichtner, G. Gebauer, and E.-D. Schulze, unpublished).

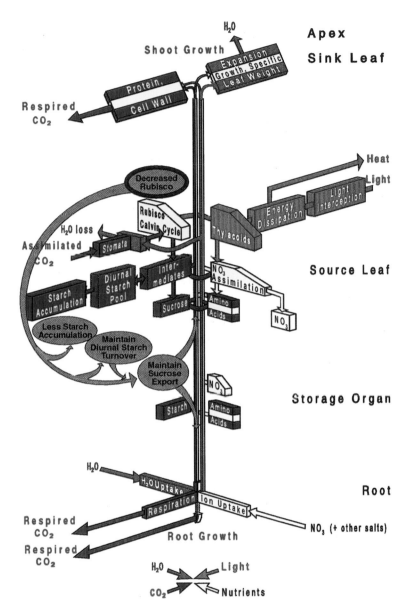

Figure 8 Influence of decreased Rubisco on accumulation of starch and export of carbohydrate. The initial perturbation causing the sequence of events is indicated by a thick black margin of the purple oval.

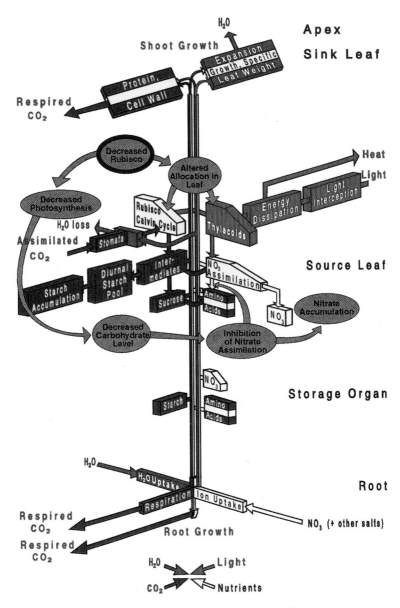

Figure 9 Influence of decreased Rubisco on the assimilation and accumulation of nitrate and on leaf expansion. The initial perturbation causing the sequence of events is indicated by a thick black margin of the purple oval.

Clearly, NO_3^- assimilation is regulated by the availability of carbohydrate. This conclusion is supported by the finding that NO_3^- assimilation is prevented in low CO_2 (Kaiser and Forster, 1989). The mechanism may involve protein phosphorylation, in the short-term (W. Kaiser, personal communication; H. W. Heldt and B. Riens, personal communication). Nitrate reductase is also highly regulated at the level of gene expression in response to light and NO_3^-, and more experiments are needed to investigate whether carbohydrate acts at this level. Although the signaling pathway and regulatory mechanisms require more research, their functional significance can already be surmised. When a plant has inadequate carbohydrate for structural (e.g., cell wall production) and energetic purposes, it would be counterproductive to continue incorporating this rare commodity into organic nitrogenous compounds. Indeed, depleting transport metabolites or the organic acids of the Krebs cycle could be fatal. Modulation of nitrogen assimilation by carbohydrate status probably plays a key role in allowing plants to grow under and adapt to varying irradiance and N fertilization regimes.

Figures 8 and 9 summarize how carbohydrate storage and N assimilation and storage are modified in response to a lower rate of photosynthesis. Less carbohydrate is "wastefully" accumulated, but the diurnal turnover of starch and the production and export of sucrose are maintained, as far as possible (Fig. 8). The decreased levels of photosynthate also lead to an inhibition of NO_3^- assimilation, and NO_3^- accumulates in the leaf (Fig. 9). As discussed in the above paragraph, a restriction of nitrate reduction will also increase the efficiency with which carbohydrate is being used in this special situation where carbohydrates are in short supply and nitrogen is available in excess. A second consequence of the decreased expression of Rubisco is a change in allocation of N between Rubisco and other proteins; this is a special case, because Rubisco represents such a large proportion of the leaf N, and would not arise if, for example, the stromal fructose-1,6-bisphosphate were to be decreased.

4. Photosynthetic Rate, Nitrogen Availability, and Biomass Allocation
Plants are thought to adapt to different growth conditions by altering the way they allocate their biomass. This allows them to invest their resources to improve acquisition of the limiting external resource. Investigation of antisense tobacco plants growing under different irradiance or nitrogen fertilization regimes has yielded the following picture of how biomass allocation is regulated in response to a changing rate of photosynthesis, or supply of N (Fig. 10).

It is well documented that the shoot/root ratio decreases when plants are grown in low N (see, e.g., Figs. 10C and 10D). This decrease is due to a preferential inhibition of shoot growth in low N. By using the antisense

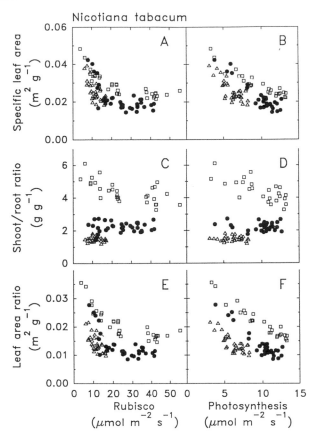

Figure 10 Specific leaf area (A, B), shoot/root ratio (C, D), and leaf area ratio (E, F) as related to changes in Rubisco and photosynthesis (after Fichtner *et al.*, 1993). The data are from the same set of plants as those shown in Fig. 8, but the symbols are changed, showing the groups of plants grown at 5 m*M* (open squares), 0.7 m*M* (closed circles), and 0.1 m*M* (open triangles) NH_4NO_3 (without differentiating between the wild-type and antisense transformants, this information can be obtained by comparison with Fig. 8).

plants, we were able to show that the shoot/root ratio decreases during growth in low N, irrespective of whether wild-type plants are compared to plants with similar amounts of Rubisco, or similar rates of photosynthesis (Fichtner *et al.*, 1993). The N supply itself, therefore, appears to be fundamental for the control of the shoot/root ratio. Chapter 5 discusses how hormonal signals interact in this regulation.

In contrast, the shoot/root ratio did not respond when the rate of photosynthesis was decreased by growing plants at low irradiance (Lauerer *et al.*, 1993) nor was it consistently altered in antisense plants with a reduced Rubisco content and rate of photosynthesis (Figs. 5C and external factors are also controlling plant growth. The two major internal factors which modulate the relation between photosynthetic rate and growth are changes in the storage of starch (Fig. 8) and changes in specific leaf area (see Figs. 10 and 11). Together, these two responses allow photosynthesis to be invested with increasing efficiency to produce new leaf area. They buffer whole-plant photosynthesis, and growth, to a remarkable extent against changes in the unit leaf area rate of photosynthesis.

B. Growth and Allocation in Starchless Mutants of *Arabidopsis thaliana*

We saw in the above section, that considerable amounts of carbohydrate are retained in mature leaves as starch. Starch formation could, in principle, represent (i) a diversion of resources away from export and growth, (ii) a contribution to growth because it buffers sucrose export against fluctuations in the environment (e.g., starch degradation could sustain sucrose export during the night), or (iii) an accumulation of surplus carbohydrate which cannot be utilized and therefore does not compete with growth. Starch accumulation might also be (iv) a mechanism to sequester carbohydrates from respiration. We have used mutants of *Arabidopsis thaliana* which have reduced or no leaf starch to take a closer look at these possibilities.

1. Significance of Leaf Starch for Plant Growth   The starting point for our experiments was the observation (Caspar *et al.,* 1986) that starchless mutants grow as fast as the wild type in continuous irradiance, more slowly in a light/dark regime, and died when the night exceeded 12–14 h. We grew wild-type and mutant plants in an intermediate light regime (14 h light/10 h dark) in saturating irradiance and N, in limiting irradiance and high N, or in limiting N at saturating irradiance. We questioned whether the daily growth increment was quantitatively related to the amount of starch moving through the leaf starch pool (i.e., being laid down during the day and remobilized at night) and whether the significance of the leaf starch for growth was influenced by the nitrogen supply (Fig. 13).

When the plants were growing with a saturating supply of N, there was a linear and near-stoichiometric correlation between the amount of carbon passing through the leaf starch pool each day and the increase in the relative growth rate (W. Schulze *et al.,* 1991). This can be interpreted as evidence that the diurnal turnover of starch contributes directly and efficiently to whole-plant growth in a day/night cycle. Presumably the rate of growth depends on the supply of carbohydrate to the growing regions, and growth is promoted when carbon is retained temporarily

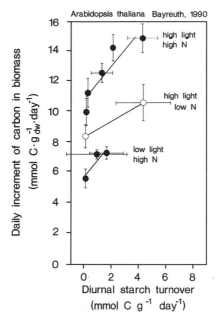

Figure 13 Comparison of the daily increment of carbon in nonstructural carbohydrates and the daily turnover of starch in wild-type *Arabidopsis* and decreased or null-starch mutants. The daily increment of carbon per unit dry weight per day is calculated from the RGR of biomass by multiplying by the fraction of carbon in dry weight (0.43) and dividing by the mmol mol wt of carbon. The daily turnover of starch is calculated after adjustment for the water content of the plants (0.2 g dry wt/g fresh wt). The variation in diurnal starch turnover was produced by using isogenic plants deficient in starch synthesis (after W. Schulze *et al.*, 1991).

in the leaves to allow export and growth to be maintained during the night as well as the day.

A different picture emerged when *Arabidopsis* was grown in low N (W. Schulze *et al.*, 1991); growth was generally decreased but the starchless mutants now grew almost as well as the wild type. Two factors were involved. First, half of the starch which was formed under these conditions accumulated in the leaves; it neither turned over nor contributed to growth (this resembles tobacco and many other wild species as discussed in this chapter). Second, the starch which was remobilized at night did not support an equivalent increase of the relative growth rate, i.e., due to N deficiency it was not efficiently used.

Evidently, the rate of growth in high N depends not only on the rate of photosynthesis (see above), but also on the orderly regulation of partitioning between export and starch formation. In contrast, plants growing in low N accumulate large amounts of excess carbohydrate, and it does not greatly matter if their assimilation rate is decreased or if they utilize their carbohydrate less effectively. Viewed in a different context, the accumulation of carbohydrate in wild-type plants in low N implies that tobacco and *Arabidopsis* do not respond "optimally" to low N; the presence of excess carbohydrate in the plants shows that they have over-invested in leaf formation and photosynthetic machinery, using N which might have been invested more productively elsewhere (e.g., in root growth to allow acquisition of more N).

2. Regulation of the Shoot/Root Ratio The experiments with starchless mutants of *Arabidopsis* provide some insights into how allocation to shoot and root growth is regulated. We found that these starchless mutants were still able to adjust their shoot/root ratio upward in response to low irradiance and downward in response to low N (W. Schulze *et al.*, 1991). It follows that partitioning between sucrose and starch cannot play an essential causal role in these adaptive responses, and that the correlation between leaf starch levels and alterations of the shoot/root ratio which have been reported previously (e.g., Geiger, 1980; Rufty *et al.*, 1984) are spurious. We have already seen that the shoot/root ratio was not affected by the rate of photosynthesis in antisense Rubisco plants. Taken together, these results show that shoot/root allocation is not directly modulated by plant carbohydrate status per se. The absence of a strong modulation of allocation by carbohydrate might explain why plants growing in low N tend to overinvest in shoot growth.

Until now we have assumed that increased allocation to root growth will actually increase N uptake. Can we use genetic manipulation to decrease the shoot/root ratio and test whether the changed allocation leads to improved competitiveness in low N, in the sense that more N is taken up? An unexpected aspect of the results with the starchless mutants provided us with an experimental system to approach this question (Fig. 14). Under a given set of growth conditions, starchless mutants always had a lower shoot/root ratio than the wild type (possibly because starch plays a more important role in shoot growth). Thus, when the starchless mutants were grown on low N, they actually produced larger roots than the wild type under similar conditions. Their leaves contained sixfold more N than the wild type, and since total biomass was only slightly reduced, N uptake per plant was increased almost sixfold compared to the wild type (this estimate omits the N in the roots). These results indicate that reallocation of biomass to root growth allows a large increase in N uptake and in the C/N balance in the plant. The conclusion is, however, preliminary, and more specific genetic changes will be needed to address this question in the future.

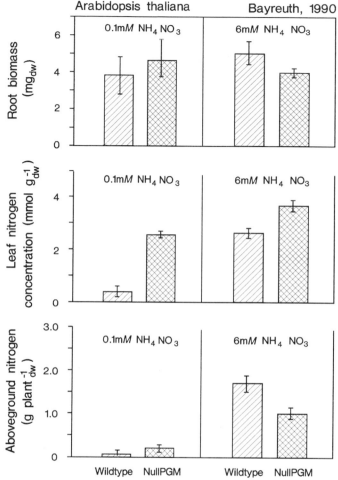

Figure 14 Root biomass (top), nitrogen concentration in leaves (middle), and whole-plant nitrogen (bottom) in *Arabidopsis* wild-type and starchless PGM mutant after 14 days of growth (after W. Schulze *et al.,* 1991).

C. Inhibition of Sucrose Export from Source Leaves in Tobacco Expressing Invertase from Yeast in Their Cell Wall

We have seen that large amounts of carbohydrate can accumulate in the leaves of plants due to an imbalance between the production and the use of photosynthate. Do plants respond to this accumulation of carbohydrate by decreasing the rate of photosynthesis; if so, how is this achieved, and what implications does this have for whole plant growth?

One possibility would be a "nonadaptive" response, in which accumulating carbohydrate feeds back to inhibit the operation of the preexisting photosynthetic machinery. For example, it has been proposed that large starch grains may disrupt the chloroplast (see references in Stitt, 1991). It has also frequently been proposed that sucrose synthesis is inhibited when soluble sugars accumulate and that higher pools of phosphorylated intermediates and decreased inorganic phosphate (P_i), in turn, lead to an inhibition of photosynthesis (for references, see Stitt, 1991). There is considerable evidence that this sequence of events operates to regulate partitioning between sucrose and starch during short-term fluctuations (see, e.g., Neuhaus *et al.*, 1989, 1990). The mechanism, which involves the deactivation of sucrose-phosphate synthase (SPS) by protein phosphorylation and the inactivation of the cytosolic fructose-1,6-bisphosphatase of a potent and specific inhibitory regulator metabolite called fructose-2,6-biphosphate (Fru2,6bisP), is summarized in Fig. 15. As a result of the lower rate of sucrose synthesis, less phosphate (P_i) is released in the cytosol and returned to the chloroplast. The resulting decline of ATP (not shown) and accumulation of glycerate-3-phosphate

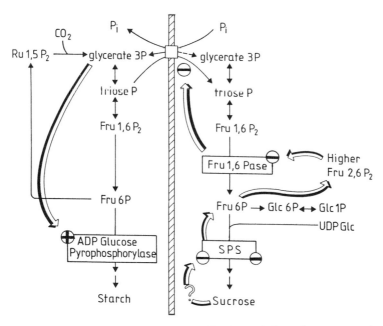

Figure 15 Scheme showing the direct feedback regulation of sucrose synthesis by accumulating sucrose in the leaf. The question mark indicates that the signal transduction path leading to deactivation (phosphorylation) of SPS has still not been elucidated.

leads to activation of ADP-glucose pyruphosphorytase and an increased rate of starch synthesis. However, there is little convincing evidence that this sequence of events leads to an inhibition of photosynthesis (see Stitt, 1991, for detailed consideration). Since phosphate is also released during starch synthesis in the chloroplast, the increased rate of starch synthesis will compensate for the inhibition of sucrose synthesis. As a result, the leaf is able to alter the partitioning of photoassimilate between export and starch accumulation without their necessarily or directly impinging on the overall rate of photosynthesis.

An alternative and very attractive possibility is that carbohydrate modulates photosynthesis via an "adaptive" process, by reducing the amount of photosynthetic machinery (rather than merely inhibiting the use of this machinery). Several experiments in the literature indicate that feedback regulation indeed changes the amount of photosynthetic proteins. For example, the amount of Rubisco protein changes in response to sink-source manipulations (von Caemmerer and Farquhar, 1984), and Rubisco often decreases when plants are exposed to enhanced CO_2 for several weeks (reviewed by Stitt, 1991). Functionally, this decrease would allow resources (e.g., N) to be invested elsewhere (recycled), to rebalance whole-plant growth (e.g., more root to allow more N uptake); for this reason it can be viewed as an adaptive response.

These possibilities have been tested in cooperative experiments with Willmitzer and co-workers (von Schaewer et al., 1990; Stitt et al., 1991). Phloem loading in the leaf is thought to involve movement of sucrose into the apoplast, followed by uptake via a proton cotransport mechanism into the phloem (see Komor, this volume, Chapter 6). To inhibit this process, Willmitzer and co-workers produced tobacco and Arabidopsis plants which expressed invertase from yeast in their cell wall (von Schaewen et al., 1990). In the transformed plants, sucrose moving through the cell wall will be hydrolyzed to glucose and fructose, which cannot be actively loaded into the phloem (Fig. 16). The young leaves of these plants appeared normal, as expected since they do not export much sucrose. As the leaves matured, a distinct visual phenotype (loss of chlorophyll) developed, first at the tip of the leaf and spreading gradually down the leaf, following the known pattern of the sink-source transition (von Schaewen et al., 1990). The visual phenotype was accompanied by a large accumulation of carbohydrate and a loss of Rubisco and other enzyme proteins which are required for photosynthesis. Measurement of metabolites showed that photosynthesis was being inhibited due to this decrease of the Calvin-cycle enzymes (Stitt et al., 1991). We have subsequently shown that this phenotype can be reproduced by feeding glucose into leaves via the transpiration stream (Krapp et al., 1991). In

Figure 16 Influence of yeast invertase on development and habit of transgenic tobacco. (A) Stunted growth of transgenic tobacco plants expressing yeast invertase in the cytosol (Cy-INV; left plant), the cell wall (Cw-INV; second plant from left), and the vacuole (V-INV; right plant) as compared with the growth of the wild type (tall plant). Leaf symptoms of mutants expressing cytosolic invertase (B), cell wall invertase (C), and vacuolar invertase (D) (after Sonnewald *et al.*, 1991).

all of these experiments, proteins involved in respiratory metabolism remained unaltered or increased after supplying glucose.

The decrease of photosynthetic proteins could be due to (a) decreased gene expression and/or (b) an increased rate of degradation of these proteins. We have recently started to investigate the potential mechanism(s) in more detail and already have evidence that carbohydrate regulates the expression of photosynthetic proteins. Measurements of steady-state mRNA levels have shown that *rbcS* and *cab* message (coding for the small nuclear-encoded subunit of Rubisco and the chlorophyll-binding protein, respectively) decrease in the invertase transformants and after glucose feeding (Krapp *et al.*, 1993). We have also observed a three- to fourfold decrease of *rbcS* message within 12 h, when phloem export was decreased by cold girdling of potato, spinach, or tobacco leaves (Krapp *et al.*, 1993). An analogous phenomenon is found in autotrophic cell suspension cultures of *Chenopodium rubrum;* after adding glucose there is a progressive inhibition of photosynthesis which is associated with loss of Rubisco and other photosynthetic proteins (Schäfer, this volume, Chapter 3). The loss of protein occurs gradually over several days but a decrease of *rbcS* message can be observed within 3 h after adding glucose (Krapp *et al.*, 1993). Independent support for the notion that high carbohydrate represses photosynthetic genes is provided by Sheen (1990), who found that the expression of reporter genes driven by promotors from several photosynthetic genes was inhibited by sugars in a maize protoplast transient expression system.

Taken together with the studies of antisense Rubisco tobacco, our results show that the supply of carbohydrate modulates the assimilation of resources in leaves. Low levels of carbohydrate inhibit N assimilation, and high levels of carbohydrate inhibit photosynthesis. This supports our initial supposition that allocation would have to be regulated by pool concentrations to allow stability. In this context, it is interesting that carbohydrate supply may also modulate gene expression in sink organs (see references in Farrar and Williams, 1991). Indeed, considerable progress has been made in analyzing the regulation of sucrose synthase expression by sucrose (Maas *et al.*, 1991). We have also seen that feeding sucrose to roots of tobacco seedlings leads to a 5- to 6-fold increase in sucrose synthase (M. Paul and M. Stitt, unpublished results) and that this enzyme decreases up to 20-fold after detaching potato tubers from the plant (P. Geigenberger, L. Merlo, and M. Stitt, unpublished results).

Studies of invertase transformants have also provided insights into species-dependent variation in the way in which plants respond to perturbations of their sink-source balance. Whereas tobacco expressing invertase responds with a large inhibition of photosynthesis, potato plants transformed with the same construction and similar invertase expression showed only a small inhibition of photosynthesis even though carbohydrates were accumulated (D. Heineke, H. W. Heldt, U. Sonnewald, and L. Willmitzer, personal communication). This contrast between tobacco and potato was even more striking in plants which had been transformed with pyrophosphatase from *Escherichia coli,* with the aim of disrupting growth processes which depend on pyrophosphate (Sonnewald *et al.,* 1991; Jellito *et al.,* 1992). Tobacco plants showed a large inhibition of growth, accompanied by accumulation of carbohydrate in the leaf and an inhibition of photosynthesis, in analogy to the phenotype in invertase transformants. Potato plants transformed with pyrophosphatase were shorter and also showed an accumulation of carbohydrate. However, photosynthesis was not so strongly inhibited: instead they developed many side shoots and initiated a large number of small tubers.

Tobacco is a plant which has been selected over the centuries for large leaf area and has few sinks and it is tempting to speculate that it has little ability to respond to an increased supply of carbohydrate. Potato has nondeterminate growth and has been selected to produce a large tuber mass. Apparently potato plants can initiate a large number of new sinks in response to increased availability of carbohydrate, or a shift in the sink/source balance brought about by inhibiting growth of the individual sinks. Significantly, it has long been known that high sucrose can promote tuber initiation in potatoes (for a comprehensive survey, see Burton, 1989), indicating that carbohydrate supply may not only be involved in modulating the growth of individual sinks, but may also act as a quasi-morphogenic factor to regulate the opening up of new sink and storage organs, as it was shown for root/shoot ratios in *Arabidopsis* (see above).

D. Summary of Experimental Results Using Genetically Manipulated Plants

If we combine the results with the *rbcS* antisense plants, starch-deficient mutants, and invertase-expressing plants we can arrive at several conclusions about how carbohydrate production and use modifies, and is modified by, plant-growth responses. In our experiments, the interaction with N plays an especially important role. The interactions are illustrated in Figs. 17 and 18 which depict how a changing carbohydrate supply alters growth and N utilization and how a changing N supply affects the production and use of carbohydrate. These interactions occur at the level of metabolism, gene expression, and allocation and whole-plant growth. They can be summarized as follows:

1. The interaction between resource acquisition and assimilation,

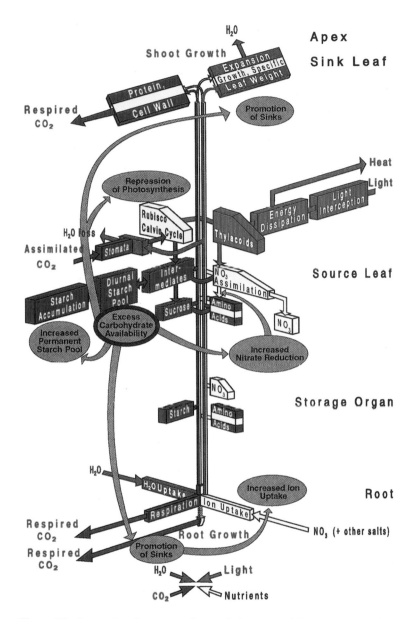

Figure 17 Interactions between carbon and nitrogen and their impact on plant growth. Effects of excess carbohydrate on storage, N uptake and assimilation, "sink" growth, and photosynthesis. The initial perturbation causing the sequence of events is indicated by a thick black margin of the purple oval.

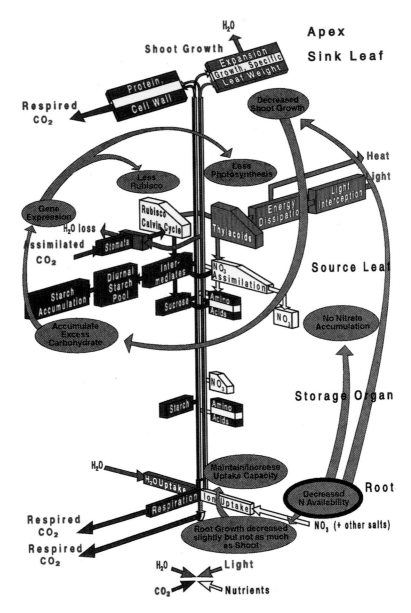

Figure 18 Interactions between carbon and nitrogen and their impact on plant growth. Impact of decreased nitrogen availability on N uptake and assimilation, carbohydrate storage, photosynthesis, and allocation. The initial perturbation causing the sequence of events is indicated by a thick black margin of the purple oval.

storage, biomass, allocation, and growth is highly integrated and is also extremely flexible.

2. In high N, the rate of growth depends on the rate of photosynthesis and on the efficient use of carbohydrate. In low N, the plants contain excess carbohydrate and growth is relatively insensitive to changes in the rate of photosynthesis and the use of carbohydrate.

3. Carbohydrate availability modulates the acquisition of further resources in several ways. When carbohydrates are in short supply, the assimilation of NO_3^- is inhibited, and allocation is changed to increase plant leaf area, via an alteration of the leaf geometry. When carbohydrates are in excess, photosynthesis genes are repressed, the assimilation of NO_3^- is increased, and sink development may be promoted (Fig. 17).

4. Carbohydrate storage is regulated to maintain a pool of starch whose diurnal turnover supports growth at night, when photosynthesis is not possible. This pool contributes directly to growth over a diurnal period. In addition, if carbohydrates are in excess, a considerable fraction of the carbohydrate in the plants can accumulate as starch in the leaves, without contributing to growth (Fig. 17). Both aspects of carbohydrate storage can be regulated independently of the photosynthetic rate per se.

5. Nitrogen availability also modulates resource acquisition. In particular, low N leads to a decreased shoot/root ratio. Preliminary results indicate these changes are a very effective way to improve N uptake by the plant (Fig. 18).

6. Provided the plant contains adequate carbohydrates, excess N may be stored in functional proteins like Rubisco. The opportunity cost may be better than when the excess N is stored as a nonfunctional protein or nitrate. This storage function of selected proteins, like nitrate accumulation, is abolished when N is limiting (Fig. 18). NO_3^- reduction also is decreased when carbohydrate is in short supply, excess N just being accumulated as NO_3^-.

7. Despite the intricate regulation of resource acquisition, storage, and allocation, it is clear that an optimum is not always achieved. For example, tobacco and *Arabidopsis* tend to overinvest in leaf formation in low N and to overinvest in Rubisco in low light.

8. Even from the limited number of species studied, it is apparent that there are important differences among species in the response to a growth imbalance.

V. Observations of Growth and Storage in Different Plant Life Forms

In the remainder of the chapter we want to investigate whether the conclusions based on transgenic plants and mutants can be generalized to a broader range of functional plant types. In the following section, we do not aim to present mechanisms. Rather, we discuss correlative evidence of growth and storage in a wider range of species. The functional types follow Monsi (1960) and Schulze (1982). The problems of classifying the plant kingdom into functional types were discussed in Schulze and Mooney (1992).

Schulze and Chapin (1987) already hypothesized that different plant life forms may regulate partitioning so that the internal pool sizes of C and N remain constant at changing resource supply (Fig. 19). When other resources were held constant, they observed that light limitation led to a slight drop in internal carbohydrate concentrations in *Urtica* and *Prunus*, which might be seen as a deviation from a "set point." The plants responded to this by producing more leaves of a reduced specific leaf weight and by reducing root growth, thereby bringing the ratio back to the set point. A similar response could operate under water or nutrient limitation; but in this case root growth would need to be stimulated and specific leaf area would increase to maintain an internal carbon-to-nutrient balance (see also Bloom *et al.,* 1985). This indicates that the information obtained from work with transgenic plants (see above) may indeed hold also when comparing species and life forms and that we can use the biochemical knowledge to better interpret responses of species in a range of habitats.

A. Annuals

Our starting point for the discussion of annuals is the work of Orians and Solbrig (1977) who proposed a cost–income model for plants growing in arid and semi-arid habitats. Their work predicted that (i) plants specialized to moist habitats with mesophylls will have higher rates of photosyn-

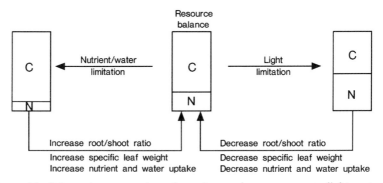

Figure 19 Schematic presentation of regulatory plant responses to light, water, and nutrient limitation which lead to a balanced resource supply within the plant (after Schulze and Chapin, 1987).

thesis when moisture is supplied than sclerophyllous plants specialized to dry conditions, but that (ii) plants from moist habitats will also lose performance in response to drought much faster than plants specialized to drought. Schulze and Chapin (1987) adopted this model for conditions of variable nutrition and generalized that plants from high-resource availability should have a higher performance than plants from low-resource availability under conditions of high supply, but will fail earlier under poor conditions. There was some evidence for this generalization, based on a comparative survey of a broad range of wild growing species representing different life forms with respect to their photosynthetic response at different leaf nitrogen concentrations (Schulze and Chapin, 1987). Since then the species comparison of Poorter (1990) has demonstrated that relative growth rates (RGR) increase with leaf area per plant weight (LAR) and with specific leaf area, supporting the idea of Orians and Solbrig (1977), at least for conditions of high nutrition. However the hypothesis had not been directly tested experimentally by growing different types of plants under variable conditions of nutrient supply.

Therefore, we ask (i) if conclusions drawn above from studies of transgenic plants can be extrapolated to a larger range of species, (ii) which parameters correlate significantly with RGR, (iii) what is the interaction of C and N storage with RGR, and (iv) what is the advantage of low-resource plants in low-resource habitats.

The growth rate of eight annuals with differing nutrient requirements (as indicated by the "nitrogen index"; Ellenberg, 1979) did not respond in the way suggested by Orians and Solbrig (1977) when they were grown under constant climatic conditions in a growth chamber at different ammonium-nitrate supplies (Fig. 20). RGRs differed between species at high supply, and this correlated with the nutrient requirement of these species in the field as expressed by the nitrogen index (Fichtner and Schulze, 1992), in that nitrophilic species had a higher RGR at high-resource supply. In contrast to earlier expectations, however, at limited N supply the low-resource plants did *not* perform better than the high-resource plants. Rather, all of the species showed a similar relative decrease in RGR when the N supply was decreased. In many cases, high-resource species still had a higher RGR than low-resource species at low N (see also Bradshaw *et al.*, 1964; Kuiper and Kuiper, 1979). In light of this result we first consider which conditions are required for maximum performance and which factors promote growth at high supply. We then consider which factors determine performance at limiting supply and try to identify which factors make low-resource plants more successful than high-resource plants in low-resource habitats.

Differences in growth rate of wild species were positively correlated with the LAR (leaf area per total biomass) independent of nutrition and

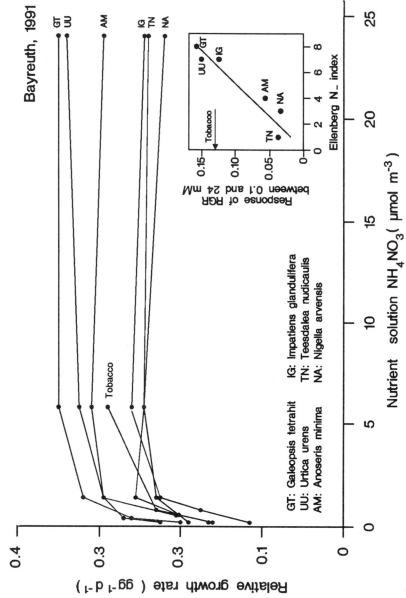

Figure 20 Relative growth rate of annuals originating from habitats of different resource availability as indicated by the Ellenberg N index as related to nitrogen supply in the growing medium.

plant species (Fichtner and Schulze, 1992; see also Lambers *et al.*, 1990; Poorter and Remkes, 1990). Also in tobacco, manipulation of Rubisco led to a change in LAR, which partly compensated for a lower rate of photosynthesis. Morphological aspects of biomass organization (structure) are obviously crucial for growth not only within a species but also when different species are compared. This will become even more important when dealing with perennial plant life forms.

In the experiment with wild annuals, specific leaf weight (SLW) had significant feedback on growth; but, opposite to the positive correlation between RGR and LAR, the correlation between RGR and SLW was negative. Similar correlations between RGR and leaf area expansion rate are known from experiments at constant nutrition but variable temperatures (Potter and Jones, 1977). Plants which expose a large leaf area of thin leaves for photosynthesis exhibited the highest RGR even though thick leaves may have higher photosynthesis per area, e.g., sun vs shade leaves. This again confirms the experiments with antisense *rbcS* plants at the level of a species comparison; tobacco plants also compensated for lower photosynthic rate by decreasing SLW, in this case due to accumulation of ions and their osmotic effect on leaf expansion.

In contrast to these morphological parameters, the correlation between RGR and net assimilation rate (NAR, growth per leaf area, which would be proportional to photosynthesis) was weak between different species (see also Poorter and Remkes, 1990). This resembles the conclusions reached from analysis of antisense *rbcS* plants, where photosynthetic rate had far less effect on growth rate than might have been expected.

At high nutrition fast-growing annual species had a higher shoot/root fresh weight ratio than slow-growing species, i.e., their success in growing fast was mainly related to exposing a larger leaf area. In addition, fast-growing annuals had fewer root tips per root biomass than slow-growing species, i.e., few big roots seem to be more efficient than many small root tips per root weight as observed in slow-growing species (Fichtner, 1991). Poorter and Remkes (1990) observed in other annual species that the highest growth rates were achieved in plants having a high leaf area per root length. This shows that not just root weight, which is often used as the only measure, but also the structure of root is important for the efficiency of nutrient uptake (see Steudle, this volume, Chapter 8) and may influence growth just as much as the photosynthetic efficiency of leaves. At least in sand culture at constant nutrient supply a thick short root (high mass per weight) seems to have a positive effect on growth. Mathematical models predicted that biomass investments in thick roots is inefficient for nutrient uptake (Boot, 1990), but such models do not take into account the effect of root structure on the efficiency with which nutrients are resorbed (see also Steudle, this volume, Chapter 8).

Maximum rates of growth result in minimal concentrations of nitrate and starch in leaves (Fig. 21), which seems to confirm the view that plants regulate partitioning so that the internal pool sizes and concentrations of C and N tend to remain constant over a broad range of conditions (see Fig. 19). Wild annuals reach maximum growth rates at similar nitrate levels as those found in tobacco (0.5 to <2.0%). However, differences exist between wild and cultivated species with respect to the level of starch which accompanies maximum growth. Starch contents at midday were lower in wild species (3 to 8%) than in tobacco (10 to 15%). The experiments with *Arabidopsis* demonstrated (see Fig. 13) that accumulation of starch during the day is necessary in order to support growth during the night hours. Complementary to our earlier interpretations, the higher starch levels in tobacco result from slower growth. In fact, the wild species actually had a higher RGR than tobacco (up to 0.35 g $g^{-1}d^{-1}$, compared to 0.30 g $g^{-1}d^{-1}$). Thus, starch in tobacco may be in part an accumulation product even when growing at high N (see also above). Only the day/night starch turnover and not the average starch level supports growth

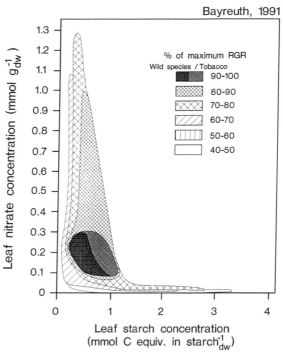

Figure 21 Relative growth rate (expressed as % of maximum) as related to leaf nitrate and starch concentrations of different wild annuals and of tobacco.

(see Fig. 8 above). It is tempting to speculate that the breeding for large leaves may have impaired the sink-source balance in the crop plant, making it inherently less effective than wild species.

Nitrate also undergoes a day/night cycle (Gebauer *et al.*, 1984), and one may hypothesize that, in analogy to starch, nitrate levels of 0.5 to 2% of dry weight are required to support nitrate reduction and growth in a day/night cycle. In contrast to the daytime accumulation of starch, nitrate reaches highest levels at night and is assimilated during the day (Schulze *et al.*, 1985). Since the plants in Fig. 21 were harvested at midday, the nitrate levels represent almost the daily minimum. An indication that nitrate levels are indeed regulated to support growth in a day/night cycle emerges from experiments using *Raphanus* (Fig. 22), in which a nitrate content of 1.5% dry wt would be sufficient to support growth for just 1 day at a RGR of 0.20 g $g^{-1}d^{-1}$. If nitrate supply were to be suddenly cut off, RGR would clearly decrease, as soon as the nitrate is depleted (Koch *et al.*, 1988). Therefore, there is a positive feedback of accumulated nitrate on

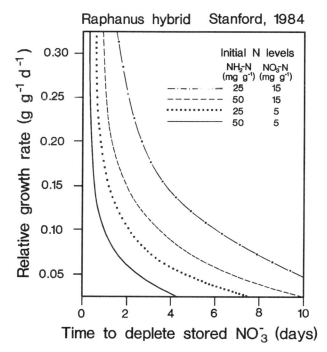

Figure 22   Time to deplete stored nitrate from different starting points of stored nitrogen and the change in relative growth rate (after Koch *et al.*, 1988).

growth in a day/night cycle or during short-term resource fluctuations. Even on a seasonal basis nitrate accumulation in rosette leaves of annuals may promote growth at a later stage of the growing season as nitrogen availability declines (Koch, 1988).

We repeatedly found in wild species and in genetically manipulated plants (see above) that if growth is reduced, either by shortage of nitrogen or by shortage of carbohydrates, the other substance accumulates beyond the level which is required to maintain growth on a daily basis (Fig. 21). When carbohydrate levels were reduced below 10 to 15% of dry weight, RGR decreased and nitrate accumulated up to 8% of dry weight. In contrast, when nitrogen supply (ammonium-nitrate) was reduced, starch accumulated up to levels of 55% of dry weight in all species. There is a very strong effect of reduced nitrogen supply on RGR below levels of 0.5% of nitrate in leaves. The effect of reduced nitrogen supply on growth appears to be stronger than the effect of reduced starch. The experiments based on a comparison of different wild species and the earlier conclusion based on transgenic plants indicate that nitrogen shortage has a stronger effect on growth than carbon shortage. This may be in part so, because the accumulated nitrate has an osmotic and compensating effect on specific leaf weight, allowing a compensatory gain in whole-plant photosynthesis, whereas (see above) plants tend to undercompensate in response to high carbohydrate and low N.

Starch and nitrate in leaves represent accumulation products which may support growth at a later stage, e.g., in a day/night or life cycle. One could in fact conceive that plants "oscillate" between situations where one or the other product accumulate, and the highest rates of growth are achieved only where long-term accumulation is at a minimum (see Fig. 21). The accumulation of products is in fact an indication that plants are not regulated in a way to optimize resource acquisition. Not only tobacco and *Arabidopsis,* but also many wild species invest too much nitrogen into photosynthetic capacity with the result that they have excess C fixation and accumulate starch under conditions when nitrogen is limiting. At this stage investment into more root would have led to much better performance at the whole-plant level.

Plants also lack optimal regulation under conditions of high N supply. They store nitrate, when investment into more leaf would alleviate the carbon shortage. We can only hypothesize that during evolution plants are adapted to short-term fluctuation which would require a buffering capacity of accumulated substances. All experiments were performed under constant conditions, which do not exist in nature, and therefore the plant does not regulate in an optimal way to overproduction under certain conditions. In nature, conditions will change so rapidly that an accumulation of nitrate or starch may be needed the next moment or day.

In conclusion we learn from annuals, with respect to "flux control" of growing structural biomass, that there is a stronger feedback of long-term investments of structure (leaf area ratio and specific leaf weight) than of short-term performance (net assimilation rate) on growth. Short-term metabolism is tightly regulated, but investment into growth contains a developmental component and cannot be so readily reversed or corrected.

Therefore, growth rate of different annual species is not strongly dependent on photosynthetic rate, while changes in structural parameters strongly influence growth. Structural parameters include also features such as the orientation of foliage in space, which is not included in the mass balance of carbon and nitrogen. The experiments with wild species also suggest that growth reaches a maximum in a dynamic process where accumulations of carbon or nitrogen storage compounds reach a minimum. The achievement of this balance appears not to be tightly regulated, but rather appears to reflect the average conditions in a naturally highly variable environment.

There are many similarities to the conclusions reached from the studies of transgenic plants, in particular the importance of morphological factors, and the occurrence and importance of mechanisms to regulate the accumulation of carbohydrate and nitrate. The maintenance of balanced pools of carbohydrate and nitrate during rapid growth can, at least in part, be understood in terms of the modulation of nitrate reduction and photosynthetic gene expression by carbohydrate and N metabolism.

We now turn to the problem of why growth of high and low resource plants at low nutrition does not follow the Orians and Solbrig model (1977). The success of nitrophilous species at low external nutrient supply can be partly explained by their generally higher seed weight (Fichtner and Schulze, 1992). However, the main difference between nitrophilous and nonnitrophilous species at low-resource supply is not RGR but, rather, the ability of low-resource species to flower and complete their full life cycle at a lower biomass and a smaller whole-plant nutrient pool than the high-resource plants. In our experiments all of the species reached a similar biomass at low-nutrient supply, but only those species originating from low-resource habitats started to flower at that low biomass. Apparently, high-resource plants require a larger pool size before flower initiation can take place. This indicates that, in addition to the strategies of resource utilization during vegetative growth, total pool size requirement for flowering may be an even more important parameter separating species in natural habitats.

B. Biennials

The 2-year life cycle of biennial species combines features of annuals and perennials, in that they store resources in the first year and remobilize them early in the second growing season and during seed filling after which they die. Therefore, biennials are an interesting model system for studying interactions of growth, storage, and reproduction (Heilmeier *et al.*, 1986; Schulze and Chapin, 1987). The central questions in the context of resource allocation are (i) does storage compete with other growth processes during the first year, (ii) what is the relative contribution of internal reallocation and cycling of substances to storage, (iii) to what extent does storage contribute to growth and seed filling at a later stage of the life cycle, (vi) what are the differences between biennial species originating from habitats of high- and low-resource availability, and (v) given its importance in annuals, how does plant structure interfere with resource gain and resource use in biennials.

Biennials exhibit large variations in their strategies for storing carbohydrates and nutrients. *Arctium tomentosum*, which is a ruderal species, harbors nitrogen and soluble carbohydrates in its storage roots at the end of the first growing season in higher concentrations than those of *Digitalis purpurea*, *Verb ascum* ssp. or *Dipsacus sylvester*, which grow in low-resource habitats. *Echium vulgare* exhibits high concentrations of carbohydrates but low levels of nitrogen, in contrast to the high levels of nitrogen and low carbohydrate concentrations found in *Melilotus officinale* (Steinlein *et al.*, 1993).

In the following we compare interactions of growth and storage in some biennial species which grow in habitats differing in nitrogen availability as indicated by the Ellenberg nitrogen index. These species are *A. tomentosum* (N index = 9), *Cirsium vulgare* (N index = 8), *D. sylvester* (N index = 6), and *Daucus carota* (N index = 4). Storage and growth of these species was investigated in an experiment in which nutrient and light availability was altered. The seasonal course of total biomass shows large differences among species (Fig. 23); in the second year *Arctium* reached four times the biomass of *D. sylvester*. This difference was maintained, irrespective of whether the plant grew under high- or low-resource supply. This resembles the conclusion drawn from the species comparison of annuals and again contradicts the original hypothesis of Orians and Solbrig (1977) of a superior performance of low-resource species at low supply. Another similarity to the response of low-resource annuals is found in the flowering response of *D. carota* (see below). This species initiated flowering in the first season at high-resource supply, but remained smaller by factor 6 or 10 than *Arctium* at low-nutrient supply. Early flowering contains the advantage of increasing the population size which will be of significance in disturbed or patchy habitats. The response is quite similar to that of low-resource annuals.

In all species the storage roots attained their maximum biomass at the end of the first season, but in *Arctium* the total weight of the storage

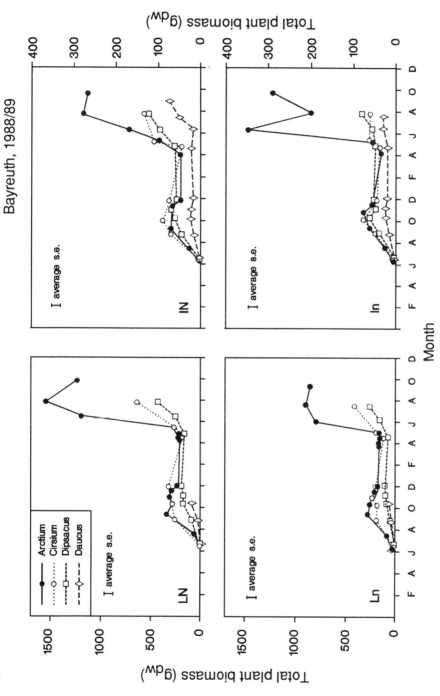

Figure 23 Seasonal change in biomass of biennial species at different light and nutrient regimes. LN, high light, low nutrition; LN, high-light and -nutrient supply; Ln, high light, low nutrition; lN, high nutrition, low light; ln: low light and nutrition (after Steinlein *et al.*, 1993).

organ was much higher than that for the other species, again irrespective of whether they grew under high- or low-nutrient supply. However, the seasonal course of total biomass does not give insight into the storage process, which actually contains two separate steps, namely growth of the structural biomass of the storage organ and filling of the storage tissue with storage products. In *Arctium* (Fig. 24) storage of carbohydrates appear to occur simultaneously with the development of the structural biomass of the storage root, whereas storage of N starts later in the season than storage of C. This is in contrast to *Cirsium* and *Daucus* where growth of structural biomass for storage may precede the actual storage process for both C and N.

With respect to the regulation of storage and growth we want to know whether growth and filling of the storage organ is an independent carbon sink which competes for resources with the development of root and shoot (i.e., a true reserve formation) or if growth and filling of the storage organ represents a deposition of surplus material. Opposite to our expectation (Figs. 25A and 25B), leaf growth precedes the development of the store in *Arctium*, which appears not to compete with growth. When leaf

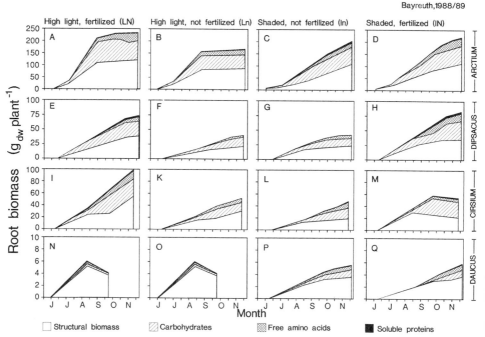

Figure 24 Seasonal change in structural biomass of the storage organ and of storage products in *Arctium tomentosum* (A to D), *Dipsacus sylvester* (E to H), *Cirsium arvense* (I to M), and *Daucus carota* (N to Q) at high and low light and nutrition (after Steinlein *et al.*, 1993).

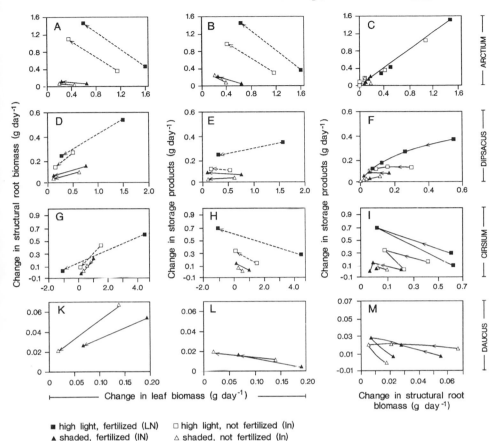

- ■ high light, fertilized (LN) □ high light, not fertilized (ln)
- ▲ shaded, fertilized (lN) △ shaded, not fertilized (ln)

Figure 25 Change in structural biomass (A, D, G, K) and of storage products (B, E, H, L) as related to changes in leaf biomass, and the relation between change in storage products and structural root biomass (C, F, I, M) of *Arctium tomentosum* (A to C), *Dipsacus sylvester* (D to F), *Cirsium vulgare* (G to I), and *Daucus carota* (K to M). The arrows indicate the change of the relations from early in the season (bottom end of arrow) to late in the season (arrowhead) for different treatments of light and fertilizer (after Steinlein *et al.*, 1993).

growth is high, growth of storage organs and storage is low. Only when leaf growth decreases does development of storage organs and accumulation of storage products increase. After leaves have developed, there is sufficient surplus carbohydrate for *Arctium* to be able to simultaneously produce the storage structures and fill the structures with storage products at the same time (Fig. 25C). This response of *Arctium* is quite different from that of low-resource plants. In *Dipsacus*, the storage structure develops simultaneously and at a constant proportion to leaf growth (Fig.

25F), and filling of the structure also occurs simultaneously with structural growth (Figs. 25D and 25E). In this case, both the laying down and the filling of the storage organ competes with growth of leaves. In *Daucus* and *Cirsium*, growth of the storage organ competes with leaf development and may be considered as part of a reserve formation which is lost as litter later on, whereas filling of the storage organ does not commence until leaf development has ceased (Figs. 25H, 25I, 25L, and 25M).

In the review of Chapin *et al.* (1990) the fact that C and N storage in *Arctium* occurred simultaneously with the growth of structural biomass was taken as evidence that it represents a process of reserve formation and competes with growth. Also the fact that the relation between storage compounds and storage tissue was independent of light or nutrient supply appeared to indicate that reserve formation was occurring. However, the relation between leaf and storage growth was not closely enough investigated at that time. The present analysis indicates that the high-resource plant *Arctium* is the only species in which growth of storage tissue and filling of storage organs are separated from leaf development. The storage process is temporarily sequestered from leaf and root growth and uses excess C and N for storage, the amount of excess C and N depending on resource availability. Once storage tissue is initiated, this tissue is filled before new storage cells are developed. At low-resource supply tissue growth is slow but filling reaches the same relative level, with the result that the concentrations of carbohydrates and N are independent of the treatment. Thus storage in *Arctium* is an accumulation of excess resources (in the sense of Chapin *et al.*, 1990), but at the same time it is a tightly regulated process with respect to growth of storage tissue and its filling with storage products.

In contrast, lower resource plants, like *Cirsium*, *Dipsacus*, and *Daucus*, seem to simultaneously maintain separate and independent sinks for growth of leaves and storage tissue. In *Dipsacus* filling also competes with leaf development. *Cirsium* and *Daucus* lay down structural components in competition with leaf growth, but start the filling process only when excess resources are available. It is difficult to determine whether the scheduling of growth events in *Arctium* is a more "advanced" trait which developed during competition with other species in high-resource environments or if storage is so essential under low-resource conditions that plants take the high opportunity cost of growing storage tissue in competition with leaf growth. These plants would probably be able to attain a higher C and N gain if investment were restricted to leaves and roots initially and if only later investments were diverted to storage structures. Thus from an economic point of view the response of *Cirsium* and *Daucus* is not optimal. We speculated that this carbon sink is a "left-over" from their evolution from woody ancesters (Steinlein *et al.*, 1993; Meusel, 1952;

Meusel and Kästner, 1990). It could be viewed as a remnant structure which only becomes functionally significant as a storage organ at a later stage of their life cycle. In either case, *Arctium* has a growth advantage under all conditions and the low-resource species appear to have no chance of competing in high-resource habitat.

Storage of carbohydrate precedes storage of nitrogen in all species even under conditions of high N availability (Fig. 24). This may indicate that the opportunity cost of storing N as amino acid in a tuber is very high and that plants regulate the flux of N in a way that incorporates N into Rubisco and produces excess C rather than storing amino acids during the growing season. Referring back to the experiments with annuals, the excess accumulation of carbohydrates would indicate that N is limiting even in the high-nutrient treatment. Additional N supply, however, would probably result in self-shading of the rosette and reduce growth as it was observed in *Plantago major* (Poorter *et al.,* 1988). N storage products accumulate only at the end of the growing season at the time of leaf senescence.

Substances which are accumulated at one time may promote growth at a later stage of the biennial life cycle. Thus the opportunity cost "deposited" during growth of structural biomass in the first year represents a "reward" in the second year. In accordance with this view, generally only a small amount of the stored carbohydrates but a large quantity of stored N are used in the second season (Heilmeier *et al.,* 1986). Plants from high-resource habitats make less use of stored substances than plants from low-resource habitats (Steinlein *et al.,* 1993), i.e., *Arctium* and *Cirsium* used 10 to 40% of the winter N store in the second season whereas *Dipsacus* and *Daucus* used 80%.

Large species-specific differences exist in the biomass development of the second season because of various feedbacks of storage on growth and because of differences in growth form. *Arctium tomentosum* develops very large leaves in the rosette stage of the second season which cover the surrounding vegetation very effectively. Because of its growth form *Arctium* is an effective competitor even in a ruderal vegetation of very aggressive grasses and herbs such as *Agropyrum* or *Urtica*. The effect of the development of a large and dense rosette is that the mineralized nitrogen of the area covered by *Arctium* leaves (about 1 m^2 ground area) is available only to this individual. The excess availability of nitrogen per plant leads to an increase of the total nitrogen pool by a factor of 3 beyond the levels of the first season (Heilmeier *et al.,* 1986). *Arctium* reaches a higher maximum biomass than the other biennial species, which do not produce a leaf rosette of similar size and density. The compound leaves of *Daucus* are the least effective in competing with other species. However, in contrast to *Arctium*, the species from low-resource habitats have the ability to switch within the first season from a biennial mode of growth into an annual mode of flowering and thus promote reproduction rather than storage, depending on resource availability (T. Steinlein, H. Heilmeier, and E.-D. Schulze, unpublished).

During the course of the second season, nitrogen is reallocated from one generation of leaves to the next younger one, into the flower stalk, and eventually into seeds. At the end of the life cycle, within about 2 weeks the plant remobilizes its resources, fills the seeds, and dies. During that period in *Arctium, Dipsacus,* and *Daucus,* about 20 to 27% of the total plant nitrogen was transported into the seeds at high nutrition and high light. This proportion decreased to 11% in shade and at low-nitrogen supply. Only in *Cirsium* was the proportion of nitrogen allocated into seed low (about 1%). Thus storage has an effect on reproductive success, but there was no obvious difference between high- and low-resource plants with respect to the nitrogen harvest index.

It is not only the total N in seed which is important for the population biology of these species. Seed number is one major determinant of reproductive success besides seed weight, and it was mentioned already that large differences exist in this respect among annual species and that seed weight explained differences in RGR. The high-resource plant *Arctium* developed seed which are about 5 times heavier than seeds of the other species, but seed number was higher in *Dipsacus* and *Daucus* (representing low-fertility habitats) than in *Cirsium* or *Arctium* (growing on sites of high N availability). Thus, in contrast to the commonly held view, the low-resource plants follow the r-strategy of reproduction, while the high-resource plants are K-strategists. This becomes even more obvious if we ask the ecologically more important question: if the biennial mode of reproduction is more successful, in terms of seed number, than an annual mode, does the accumulation of storage products and the evolution of storage tissues have a positive effect on the maintenance of the population? This can be tested using the facultative biennial species *Daucus* and *Cirsium* as an example. Under high-light conditions all *Daucus* plants flowered in the first year and produced three to six times more seeds than those in the shade, where the same species produced about twice as many seeds in the biennial mode as in the annual mode. Nevertheless, in terms of increasing population size, *Daucus* would do better in the annual mode. Thus, the low-resource biennial is r-selected, but uses the rapid reproduction capability if it experiences high-resource availability (T. Steinlein, H. Heilmeier, and E.-D. Schulze, unpublished).

In addition to seed numbers, population biology is also dependent on seedling establishment. In this respect, the structure of *Arctium* becomes important again. The large area which was covered by the rosette leaves allows seedlings of *Arctium* to become established while competition of other species is still low because they were outshaded in the previous season. In contrast to *Arctium* the other species cannot rely on the competitive structure of the parent plant, and a high seed number is needed to facilitate establishment on naturally disturbed sites, such as gopher mounds.

In terms of whole-plant regulation of carbon and nitrogen fluxes we may summarize the results obtained from biennial plants. (i) In *Arctium,* growth and filling of the storage organ only follow leaf formation. In this species, storage appears to represent accumulation even under conditions of low supply. (ii) Growth of the storage organ competes in most species with growth of leaves (equivalent to reserve formation). The filling of the storage tissue is an accumulation process (except in *Dipsacus*), as the storage products from reallocation and uptake build up at a time when seasonal growth has terminated. (iii) Storage of N in leaves (as Rubisco) dominates during the growing season and N products do not accumulate in the storage organ until after termination of leaf growth. (iv) The accumulated N substances promote growth in the second season. Thus, the negative feedback of partitioning resources for structural growth in the first season has a promoting (feedforward) effect on seed set in the second season. (v) Structural aspects of the rosette and seed size and numbers affect the competitive ability of the species. This has consequences in high-resource habitats, in which the fixed biennial mode of *Arctium* and its production of a small number of heavy seeds is an advantage in competing with other species. In low-resource habitats competition for light is not the primary constraint. Therefore biennials of low-resource habitats may switch between annual and biennial modes of reproduction and produce many but light seeds. The annual system would be more effective if viewed in terms of the rate of population increase and is indeed used if resources are available. Thus we have several reproductive strategies in the low-resource species—some (obligate annuals) flower at low-pool size and others are facultative biennial using storage and a second growth season as an option to survive if poor resources are encountered.

C. Perennial Herbaceous Species

Growth and storage in perennial herbaceous species follows similar general patterns as those in annuals. In a comparison of annual and perennial species including woody plants, Poorter (1990) concluded that leaf area per plant weight was the most important factor in explaining inherent differences in RGR between species, while net assimilation rate (growth increment per leaf area and time) was only of secondary importance (Hirose, 1988). In an additional comparison of 24 perennial herbaceous European species Poorter and Bergkotte (1992) demonstrated that the carbon content of leaves decreased while organic nitrogen compounds, organic acids, and mineral and water content increased with species-specific increases in RGR. In a productive environment a high leaf biomass and a high specific leaf area will be of paramount importance for survival, because these traits increase the competitive capability for light (Schulze and Chapin, 1987). This contrasts with low-resource environments in which allocation to roots and additional characteristics related to herbivore defense and leaf longevity result in low specific leaf weight but promote survival (Chapin and Shaver, 1985). These structural characteristics explain variations in RGR of perennial European grasses over a range of nutrient supply (Bradshaw *et al.*, 1964). High-resource plants maintain higher biomass than low-resource plants under a whole range of conditions. This implies that it is not the potential RGR of a species that determines its success in a competitive situation, but rather factors determining specific leaf area and allocation for leaf growth (Poorter, 1989), as it was demonstrated already for biennials and annuals (Chapter 5, this volume; Heilmeier *et al.*, 1993; Steinlein *et al.*, 1993). Structural features have very strong feedback not only on RGR but also on the competitive ability of plants, and this latter interaction becomes even more important when dealing with woody species.

In the context of storage, growth, and competition we ask (i) is the storage process of perennial herbaceous species different from what we learned in annuals, (ii) how and when does storage promote growth and what is the function of a permanent storage organ, and (iii) how do structural components contribute to the success of perennial herbaceous species in stable but extreme environments (Schulze, 1982).

Growth analysis of the bulk biomass reveals general relations between biomass partitioning and growth (Poorter, 1990), but it does not provide insights into internal processes, longevity of plant organs, and total plant reallocation, which are the basis for the success of perennials in stable environments (Schulze, 1982; Millard, 1988; Chapin *et al.*, 1990). Structural biomass may only be a small proportion of total biomass and stored or accumulated substances may or may not support growth at a later stage of development, depending on the resource supply. However, this reallocation of stored substances is a prerequisite for the success of perennial species in a competitive situation. It may be hypothesized that, as in annuals, in perennial herbaceous species storage also indicates an imbalance between supply and demand, but the opportunity cost for storage is lower than that in annuals since the storage organ already exists as a permanent buffer to temporarily accept the spillover of surplus.

In the following we investigate storage processes, taking *Urtica dioica* as an example, because this species inhabits a large range of habitats of different resource availability (Reif *et al.*, 1985; Beck, this volume, Chapter 5). We investigate whole-plant stands at a steady state of performance (rather than the seedling stage as was done in most species comparisons; Poorter and Bergkotte, 1992), because only then the "evolutionary advantage" of permanent storage organs and of interactions between storage and growth can be demonstrated. In this case the results are based on observations in different habitats, namely a site at high nutrition and full sunlight, a small forest clearing at a river bank, and a peat site under a dense forest canopy.

The perennial part of *U. dioica* is a rhizome, which contains high concentrations of organic nitrogen in winter (mainly arginine and asparagine; Rosnitschek-Schimmel, 1985a,b). The concentration of organic nitrogen in spring is surprisingly independent of the habitat conditions. In 1-year-old rhizomes the organic nitrogen ranges from 3.9 mmoleq N g^{-1} dry wt in sun-exposed habitats to 3 mmoleq N g^{-1} dry wt at shady sites. The organic nitrogen of the rhizome decreases rapidly in spring (Fig. 26), the current- and 1-year-old parts being the main storage organ. The nitrate concentrations are also high before onset of spring growth and do not decrease until after leaf development. In fact, the nitrate reductase capacity of leaves in *Urtica* is very high, and this may be a physiological trait which differentiates between species of different resource environments (Gebauer *et al.*, 1988), because this step may determine growth more directly than the availability of carbohydrates. The remobilized nitrogen is invested into aboveground growth in the spring. However, the rhizome does not lose its storage function in spring. Instead it accumulates high concentrations of starch instead of nitrogen during the summer. At that time of year about 1/3 of total rhizome biomass may be starch. In autumn the storage mode changes again. When aboveground parts die back and belowground growth activity of new rhizomes is initiated, the starch content decreases within few weeks. At the same time as starch concentrations decrease, the amino acids start to increase again. This organic nitrogen originates in part from reallocation from the dying aboveground biomass.

This seasonal course of storage products is remarkable because the same structure, e.g., current- and 1-year-old rhizomes, may act as nitrogen or as carbon store depending on season. Few examples of this response type have been documented. The rhizome of the sedge *Eriophorum* was shown to have a similar alternating storage function (Chapin *et al.*, 1986), and other perennial graminoids, such as *Carex* and *Calamagrostis*, show a pronounced seasonal cycle in nitrogen and phosphorous, but not in carbohydrates (Chapin and Shaver, 1989). Also Brocklebank and Hendry (1989) observed highest starch concentrations in late summer or autumn and the lowest in spring in several perennial European herbaceous species. With increasing rhizome age the physiological ability to

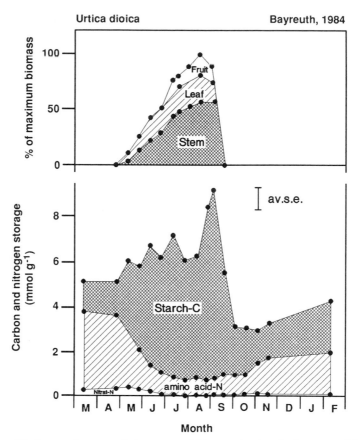

Figure 26 Seasonal course of aboveground biomass (top) and of storage of N and of starch in rhizomes of *Urtica dioica* (after Teckelmann, 1987).

perform these transport and storage processes appears to be lost. Old rhizomes maintain a high storage level of N and carbohydrates without the pronounced seasonal fluctuation (Teckelmann and Schulze, 1992). A similar aging process was observed in gramineoid rhizoms, e.g., in *Dupontia fisheri* (Chapin *et al.*, 1980).

The very broad species comparison of Poorter (1990) already points to the importance of structural characteristics which differentiate between species of high and low RGR. In *U. dioica* there are additional structural features, which have a strong feedback on its performance and which are a basis for the success of this species in a competitive environment. The rhizome of *Urtica* initiates a great number of tillers in autumn which start growth very early in the season and which cover the ground with a low leaf area index before other species even initiate growth (Teckelmann, 1987). However, from very early on in the season the large number of tillers suffer from high intraspecific competition because of increasing leaf area (see also Schulze and Zwölfer, this volume, Chapter 12). This leads to a process of self-thinning in which tillers operate completely independent of each other as indicated by the fact that the "3/2 power law" of self-thinning can be applied which describes self-thinning in a stand of competing individuals (Westoby, 1984). The process of self-thinning of *Urtica* tillers therefore operates as if it was a stand of individual plants not connected by rhizome. The tiller number decreases from about 600 m^{-2} in April to about 50 m^{-2} in June. The resources in the dying tillers are either shed as litter or they are reallocated into the rhizome and promote growth of the dominant tillers. Thus, not only is tillering an effective competitive mechanism which serves to suppress growth of other species, but tillers are also temporary storage organs which promote the input of photosynthates.

Following self-thinning, *U. dioica* will continue height growth of the main stem until September. This causes additional competition for light

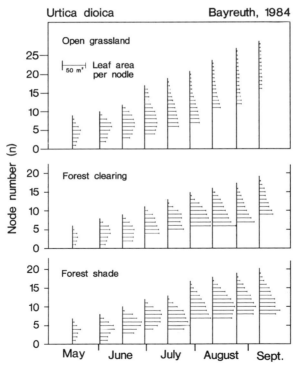

Figure 27   Seasonal course of leaf development in *Urtica dioica* at different levels of shade (after Teckelmann, 1987).

between leaves along the same tiller (Fig. 27). Newly grown leaves cause self-shading and leaf abscission at the base of the same tiller. This ability of leaf replacement does not exist in a similar way in most other species, where RGR decreases at optimal resource supply because of self-shading (Poorter *et al.*, 1988; Hirose *et al.*, 1988). The process of leaf replacement is very rapid in *U. dioica* with growth of one node carrying two leaves per week. Thus the complete cohort of five to eight leaves on a tiller at any one time are replaced two to three times per year. Each time the resources from old leaves are reallocated to promote new apical growth and part of the nitrogen and carbon is lost as litter. If we consider the processes of tiller formation, self-thinning, and leaf replacement, we can estimate that the same molecule of amino nitrogen may function intermittently as Rubisco or free amino acid four times during a seasonal life cycle before it is shed as litter or incorporated into seed.

With respect to the regulation of nutrient cycling and reallocation, *Urtica* is not different from low-resource plants in general. *Eriophorum vaginatum* also exhibits a circulation of resources during sequential leaf development, however, at a much lower rate. In *Eriophorum* each leaf is active for two seasons. Similar to *Urtica*, in *Eriophorum* about six cohorts of leaves are present (Jonasson and Chapin, 1985). Thus a cycle which takes place within 1 month in *Urtica* is spread over a whole season in *Eriophorum*. Despite this, the short-term RGR of aboveground parts in *Eriophorum* reaches 0.6 and thus exceeds that of *Urtica* (0.45). Jonasson and Chapin (1985) suggested that sequential leaf development paired with highly efficient remobilization of nutrients from senescing leaves enables plants to recycle resources within the shoot and minimize dependence upon soil nutrition.

The high growth activity of *Urtica* and the unavoidable losses of litter cause a great demand for nitrogen in the whole plant, which leads to a depletion of the nitrogen store in the rhizome even in a habitat of high-resource availability. Nitrogen appears to limit growth in summer, and we may take the accumulation of starch in the rhizome as a very sensitive indicator for unbalanced C/N status (see Fig. 19). Similar stimulation is also seen if other minerals such as phosphorous are limiting (Ariovich and Cresswell, 1983). The available N is invested into new leaves of high photosynthetic performance. These apparently produce an excess of carbohydrates which are stored in the rhizome and serve to support growth of new rhizomes in autumn when photosynthetic input is declining. Photosynthetic capacity of the new leaves is maintained at maximum performance, because nitrogen is reallocated from the shaded to the young leaves such that the uppermost sun-exposed leaves contain the highest concentration of leaf nitrogen (Fig. 28) which results in an increase of total productivity (Field, 1983; Hirose *et al.*, 1988).

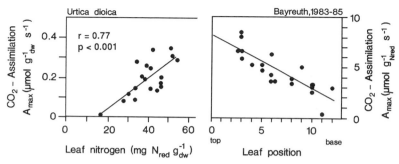

Figure 28 CO_2 assimilation as related to leaf nitrogen concentration (left) and to leaf position (right) (after Teckelmann, 1987).

The high apical activity of *Urtica*, which may be stimulated by high root activity and hormonal feedback of roots (see Beck, this volume, Chapter 5), in a high-resource environment results in a closed canopy which increases canopy height continually and outshades all competing lower growing species. Thus, *U. dioica* has a mechanical advantage over most other forest herbs (Givnish, 1986) and builds monospecific stands of high community stability. Only winter annual climbers, such as *Galium aparine*, are able to cope with the growth activity of *Urtica*. They do this because they can save on structural biomass by using *Urtica* as a support. The effect of height growth as it was demonstrated for *Urtica* is significant also in semi-natural grasslands where nutrient input through fertilizers allows a small number of tall growing species to dominate. This results in a reduction of biodiversity, caused by competition for light (Schulze and Chapin, 1987).

In contrast to annual or biennial species, *U. dioica* does not depend on fruit set, but flowering and fruiting is proportional to the stored biomass and may be suppressed under conditions of low light or nitrogen input.

From the *Urtica* experiments, we may generalize the following: (i) As in annuals and biennials, nitrogen and carbohydrates are accumulated whenever another resource is limiting. This means that low N leads to accumulation of carbohydrate and vice versa. (ii) The storage organ acts as a highly efficient buffer system, which reduces the loss of resources due to litter formation and mineralization, and which also supports above- or underground growth whenever conditions change and a different set of resources becomes limiting. (iii) In addition to the seasonal storage, there is a remarkable internal cycling of nitrogen between various functions, such that productivity is maximized and the opportunity cost for storage is kept low. (iv) *Urtica* also demonstrates the strong feedback effects of structural features which not only promote growth but which are the basis for successful competition with other species in a high-resource environment.

D. Woody Plants

Woody species have a lower respiratory loss per plant weight than herbaceous species due to the formation of conducting but nonrespiring wood tissue which is important as structure, especially upon competition for light. Schulze and Chapin (1987) concluded, that whenever the resource supply over time and the competitive ability increases, evergreen or deciduous woody species gain dominance over herbaceous plants (Fig. 29). More than in any other functional group of plants, in woody species the deciduous and evergreen foliage has ecological significance. Evergreen foliage gains dominance whenever nutrient supply is low (tropical rain forest) and conditions for photosynthesis are poor (boreal forest), or in winter rainfall climates, where the period of drought is not too extreme (mediterranean climates), or where drought is extreme (deserts). In contrast, deciduous growth is favored by seasonal climates of summer rainfall (temperate zone, savanna, subtropical forest) and where nutrient supply is high (river banks in boreal climates). In many situations deciduous and evergreen woody species coexist, but in these cases, deciduous plants occupy the more favorable microhabitats (Schulze, 1982).

In the following, we discuss (i) interactions between storage and growth in woody species (reviewed by Schulze, 1982). In addition we concentrate on resource allocation and growth. The paramount question is: (ii) is there a regulation of root/shoot ratios similar to what is known in herba-

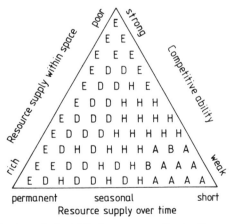

Figure 29 Distribution of plant life forms on gradients of resource supply over time and space and competitive ability for light. A, annual; B, biennial; H, herbaceous perennial; D, deciduous woody; E, evergreen woody (after Schulze and Chapin, 1987).

ceous plants? In woody species the root has important supporting functions in addition to maintaining a functional equilibrium of source-sink relations. It is quite unclear how the permanent structure of 1 year has a compounding effect on future years. (iii) We therefore investigate how the permanent structure contributes to the competitive ability in woody species. For obvious reasons the results on woody species are much more "observational" than results obtained by experiments using herbaceous species, and experimental manipulations in older forest stands are badly needed.

There is very limited quantitative information about root growth and biomass, because tree roots are generally too deep to be excavated or so extended and intermingled with other woody species that quantitative information is impossible to obtain. Because of this inherent lack of information about tree roots, a lysimeter experiment was conducted in which plants received different soil volumes and different annual supplies of water (Heilmeier *et al.*, 1993), namely 2, 4, and 6 m^3 water/plant and season in 7-, 14-, and 21-m^3 pots. Plants were allowed to grow up to 4 years. We wanted to see if perennial woody biomass contributes to the performance in the next season and if there is a carryover effect which leads to different root/shoot ratios, depending on the total water supply. Figure 30 shows that this is not the case; the root/shoot ratio is constant for different water supplies and tree ages. There is also a linear relation between biomass of fine roots and fine root length, which under these conditions may reach 60 km in a 4-year-old almond plant growing in a 21-m^3 pot. Microscopic inspection showed that we may have underestimated root length by a factor of 2. The result shows a clear root/shoot interaction in a case in which water supply determined the performance (see Schulze, this volume, Chapter 7). The plants terminate growth as soon as a critical water status in the soil is reached and therefore restrict

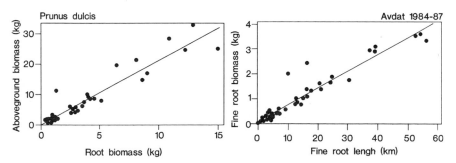

Figure 30 Shoot biomass as related to root biomass (left) and fineroot biomass as related to root length (right) in *Prunas dulcis*. The variation is caused by 1- to 4-year-old plants grown in pots of 1, 2, and 3 m depth and 3 m diameter (after Heilmeier *et al.*, 1993).

their growth to an increasingly short period when the soil is moist. It is demonstrated in Chapter 8 (this volume) that, in the case of water supply, the root/shoot interaction is regulated by a hormonal root signal, namely abscisic acid.

Considering interactions of carbon and nutrient relations trees appear to have a lower rate of photosynthesis than herbaceous species and exhibit a much lower capacity for nutrient uptake on a fine root basis. Uptake of [^{15}N]nitrate and [^{15}N]ammonium was lower in *Picea abies* than in dwarf shrubs or grasses (Buchmann, 1993). Not only is the uptake capacity for nitrogen and carbon low, but the transport processes is also low, including xylem and phloem flow (Ziegler, 1990). For example, when photosynthates were labeled with ^{14}C in *Pinus,* only 10 to 15% of the newly assimilated carbohydrates were transported from the needle to the stem and only 0.1% reached the root (Hansen, 1992). This may explain the low levels of nonstructural carbohydrates which were found in tree roots.

In a seasonal course, nonstructural carbohydrate levels reached a maximum just before onset of needle growth and remained at a low level during summer, at a time when stem growth is the major carbon sink (Oren *et al.,* 1988). In contrast to the seasonal course of carbohydrates, needle nitrogen content of *P. abies* is remarkably constant throughout the season. A seasonal decrease in the nitrogen content is only apparent in 1-year-old twigs and in large roots, but this may be due to the change in dry weight during incremental growth. Because of this interaction between concentrations and pool size in trees Timmer and Morrow (1984) used a diagram, in which element concentration and content per leaf or needle are compared at different weight of the needles (see also Oren and Schulze, 1989). If element concentration and element content increase at a constant weight per leaf, this is defined as accumulation. If uptake is invested into new growth and the concentration is maintained constantly, this element is available at a rate-limiting concentration. This analysis enabled Oren and Schulze (1989) to identify cation–nitrogen interactions which play an important role in nutritional forest decline, especially the deficiency of magnesium (Schulze *et al.,* 1989; Schulze, 1989).

The case of magnesium deficiency in *Picea* illustrates another important interaction between resource supply, accumulation, and growth. In spring *Picea* initiates needle and root growth by using nitrogen resources stored mainly in mature tissue and gained via net uptake of nitrogen by the canopy (20 to 25% of N requirement). Root uptake of N is low, because soils are cold and root growth has not been fully initiated. The cation requirement to support this growth originates mainly from reallocation from mature tissues. This seems to be a very general mechanism in woody species (Chapin and Kedrowski, 1983; Millard and Thomson,

1989; Jonasson, 1989). Cation uptake from roots is also low in spring in *P. abies;* but in addition to supporting growth of mature tissues, cation uptake and mobilization must also equilibrate cation losses due to leaching from the canopy as a result of ammonium uptake from aerosols and acid rain in the German climate. The discrepancy between the storage pool, requirement for growth, and the leaching loss results in severe magnesium deficiency in those stands which are not able to recharge the cation pools in their mature tissues during the winter. In *Picea*, carbon supply appears to limit growth in addition to nitrogen, since stem growth follows needle growth, but requires less cations and nitrogen. Therefore, the magnesium deficiency symptom of forest decline may partly disappear later during the season, if uptake again exceeds the requirements for stem growth.

Although the mass balance illustrates a nitrogen–cation interaction, the physiological basis of reallocation, transport and uptake are not at all clear. In spruce about 100 mmol of Mg and about 1500 mmol of N are transported from the root via the xylem to the top of the tree, but only about 15 mmol Mg and about 400 mmol N are used for crown growth (Schulze, 1991). There must be a very rapid futile cycle of Mg, N, and other elements via xylem and phloem. In fact, the same molecule of Mg may enter into the endodermis of a needle 10 times during the course of a season, but it will not be transported into the mesophyll even if this mesophyll exhibits symptoms of Mg deficiency. It is circulated back into the phloem. One may hypothesize that the element cycle of xylem and phloem is crucial for the functioning of the whole plant and thus has precedence (greater sink strength) over the requirement of an old needle, which could easily be shed or replaced if necessary without endangering the survival of the whole plant. Only newly growing shoots have access to phloem Mg and accumulate it in excess even though old needles loose Mg and turn yellow (Schulze, 1991). Thus, the sink strength of the plant internal cycle is the regulating mechanism that reallocates Mg from old needles and maintains the phloem–xylem cycle. In the case of *P. abies* the sink strength of growing tissue appears to be regulated by the availability of nitrogen rather than carbohydrate levels. Nevertheless, reduction of nitrate again depends on the carbohydrate cycle (Gebauer and Stadler, 1990) as we have already seen in the transgenic plants. Highest nitrate reductase activity is found in the leaves, but since the whole biomass of wood is very large, the total amount being reduced is higher in the stem than in leaves.

Under stable environmental conditions, woody species will eventually outcompete herbaceous plants because of their structural advantage in height growth. This even allows late successional species to "downregulate" photosynthesis. Schulze *et al.* (1986) demonstrated that CO_2

assimilation and water use decreased in late successional species as compared to early successional species by factor 3 from 9 to 12 μmol CO_2 m^{-2} s^{-1} in *Prunus spinosa* to only 3 to 4 μmol CO_2 m^{-2} s^{-1} in *Fagus sylvatica*. However, the volume of living space gained by the dry matter investment increased 10-fold from 550 liter kg^{-1} in late successional plants as compared to 38 liter kg^{-1} in early successional plants. This occurs among other factors together with changes in shade tolerance and leaf area index (Küppers, 1989). The differences in growth of woody species can to a large extent be explained by the rules of bud development, i.e., by the dominance of a longitudinal or lateral symmetry (Troll, 1937; Küppers, 1989). The importance of structure in promoting growth can be impressively demonstrated using crossings of forest trees. The common crossing of *Larix decidua* and *Larix leptolepis* exhibits higher growth than that of the parent plants at similar CO_2 assimilation rates and nitrogen contents of the needles (Matyssek and Schulze, 1987a,b). The increased growth is related mainly to parameters affecting the leaf area ratio. The hybrid has a length of main branches characteristic of the parent *L. decidua* and exhibits needle density on these branches which is typical for *L. leptolepis*. Both factors result in a foliage biomass which exceeds that of both parent trees and suggest a classical heterosis effect.

E. Summary of Observations on Different Plant Life Forms

We may summarize the observations in different species representing a range of plant life forms:

- There is a clear interaction between storage and growth, which is in fact quite similar to what was observed in transgenic plants.
- Nitrogen seems to be a much stronger regulator of growth than carbohydrates.
- There is a functional equilibrium between root and shoot biomass even in perennial woody plants, i.e., the compounding effect of perennial wood and growth of new structure is tightly regulated.
- Structural aspects of carbon investment become of paramount importance during competition for light in perennial species, and structural parameters explain succession and heterosis effects.

VI. Conclusions

The present study has attempted to integrate factors which regulate assimilation of carbon and nitrogen, storage, and growth at the biochemical and the morphological level using transgenic plants and species comparisons.

1. A strong regulatory interaction exists between the availability of carbohydrates and the further assimilation of carbon and nitrogen, which affects the use of these resources for growth. When photosynthesis is in excess, large amounts of starch are accumulated which do not contribute to growth, and, in the mid- to long-term, the rate of photosynthesis is decreased, possibly via regulation of gene expression. When carbohydrate is in short supply, nitrogen assimilation is decreased with the result that nitrate accumulates in the leaf.

2. Nitrogen strongly regulated biomass allocation to shoot and root growth. Within an organ, excess nitrogen may accumulate as nitrate. However, provided enough carbon is available, vegetatively growing plants seem to prefer to allocate the excess N to functional proteins like Rubisco. As discussed for this example, the opportunity cost is lower and several marginal gains may be made. Interestingly, biennials also delayed the formation of storage organs sensu strictu until late in the season, to allow N to be invested in "productive" proteins.

3. Morphological and structural features are very important for plant growth. For example, we have seen that a decreased rate of photosynthesis leads almost automatically to an increased LAR due to changes in leaf morphology in transgenic plants. We have also seen that RGR is widely correlated with LAR (and negatively correlated with NAR) in a wide variety of wild species and conditions. Structural features become even more important when comparing woody species and interpreting the succession of plant life forms. These are genetically fixed traits which are, at first sight, independent of the metabolic processes (e.g., dominance of buds in woody species).

4. Storage is an important buffer in all plant life forms, and those substances accumulate which are assimilated by excess capacity. Obviously plants do not completely down-regulate the respective processes, probably because the opportunity cost for reducing the assimilatory or uptake capacity in the short-term is higher than the cost for storage. Situations were demonstrated where obviously the regulation is not optimal. However, it is quite likely that such response is an advantage under the natural and thus highly variable environmental conditions.

5. Reserve formation, defined as a storage process which competes with vegetative growth, is rare in the plants investigated. Very likely it is linked to the evolution of that species. For instance the growth of storage structure before filling of the storage organ could potentially result from the fact that these species had woody ancestors (biennials). Reserve starch formation in leaves (for diurnal turnover) actually increased growth over the entire day/night cycle and is obviously

regulated so that enough reserves are laid down to support growth during the night.

6. We have also learned that single gene manipulations (antisense *rbcS*, starch synthesis, invertase) can cause massive changes not only in metabolism, but also in structural components. Thus, if effects of structural parameters, such as specific leaf weight, are significant for growth, this might sometimes be based on single gene changes in the photosynthetic apparatus, e.g., the often-observed negative relation between photosynthetic rate and specific leaf area can actually be reproduced in tobacco by reducing the rate of photosynthesis by decreasing expression of Rubisco.

7. There are numerous "side effects" caused by single gene changes. For instance if Rubisco is reduced, nitrate may accumulate. This has feedback on specific leaf weight and leaf area expansion, which in turn compensates for the loss in photosynthesis. We therefore learn from the genetic manipulations that the system is not only highly flexible (well regulated), but it is also highly interactive and that numerous "surprises" emerge, which were not foreseen in the primary hypothesis.

8. We find that all major control types which were identified at the cellular level interact, when viewed at the level of a complex phenomena like whole-plant growth. Numerous feedback and feedforward effects act together in regulating the same process and shared control exists. In addition, the plant internal circulation between phloem and xylem and the reallocation of resources during plant development represent a major cycle from which various sinks tap their resources in a very competitive way.

9. Secondary factors, such as shading, or possible inbalances between root and shoot create a very dynamic equilibrium, which is not tightly regulated, but which may oscillate between several quasi-stable conditions. The buffering capacity of storage may be one property which is important in coping with this dynamic process.

10. Although regulation often improves performance, we have found cases where an optimum is clearly not reached, and in some cases the regulation can even decrease performance. For example, all the annual species studied tended to overinvest in photosynthetic apparatus with the result that they tended to overproduce carbohydrate. In tobacco in low light there was overinvestment in Rubisco. This trait, however, may be of advantage in a competitive situation at the plant community level.

11. The question of whether regulation increases or decreases performance (see Scheibe and Beck, this volume, Chapter 1) is hardly applicable in such a complex system. We already saw in Chapter 2 that an enzyme, once integrated into a metabolic path, necessarily cannot

realize maximal activity, for the simple reason that the rest of the pathway could not operate under conditions perfect to one enzyme. We have seen in this chapter that carbon cannot be allocated just for the purpose of maximizing photosynthetic apparatus, because this process itself depends on water and nitrogen supply. An absence of regulation at this level would lead to chaos.

It is quite clear that this chapter is focused on the use of carbon and nitrogen, and additional mechanisms of regulation become important when dealing with other limiting resources such as water (Schulze, this volume, Chapter 7). We also emphasize that we were investigating the long-term allocation of carbon. The regulation of short-term allocation may involve additional factors such as hormonal interactions (Beck, this volume, Chapter 5).

Finally, we are always inclined to explain competitiveness in terms of high growth rate and maximum performance, but it is often very difficult to explain the performance of low-resource species and their persistence in nature. Our results indicate that additional components such as seed number and seed weight become important in explaining the survival of these species in the real world, even though high-resource plants may perform better over the whole range of conditions.

Acknowledgments

A large number of guest scientists, postdoctoral students, doctoral students, and master's degree students have joined to complete this 12-year project. We acknowledge their cooperation and help. Especially we thank our scientific guests Dr. C. Labate, Dr. R. K. Monson, Dr. H. A. Mooney, Dr. R. Oren, Dr. D. Pankovic, Dr. P. Quick, Dr. R. Wender, Dr. K. Werk, and Dr. D. Whale, as well as our doctoral students M. Brauer, N. Buchmann, K. Fichtner, H. Heilmeier, T. Jelitto, A. Krapp, R. Matyssek, T. Steinlein, M. Teckelmann, and A. Wartinger. For technical assistance we thank A. Dominikus, K. Jäschke, B. Probst, B. Scheitler, B. Seewald, and M. Wartinger. For help in drawing the "functional plant" of Figs. 1, 8, 9, 11, 17, and 18, we thank L. Badewitz, G. Grimm, and R. Asche.

References

Ariovich, D., and Cresswell, C. F. (1983). The effect of nitrogen and phosphorus on starch accumulation and net photosynthesis in two variants of *Panicum maximum* Jacq. *Plant, Cell Environ.* **6,** 657–664.

Björkman, O. (1981). Responses to different quantum flux densities. *Encycl. Plant Physiol., New Ser.* **12,** 57–107.

Bloom, A. J., Chapin, F. S., III, and Mooney, H. A. (1985). Resource limitation in plants—an economic analogy. *Annu. Rev. Ecol. Syst.* **16,** 363–392.

Boot, R. (1990). The significance of size and morphology of root systems for nutrient acquisition and competition. In "Causes and Consequences of Variation in Growth Rate and Productivity of Higher Plants" (H. Lambers, M. L. Cambridge, H. Konings, and T. L. Pons, eds.), pp. 299–312. SPB Academic Publishing, The Hague, The Netherlands.

Bradshaw, A. D., Chadwick, M. J., Jowett, D., and Snaydon, R. W. (1964). Experimental investigations into the mineral nutrition of several grass species. *J. Ecol.* **52,** 665–676.

Brocklebank, J., and Hendry, G. A. F. (1989). Characteristics of plant species which store different types of reserve carbohydrates. *New Phytol.* **112,** 255–260.

Buchmann, N. (1993). Wege and Umsetzungen von ^{15}N-Ammonium and ^{15}N-Nitrat in einem Fichtenjungbestand (Picea abies (L.) Karst.). Doctoral Thesis, University of Bayreuth.

Burton, W. G. (1989). "The Potato," 3rd ed. Longman, New York.

Caspar, T., Huber, S. C., and Sommerville, C. R. (1986). Alterations in growth, photosynthesis, and respiration in a starchless mutant of *Arabidopsis thaliana* deficient in chloroplast phosphoglucomutase activity. *Plant Physiol.* **79,** 1–7.

Chapin, F. S., III, and Kedrowski, R. A. (1983). Seasonal changes in nitrogen and phosphorous fractions and autumn retranslocation in evergreen and deciduous trees. *Ecology* **64,** 376–391.

Chapin, F. S., III, and Shaver, G. R. (1985). Individualistic growth response of tundra plant species to environmental manipulations in the field. *Ecology* **66,** 564–576.

Chapin, F. S., III, and Shaver, G. R. (1989). Lack of latitudinal variations in graminoid storage reserves. *Ecology* **70,** 269–272.

Chapin, F. S., III, Tieszen, L. L., Lewis, M. C., Miller, P. C., and McCowen, B. H. (1980). Control of tundra plant allocation patterns and growth. In "An Arctic Ecosystem: The Coastal Tundra at Barrow, Alaska" (J. Brown, P. C. Miller, L. L. Tieszen, and F. L. Bunnell, eds.), pp. 140–193. Dowden, Hutchinson & Ross, Stroudsburg, PA.

Chapin, F. S., III, Shaver, G. R., and Kedrowski, R. A. (1986). Environmental controls over carbon, nitrogen and phosphorous fractions in *Eriophorum vaginatum* in Alaskan Tussock tundra. *J. Ecol.* **74,** 167–195.

Chapin, F. S., III, Schulze, E. D., and Mooney, H. A. (1990). The ecology and economics of storage in plants. *Annu. Rev. Ecol. Syst.* **21,** 423–447.

Ellenberg, H. (1979). Zeigerwerte der Gefässpflanzen Mitteleuropas. *Scr. Geobot.* **9.** 22 pp.

Farrar, J. F., and Williams, J. H. H. (1990). Control of barley root respiration. *Physiol. Plant.* **79,** 254–266

Fichtner, K. (1991). Dea Einfluβ des Stickstoffangebots auf das Wachstum unterschiedlich nitrophiler annueller Pflanzen und die Wechselwirkung zwischen Stickstoffhaushalt und Photosynthese bei Phaseolus lunatus und transgenem *Nicotiana tabacum.* Doctoral Thesis, University of Bayreuth.

Fichtner, K., and Schulze, E.-D. (1992). The effect of nitrogen nutrition on growth and biomass partitioning of annual plants originating from habitats of different nitrogen availability. *Oecologia* **92,** 236–241.

Fichtner, K., Quick, W. P., Schulze, E.-D., Mooney, H. A., Rodermel, S. R. Bogorad, L., and Stitt, M. (1993). Decreased ribulose-1.5-bisphosphate carboxylase-oxygenase in transgenic tobacco transformed with "antisense" rbcS. V. Relationship between photosynthetic rate, storage strategy, biomass allocation and vegetative plant growth at three different nitrogen supplies. *Planta* **190,** 1–9.

Field, C. (1983). Allocating leaf nitrogen for the maximization of carbon: Leaf age as a control on the allocation program. *Oecologia* **56,** 341–347.

Gebauer, G., and Stadler, J. (1990). Nitrate assimilation and nitrate content in different organs of ash trees (*Fraxinus excelsior*). In "Plant Nutrition: Physiology and Application"

(M. L. van Beusichem, ed.), pp. 101–106. Kluwer Acad. Publ., Dordrecht, The Netherlands.

Gebauer, G., Melzer, A., and Rehder, H. (1984). Nitrate content and nitrate reductase activity in *Rumex obtusifolius* L. I. Differences in organs and diurnal changes. *Oecologia* **63,** 136–142.

Gebauer, G., Rehder, H., and Wollenweber, B. (1988). Nitrate, nitrate reduction and organic nitrogen in plants from different ecological and taxonomic groups of Central Europe. *Oecologia* **75,** 371–385.

Geiger, D. R. (1980). Processes affecting carbon allocation and partitioning among sinks. *In* "Phloem Transport" (J. Cronshaw, W. T. Lucas, and R. T. Giaquinta, eds.), pp. 375–388. Alan R. Liss, New York.

Givnish, T. J. (1986). Biomechanical constraints on crown geometry forest herbs. *In* "On the Economy of Plant Form and Function" (T. J. Givnish, ed.), pp. 525–584. Cambridge Univ. Press, Cambridge.

Hansen, J. (1992). Jahreszeitliche Dynamik des Kohlenhydratmetabolismus der Waldkiefer (*Pinus sylvestris* L.). Doctoral Thesis, University of Bayreuth.

Heilmeier, H., Schulze, E.-D., and Whale, D. M. (1986). Carbon and nitrogen partitioning in the biennial monocarp Arctium tomentosum Mill. *Oecologia* **70,** 466–474.

Heilmeier, H., Wartinger, A., Horn, R., and Schulze, E.-D. (1993). Biomass partitioning in *Prunus aremniaca*. *Oecologia* (to be published).

Hirose, T. (1988). Nitrogen availability, optimal shoot/root ratios and plant growth. *In* "Plant Form and Vegetation Structure" (M. J. A. Werger, P. J. M. van der Aart, H. J. During, and J. T. A. Verhoeven, eds.), pp. 135–145. SPB Academic Publishing, The Hague, The Netherlands.

Hirose, T., and Werger, M. J. A. (1987). Maximizing daily canopy photosynthesis with respect to the leaf nitrogen allocation pattern in the canopy. *Oecologia* **72,** 520–526.

Hirose, T., Werger, M. J. A., Pons, T. L., and van Rheenen, J. W. A. (1988). Canopy structure and leaf nitrogen distribution in a stand of Lysimachia vulgaris L. as influenced by stand density. *Oecologia* **77,** 145–150.

Jelitto, T., Sonnewald, V., Willmitzer, L., Hajirezaei, M., and Stitt, M. (1992). Inorganic pyrophosphate content and metabolites in leaves and tubers of potato and tobacco plants expressing *E. coli* pyrophosphatase in the cytosol: Biochemical evidence that sucrose metabolism has been manipulated. *Planta* **788,** 238–244.

Jonasson, S. (1989). Implications of leaf longevity, leaf nutrient re-adsorption and translocation for the resource economy of five evergreen plant species. *Oikos* **56,** 121–131.

Jonasson, S., and Chapin, F. S., III (1985). Significance of sequential leaf development for nutrient balance of the cotton sedge, *Eriophorum vaginatum* L. *Oecologia* **67,** 511–518.

Kaiser, W., and Forster, I. (1989). How CO_2 prevents nitrate reductase assimilation in leaves. *Plant Physiol.* **91,** 970–975.

Koch, G. W. (1988). Acquisition and allocation of carbon and nitrogen in the wild radish, Raphanus sativa x raphanistrum (Brassicaceae). Ph.D. Dissertation, Stanford University, Stanford, CA.

Koch, G. W., Schulze, E.-D., Percival, F., Mooney, H. A., and Chu, C. (1988). The nitrogen balance of Raphanus sativus x raphanistrum plants. Growth, nitrogen redistribution and photosynthesis under NO_3^- deprivation. *Plant, Cell Environ.* **11,** 755–767.

Krapp, A., Quick, W. P., and Stitt, M. (1991). There is a dramatic loss of *Rubisco*, other Calvin cycle enzymes and chlorophyll when glucose is supplied to mature spinach leaves via the transpiration stream. *Planta* **186,** 58–59.

Krapp, A., Hofmann, G., Schäfer, C., and Stitt, M. (1993). Regulation of the expression of *rbcS* and other photosynthetic genes by carbohydrates: A mechanism for the "sink regulation" of photosynthesis. *Plant J.* **3,** 817–828.

Kuiper, D., and Kuiper, P. J. C. (1979). Comparison of *Plantago* species from nutrient-rich and nutrient-poor conditions: Growth response, ATPases and lipids of the roots, as affected by the level of mineral nutrition. *Physiol. Plant.* **45**, 489–491.

Küppers, M. (1989). Ecological significance of aboveground architectural patterns in woody plants: A question of cost-benefit relationships. *Trees* **4**, 375–379.

Lambers, H., Freijsen, H., Poorter, H., Hirose, T., and van der Werf, A. (1990). Analysis of growth based on net assimilation rate and nitrogen productivity. Their physiological background. *In* "Causes and Consequences of Variation in Growth Rate and Productivity of Higher Plants" (H. Lambers, M. L. Cambridge, H. Konings, and T. L. Pons, eds.), pp. 1–17. SPB Academic Publishing, The Hague, The Netherlands.

Lauerer, M. (1992). der Einfluss der Lichtintensität und Rubisco-1,5-bisphosphat-Carboxilase/oxigenase auf die Photosynthese, Wachstum und Blattanatomie bei Tabak (Nicotiana tabacum L.) transformiert mit "antisense" rbcS. Diploma Thesis, University of Bayreuth.

Lauerer, M., Saftic, D., Quick, W. P., Labate, C., Fichtner, K., Schulze, E.-D., Rodermel, S. R., Bogorad, L, and Stitt, M. (1993). Decreased ribulose-1.5-bisphosphate carobxylase-oxygenase in transgenic tobacco transformed "antisense" rbcS. *VI*. Effect on photosynthesis in plants grown at different irradiance. *Planta* **190**, 332–345.

Maas, C., Schaal, S., and Werr, W. (1990). A feedback control element near the transcription start site of maize shrunken gene determines promotor activity. *EMBO J* **9**, 3447–3452.

Matyssek, R., and Schulze, E.-D., (1987a). Heterosis in hybrid larch (*Larix decidua* x leptolepis). I. The role of leaf characteristics. *Trees* **1**, 219–224.

Matyssek, R., and Schulze, E.-D. (1987b). Heterosis in hybrid larch (*Larix decidua* x leptolepis). II. Growth characteristics. *Trees* **1**, 225–231.

Meusel, H. (1952). Über Wuchsformen, Verbreitung und Phylogenie einiger mediterran-mitteleuropäischer Angiospermen-Gattungen. *Flora (Jena)* **139**, 333–393.

Meusel, H., and Kästner, A. (1990). Lebensgeschichte der Gold- und Silberdisteln. *Öesterr. Akad. Wiss. Math.-Naturwiss.Kl.* **127**, 1–294.

Millard, P. (1988). The accumulation and storage of nitrogen by herbaceous plants. *Plant, Cell Environ.* **11**, 1–8.

Millard, P., and Thomson, C. M. (1989). The effect of the autumn senescence of leaves on the internal cycling of nitrogen for the spring growth of apple trees. *J. Exp. Bot.* **40**, 1285–1289.

Monsi, M. (1960). Dry matter production in plants. I. Schemata of dry-matter reproduction. *Bot. Mag.* **73**, 81–90.

Neuhaus, H. E., Kruckeberg, A. L., Feil, R., and Stitt, M. (1989). Reduced activity mutants of phosphoglucose isomerase in the cytosol and chloroplast of *Clarkia xantiana*. Study of the mechanisms which regulate photosynthetic partitioning. *Planta* **178**, 110–122.

Neuhaus, H. E., Quick, W. P., Siegl, G., and Stitt, M. (1990). Control of photosynthate partitioning in spinach leaves. Analysis of the interaction between feedforward and feedback regulation of sucrose synthesis. *Planta* **181**, 583–592.

Oren, R., and Schulze, E.-D., (1989). Nutritional disharmony and forest decline: A conceptual model. *Ecol. Stud.* **77**, 425–443.

Oren, R., Schulze, E.-D., Werk, K., Meyer, J., Schneider, B. U., and Heilmeier, H. (1988). Performance of two *Picea abies* (L.) Karst. stands at different stages of decline. I. Carbon relations and stand growth. *Oecologia* **75**, 25–37.

Orians, G. H., and Solbrig, O. T. (1977). A cost-income model of leaves and roots with special reference to arid and semi-arid areas. *Am. Nat.* **111**, 677–690.

Poorter, H. (1990). Interspecific variation in relative growth rate: On ecological causes and physiological consequences. *In* "Causes and Consequences of Variation in Growth Rate and Productivity of Higher Plants" (H. Lambers, M. L. Cambridge, H. Konings, and T. L. Pons, eds.), pp. 45–68. SPB Academic Publishing, The Hague, The Netherlands.

Poorter, H., and Bergkotte, M. (1992). Chemical composition of 24 wild species differing in relative growth rate. *Plant, Cell Environ.* **15,** 221–229.

Poorter, H., and Remkes, C. (1990). Leaf area ratio and net assimilation rate of 24 wild species differing in relative growth rate. *Oecologia* **83,** 553–559.

Poorter, H., Pot, S., and Lambers, H. (1988). The effect of an elevated atmospheric CO_2 concentration on growth, photosynthesis and respiration of *Plantago major. Physiol. Plant.* **73,** 553–559.

Potter, J. R., and Jones, J. W. (1977). Leaf area partitioning as an important factor in growth. *Plant Physiol.* **59,** 10–14.

Quick, W. P., Schurr, U., Scheibe R., Schulze, E.-D., Rodermel, S. R., Bogorad, L., and Stitt, M. (1991a). Decreased ribulose-1,5-bisphosphate carboxylase-oxygenase in transgenic plants transformed with antisense rbcS. I. Impact on photosynthesis in ambient growth conditions. *Planta* **183,** 542–554.

Quick, W. P., Schurr, U., Fichtner, K., Schulze, E.-D., Rodermel, S. R., Bogorad, L., and Stitt, M. (1991b). The impact of decreased Rubisco on photosynthesis, growth, allocation and storage in tobacco plants which have been transformed with antisense rbcS. *Plant J.* **1,** 51–58.

Quick, W. P., *et al.*, (1992). Decreased ribulose-1,5-bisphosphate carboxylase-oxygenase in transgenic tobacco transformed with antisense *rbcS*. IV. Impact on photosynthesis in condition of altered nitrogen supply. *Planta* **188,** 522–531.

Reif, A., Teckelmann, M., and Schulze, E.-D. (1985). Die Standortamplitude der Grossen Brennessel (Urtica dioica L.) eine Auswertung vegetationskundlicher Aufnahmen auf der Grundlage der Ellenbergschen Zeigerwerte. *Flora (Jena)* **176,** 365–382.

Rosnitschek-Schimmel, I. (1985a). Seasonal dynamics of nitrogenous compounds in a nitrophilic weed. I. Changes in inorganic and organic nitrogen fractions of the different plant parts of Urtica dioica. *Plant Cell Physiol.* **26,** 169–176.

Rosnitschek-Schimmel, I. (1985b) Seasonal dynamics of nitrogenous compounds in a nitrophilic weed. II. The role of free amino acids and proteins as nitrogen store in Urtica dioica. *Plant Cell Physiol.* **26,** 177–183.

Rufty, T. W., Huber, S. C., and Volk, R. J. (1984). Alterations in leaf carbohydrate metabolism in response to nitrogen stress. *Plant Physiol.* **88,** 725–730.

Schulze, E.-D. (1982). Plant life forms and their carbon, water and nutrient relations. *Encycl. Plant Physiol. New Ser.* **12B,** 615–676.

Schulze, E.-D. (1989). Air pollution and forest decline in a spruce (*Picea abies*) forest. *Science* **244,** 776–783.

Schulze, E.-D. (1991). Water and nutrient interactions with plant water stress. *In* "Response of Plants to Multiple Stresses" (H. A. Mooney, W. E. Winner, and E. J. Pell, eds.), pp. 89–103. Academic Press, San Diego.

Schulze, E.-D., and Chapin, F. S., III (1987). Plant specialization to environments of different resource availability. *Ecol. Stud.* **61,** 120–148.

Schulze, E.-D., and Mooney, H. A. (1992). Ecosystem function of biodiversity. *Ecol. Stud.* **99,** 520.

Schulze, E.-D., Koch, G., Percival, F., Mooney, H. A., and Chu, C. (1985). The nitrogen balance of Raphanus sativus x raphanistrum plants. I. Daily nitrogen use under high nitrate supply. *Plant, Cell Environ.* **8,** 713–720.

Schulze, E.-D., Küppers, M., and Matyssek, R. (1986). The role of carbon balance and branching pattern in the growth of woody species. *In* "On the Economy of Plant Form and Function" (T. J. Givnish, ed.), pp. 585–602. Cambridge Univ. Press, Cambridge.

Schulze, E.-D., Oren, R., and Lange, O. L. (1989). Nutrient relations of trees in healthy and declining Norway spruce stands. *Ecol. Stud.* **77,** 392–417.

Schulze, W., Stitt, M., Schulze, E.-D., Neuhaus, H. E., and Fichtner, K. (1991). A quantification of the significance of assimilatory strach for growth of *Arabidopsis thaliana* L. Heynh. *Plant Physiol.* **95**, 890–895.

Sheen, J. (1990). Metabolic repression of transcription in higher plants. *Plant Cell* **2**, 1027–1038.

Sonnewald, U., Brauer, M., von Schaewen, A., Stitt, M., and Willmitzer, L. (1991). Transgenic tobacco plants expressing yeast-derived invertase in either the cytosol, vacuole or apoplast: A powerful tool for studying sucrose metabolism and sink/source interactions. *Plant J.* **1**, 95–106.

Steinlein, T., Heilmeier, H., and Schulze, E.-D., (1993). Nitrogen and carbohydrate storage in biennials originating from habitats of different resource availability. *Oecologia* **93**, 374–382.

Stitt, M. (1991). Rising CO_2 levels and the potential significance for carbon flow in photosynthetic cells. *Plant, Cell Environ.* **14**, 741–762.

Stitt, M., von Schaewen, A., and Willmitzer, L. (1991). "Sink" regulation of photosynthetic metabolism in transgenic tobacco plants expressing yeast invertase in the cell wall involves a decrease of the Calvin cycle enzymes. *Planta* **183**, 40–50.

Teckelmann, M. (1987). Kohlenstoff-, Wasser- und Stickstoffhaushalt von Urtica dioica L. an natürlichen Standorten. Doctoral Thesis, University of Bayreuth.

Teckelmann, M., and Schulze, E.-D. (1992). Growth and storage in the perennial herbaceous plant *Urtica dioica. Oecologia* (to be published).

Timmer, V. R., and Morrow, L. D. (1984). Predicting fertilizer growth response and nutrient status of jack pine by foliar diagnosis. *In* "Forest Soils and Treatment Impact" (E. L. Stone, ed.), pp. 335–351. University of Tennessee, Knoxville.

Troll, W. (1937). "Vergleichende Morphologie der höheren Pflanze," Vol. "Vegetationsorgane", Part 1, "Der Sproß" Bornträger, Berlin.

von Caemmerer, S., and Farquhar, A. D. (1984). Effects of partial defoliation, changes of irradiances during growth, short-term water stress, and growth at enchanced CO_2 on the photosynthetic capacity of leaves of *Phaseolus vulgaris* L. *Planta* **160**, 320–329.

von Schaewen, A., Stitt, M., Schmidt, R., Sonnewald, U., and Willmitzer, L. (1990). Expression of yeast-derived invertase in the cell wall of *Arabidopsis* and tobacco leads to inhibition of sucrose export, accumulation of carbohydrate and strongly influences the growth and habitus of transgenic tobacco plants. *EMBO J.* **9**, 3033–3044.

Westoby, M. (1984). The self-thinning rule. *Adv. Ecol. Res.* **14**, 167–225.

Woodrow, I. E., and Berry, J. A. (1987). Enzymatic regulation of photosynthetic CO_2 fixation in C-3plants. *Annu. Rev. Plant Physiol. Plant Mol. Biol.* **39**, 533–594.

Ziegler, H. (1990). Flüssigkeitsströme in Pflanzen. *In* "Biophysik" (W. Hoppe, W. Lohmann, H. Markl, and H. Ziegler, eds.), pp. 561–570. Springer-Verlag, Berlin.

5

The Morphogenic Response of Plants to Soil Nitrogen: Adaptive Regulation of Biomass Distribution and Nitrogen Metabolism by Phytohormones

E. Beck

I. Introduction: Nitrogen-Induced Changes of Root Growth

Distribution of photosynthetically produced biomass between photosynthesizing and nonphotosynthesizing, water- and nutrient-acquiring plant parts is highly variable and obviously corresponds to economic principles (Bloom *et al.*, 1985). A plant, growing at a limited supply of water or nutrients will profit from investing a higher portion of its biomass into the root system in order to exploit more of the limiting soil resources. On the other hand, under optimal conditions, promotion of the shoot increases the plant's capacity for photosynthesis and consequently increases biomass production and growth (see also this volume, Chapter 7). Production of a larger root system at the cost of shoot growth has been observed with numerous herbaceous (Brouwer, 1962; Barta, 1975; Pate *et al.*, 1979; Rufty *et al.*, 1984; Kuchenbuch *et al.*, 1988) and woody plants (Ericsson, 1981), especially in response to an insufficient nitrogen supply and has been described as "N-etiolement" of the root (Lundegårdh, 1954). Various mathematic models for that phenomenon, in particular with respect to crop plants, have been elaborated (Thornley, 1972; Chung *et al.*, 1982; Johnson, 1985; Spek and Van Oijen, 1988). Despite numerous studies concerning sink activities and phloem transport (Patrick, 1991), source–sink relations (Geiger and Fondy, 1991), the response of carbohydrate metabolism to the nitrogen supply (Rufty *et al.*, 1988), photosynthetic or growth rates (Ågren and Ingestad, 1987), the biochemical mechanisms controlling carbon fluxes, and biomass distribution in the plants are still unknown (see also Komor, this volume, Chapter 6). In the following I try to address these biochemical control mechanisms, using the nitrophilic herbs *Urtica dioica* L. and *Chenopodium rubrum* L. as experimental plants (see also Stitt and Schulze, this volume, Chapter 4).

Urtica dioica tolerates a wide range of N supply (Rosnitschek-Schimmel, 1982; Hofstra *et al.*, 1985) while growing well only in a narrow range of neutral soil pH (Reif *et al.*, 1985). Depending on water availability and soil type (Olsen, 1921) its subterraneous parts are highly variable, ranging from a poorly branched, slender rhizome system with only a few roots to a pronounced tap root system with secondary growth (Teckelmann, 1987). Stinging nettles can easily be raised from seeds or from rhizome cuttings of at least 500 mg fresh weight (Dauberschmidt, 1992). In the experiments reported here, plants were grown from seeds. The seedlings were transferred to modified Kick-Brauckmann pots filled with approximately 1 liter of washed quartz sand. The pots were continuously percolated with Knoop nutrient solutions of various concentrations of NO_3^- and NH_4^+, respectively, ranging from 1 to 22 mM (Rosnitschek-Schimmel, 1982). The nettles were grown in the greenhouse under short-day conditions (light period of 10 h) to keep them in the vegetative stage.

In addition to the perennial stinging nettle, *C. rubrum*, an annual nitrophilic weed, was also used, especially for biochemical studies of the nitrogen and cytokinin metabolism. *Chenopodium rubrum* was raised from seeds (a gift from Professor E. Schäfer, Freiburg) and grown in the greenhouse for 10 days in normal soil under long-day conditions. Plants with shoots of about 25 cm height were used. Two to three weeks prior to the uptake experiments, they were transferred to a hydroponic culture in 10-fold-diluted, continuously aerated Knoop nutrient solution. In addition, autotrophic cell suspension cultures of this species (Hüsemann and Barz, 1977), supplied by Professor W. Barz (Münster), were grown under a light/dark regime of 16/8 h. These allowed the investigation of the nigrogen (Beck and Renner, 1990) and cytokinin (Fusseder and Ziegler, 1988; Fusseder *et al.*, 1989; Peters and Beck, 1991) metabolism at the cellular level.

II. The Response of Nitrogen Uptake to Nitrogen Supply

One way to cope with nitrogen deficiency is to increase the absorbing surface by an increase of the root-to-shoot (R/S) biomass ratio. The other strategy of a plant is to increase the uptake capacity per unit root weight

(Schulze and Chapin, 1987). Studies with several plant species have shown that net nitrate uptake appears to be mediated by a dual system, termed mechanisms I and II, as is indicated by biphasic uptake kinetics (Siddiqi *et al.*, 1990; Goyal and Huffaker, 1986; Warner and Huffaker, 1989). In the low-concentration range, net influx is usually saturable (mechanism I) while, in the range beyond 1 mM, saturation of uptake by mechanism II is frequently obscured by a third type of influx that is linearly dependent on the nitrate concentration (Doddema and Telkamp, 1979). Nissen (1987) reexamined the published uptake kinetics for inorganic nutrients with respect to being adequately represented by two Michaelis-Menten terms and a linear term. He came to the conclusion that uptake occurs by a single, multiphasic mechanism rather than by a dual or a threefold system. While the characteristic of the biphasic system is a curvilinear Lineweaver-Burk plot, the multiphasic interpretation relies on a multistep characteristic. Both types of curves do not allow clear-cut kinetic parameters such as K_m and V_{max} (Q_{max}) values. Strangely, Nissen (1987) did not consider nitrogen uptake despite the large body of existing data. These have been gained predominantly with excised roots while intact and mature plants have been used only to a minor extent (Schulze and Bloom, 1984; Schulze *et al.*, 1986; Oscarson *et al.*, 1989). Such investigations are presumed to stress the unsaturable term because of the mass-flow effect in connection with transpiration. However, up to now, such a phenomenon has not been observed. On the contrary, the effects of mass flow were seen in the concentration of the xylem fluid and with those parameters which depend on that concentration (Schulze and Bloom, 1984). Thus, while nitrate supply to the roots depends on mass flow in the soil (Strebel and Duynisveld, 1989), uptake of nitrate by the roots obviously is independent of the rate of water uptake.

For growing plants in sand culture, a 10-fold more concentrated nutrient solution must be supplied as compared to a hydroponic culture in order to achieve optimal growth (H. Marschner, personal communication). Even when the nutrient solution is continuously percolating through the substrate, a steep concentration gradient obviously develops around the roots. Kinetic data of ion uptake can therefore only be determined in hydroponic culture.

Because of the complexity of the subterraneous parts of *U. dioica* which consist of two types of organs of various age classes (Teckelmann, 1987), namely roots and rhizomes, the annual nitrophilic weed *C. rubrum* was used, which has a simple root system, for studying the uptake of NO_3^- and NH_4^+. Two to three weeks prior to the experiments the plants were transferred from soil to nutrient solution which was 10-fold diluted as compared to that used in sand culture. For induction of nitrogen deficiency nitrate or ammonium nitrate were omitted from the hydroponic nutrient solution. These plants, like those of *U. diocia*, responded with a significant increase of the R/S ratio ($+$N, 0.35 ± 0.005; $-$N, 0.56 ± 0.13).

The uptake kinetics were determined with the nutrient-flow system for monitoring NH_4^+ and NO_3^-, as described by Bloom and Chapin (1981). Subsequent to each change of the nitrogen concentration in the nutrient solution, several hours were required by the plant to attain a new constant uptake rate. All data result from measurements performed during the daily illumination period.

In the Michaelis-Menten plot, a two-phase characteristic was observed for NO_3^- uptake which did not show an additional, unsaturable and therefore linear term. However, the Lineweaver-Burk plot revealed the typical steps as described by Nissen (1987) for the multiphasic interpretation. For reasons of convenience, however, the K_m and Q_{max} data were determined from the nearly linear sections of the curves (Table I). A K_m of about 15–30 μM and a maximal rate of net influx of 7 μmol g^{-1} dry wt h^{-1} were calculated for the low-concentration uptake system (I) while the high-concentration system II exhibited about a 12-fold higher half-saturation and a 22-fold greater capacity. The characteristics of the latter system are in the same range as described for barley seedlings (Goyal and Huffaker, 1986; Warner and Huffaker, 1989). The Michaelis-Menten plot of nitrogen-deficient plants was slightly sigmoid but did not clearly show system I. The uptake capacity in the low-concentration range was significantly increased as it has been shown also for mutants of *Arabidopsis* (Doddema and Telkamp, 1979). This was presumably due to a substantially increased affinity of system II, the K_m of which was only one-third of that of optimally supplied plants. Q_{max} of system II, on the other hand, was not increased. The K_m and Q_{max} data of system I for NH_4^+ uptake are in good agreement with those reported for wheat seedlings (Goyal and Huffaker, 1986).

When *C. rubrum* was grown on NH_4NO_3 instead of on KNO_3, system I was not significantly affected, neither in plants grown at optimal nitrogen

Table I   Uptake Kinetic Parameters for NO_3^-, NH_4^+ and NH_4NO_3 of *Chenopodium rubrum* plants

		System I		System II	
Treatment	Ion	$K_m(\mu M)$	Q_{max}	$K_m(\mu M)$	Q_{max}
$+$N	NO_3^-	23	7	360	150
$-$N	NO_3^-	nd	nd	120	130
$+$N	NH_4^+	58	4	230	60
$+NH_4NO_3$	NO_3^-	15	4	690	77
$-NH_4NO_3$	NO_3^-	80	11	640	110

Q_{max} refers to μmol x g^{-1} dry wt h^{-1}, nd, not to be determined.

supply nor in N-deprived plants. However, system II was less sensitive to nitrate (considerably higher K_m values) and less effective as well. Similar to the results with radish (Schulze and Bloom, 1984), and increase of nitrate uptake that was attributable to transpiration could not be observed. The reviewed uptake studies until now show that a plant, in addition to increasing the absorbing surface, is also capable of increasing the uptake potential per unit root weight. Interestingly, mechanism II (or phase II) represents the variable part of the uptake system and not, as might have been expected from the lower K_m, system I.

III. Changes in Growth Reduction and in Root-to-Shoot Ratios by Nitrogen Deficiency

When growing at an insufficient nutrient supply, plants invest more of their photosynthetic carbon gain into the root and thus increase the root/shoot ratio of biomass (*R/S* ratio). Producing less photosynthetically active biomass in turn decreases its growth (Schulze and Chapin, 1987). Principally, growth limitation due to a suboptimal supply of a macronutrient and increase of the *R/S* ratio should be inseparable processes. Nevertheless, when investigating the biochemical control of biomass distribution, we tried to separate both phenomena to simplify the system. By carefully selecting the nitrogen concentrations of the nutrient solutions we were able to shift, though not very dramatically, the *R/S* ratio of *U. dioica* without affecting biomass production (Fig. 1).

Optimal biomass production was observed at a nitrate supply between 3 and 15 mM. The *R/S* ratio on the other hand decreased as NO_3^- increased from 1 to 15 mM NO_3^- and remained constant at high-nitrate supply (15 to 22 mM). The separation of the growth and *R/S* ratio effects can be explained by the fact that under optimal nitrogen supply, self-shading is substantial. The smaller leaf area of the low-nitrate plants (3 mM) resulted in a considerably better illumination of all leaves and, in turn, in a higher photosynthetic carbon gain by the lower leaves (Fetene *et al.*, 1993).

IV. Analysis of Sink–Source Relations of Plants with Different Nitrogen Status

A specific *R/S* ratio results from the pattern of assimilate transport from source leaves to sinks. *Urtica dioica* typically bears 8–12 pairs of opposite leaves, the number of which is maintained by shedding the old leaves when new ones emerge (Teckelmann, 1987; Stitt and Schulze, this vol-

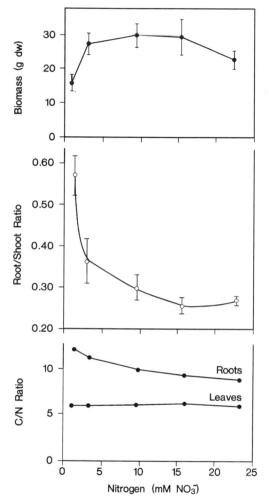

Figure 1 Influence of the nitrogen supply on total dry weight (top), root/shoot ratio of biomass (middle), and the C/N ratio of the shoots (bottom) of 3-month-old *Urtica dioica* plants grown at five different concentrations of nitrate in the nutrient solution (n = 4).

ume, Chapter 4). Nettles in the vegetative developmental stage were considered to sustain only two gross carbon sinks: The shoot apex (the apical meristem plus developing leaves) and the belowground parts consisting of rhizomes and roots. For simplicity, roots and rhizomes are treated together as root fraction. The detailed source–sink relations of *U. dioica* plants, growing at low (3 mM)-, optimal (15 mM)-, and high (22 mM)-nitrate levels were analyzed by ^{14}C-labeling of the photosynthates and a subsequent chase during the following night (Fetene *et al.*, 1993). Only the youngest two leaf pairs represented pure sinks. Older expanding leaves imported assimilates from the source leaves, but simultaneously exported a minor portion of their own photosynthates. The fifth leaf pair (as numbered from the top) and the older leaves were pure source leaves which typically did not import any carbon from other leaves. In principle each sink attracts assimilates from the closest source (Kursanov, 1984; Wardlaw, 1990). However, the relative sink strengths of the shoot apex and of the belowground parts, as measured at the individual nodes of the stem, are subjected to modulation by environment-caused factors such as the N status of a plant. The effect of such factors may largely override the principle of sink–source distances. Thus, the transition point from a mainly acropetal to a predominantly basipetal transport of ^{14}C-labeled assimilates was strongly influenced by the N status of the plants: At high-nitrate supply, a relative sink strength of 1 (i.e., an equal sink strength of both sinks) was found at the node where the fifth leaf pair feeds into the stem (sixth node); at low-nitrate supply, equivalence of both sink strengths was found with the fourth leaf pair, i.e., at the fifth node (Fig. 2). Nitrogen supply can thus at least partly change the direction of carbon flow in the plant (see also this volume, Chapter 4).

The normalized rates of net CO_2 uptake under optimal illumination and at ambient CO_2 concentration were similar with all leaves, except

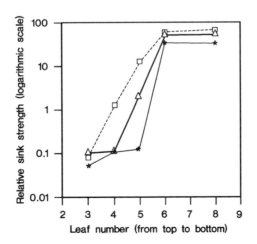

Figure 2 Influence of the level of the nitrate supply (□, low; *, medium; △, high) on relative sink strength (transport of ^{14}C into the plant parts below the $^{14}CO_2$-fixing leaf/ transport into parts above that leaf). Values are averages of two experiments. Leaves are counted basipetally.

the expanding ones. These fixed CO_2 at a rate almost twice as high. This phenomenon was interpreted to indicate sink limitation of photosynthesis in the older leave (Wareing *et al.*, 1968). Sink limitation was, in addition, suggested by the finding that removal of all except one source leaf resulted in a significant stimulation of the rate of photosynthetic net CO_2 uptake (Fig. 3). However, increasing the sink strength of the roots in the low-nitrogen plants did not result in a higher rate of net photosynthesis in the lowest (closest) pair of source leaves. It was concluded that the effect of the N status of the plant on its sink strengths was not sufficient to relieve the sink limitation of photosynthesis in the associated source leaves. Consequently, the different patterns of carbon allocation resulting from growth at different nitrogen levels reflected a rearrangement of the relative sink strengths rather than an increase of the plant's total sink capacity. Otherwise, the biomass production would have responded to the nitrogen status, as it was observed in the low-nitrogen range between 1 and 3 mM NO_3^- (Fig. 1).

V. Are Ammonium-Induced Carbon Sinks Morphogenically Effective?

The simplest and most direct hypothesis of how the level of N supply could mediate a plant's biomass distribution between root and shoot is based on the fact that nitrogen assimilation requires carbon skeletons, i.e., that ammonium-producing tissues attract carbon (see also Komor, this volume, Chapter 6). While nitrate reduction in *U. dioica* is confined more or less exclusively to growing leaves and stalks (less than 5% of the nitrate reductase activity was found in the belowground tissues), the

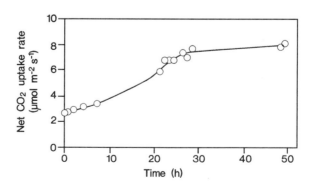

Figure 3   Increase of the rate of net CO_2 uptake of a single source leaf (No. 5) following removal of all other source leaves at time zero of a nonflowering *Urtica dioica* plant.

enzymes involved in ammonium assimilation are present in all parts of the nettles (Ludwig, 1981; Rosnitschek-Schimmel, 1983). *In vivo* operation of the ammonium assimilation pathway in the roots is evidenced by the composition of the root pressure exudates of *Urtica* plants which were grown on 3, 15, and 22 mM NO_3^- or NH_4^+, respectively (Rosnitschek-Schimmel, 1985). When the plants were grown on nitrate the NO_3^- concentration in the root pressure exudate corresponded closely to the supplied concentration (Fig. 4). By contrast the concentration of amino acids in the xylem sap was low and independent of the supplied level of NO_3^-. However, high concentrations of amino acids were obtained in the root pressure exudates of ammonium-fed nettles and the levels of these compounds (mainly asparagine and glutamine) clearly corresponded to the level of NH_4^+ supply (Fig. 4). So did the concentrations of free ammonium, but on a much lower level. Since in ammonium-fed nettles, NH_4^+-assimilation takes place predominantly in roots and rhizomes, the belowground parts of these nettles should attract carbon skeletons at a higher rate than those of *Urtica* plants that were grown on nitrate. If this carbon attraction in the long-term results in a higher biomass accumulation in roots and rhizomes, this effect would be most pronounced in plants growing at a high level (22 mM) of ammonium.

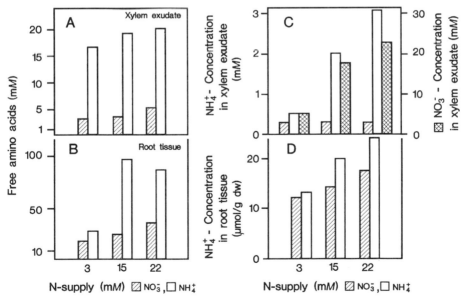

Figure 4 Comparison of the concentrations of NH_4^+, NO_3^-, and amino acids in the root tissue (B, D) and root pressure exudate (A, C) of an *Urtica dioica* plant which had been grown at various levels of nitrogen supply.

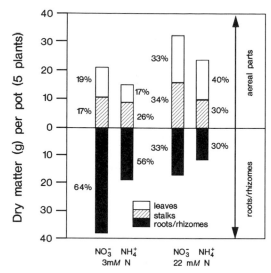

Figure 5 Biomass production and distribution in *Urtica dioica* as influenced by the nitrogen nutrition (3 and 22 m*M* nitrate or ammonium). The percentage values represent percent of total biomass of the various plant parts (after Rosnitschek-Schimmel, 1982).

However, Fig. 5 shows an opposite effect (Rosnitschek-Schimmel, 1982): Considerably less biomass was found in the belowground parts of high-ammonium nettles as compared to the low NH_4^+ plants. The comparison of the biomass distribution of low-ammonium and low-nitrate plants, likewise, argues against the ammonogenic carbon sink hypothesis (Fig. 5). In both cases the bulk of the biomass was found with the fraction of subterraneous parts. It was therefore concluded that an ammonogenic activity of a sink does not result in long-term biomass accumulation. Such a principle would favor growth of roots and rhizomes in the plants growing on ammonium and of the shoots when nitrate is the nitrogen source. Hence, nitrogen metabolism as such cannot be considered to control biomass distribution in the whole plant. Consequently, special patterns of photosynthate distribution must be induced by signals that correspond to the N status of root and shoot and give rise to enhanced growth activities or increased phloem unloading.

VI. Cytokinins Control the Growth Patterns of *Urtica dioica*

It is well known that cytokinins represent a class of phytohormones which directly responds to nitrogen status. This has been shown for a number of annual or perennial species (Wagner and Michael, 1971; Sattelmacher and Marschner, 1978; Salama and Wareing, 1979; Kuiper *et al.*, 1988, 1989; Darral and Wareing, 1981). Due to their predominant production in the roots, they have been discussed as a root signal transferred by the xylem sap to the shoot (Van Staden and Davey, 1979; Neumann *et al.*, 1990). However, not only the molecular mechanism by which cytokinins mediate growth (Romanov, 1990; Trewavas, 1991) but also the cytokinin relations of a plant are still completely unknown. In addition, knowledge of cytokinin metabolism is also incomplete (Letham and Palni, 1983; Koshimizu and Iwamura, 1986; McGaw, 1988), due to the fact that until the 1980s these phytohormones could only be measured with overall bioassays. Today, HPLC combined with the ELISA technique provides a powerful analytical procedure to identify cytokinin patterns of plant tissues and to quantify the individual constituents (Strnad *et al.*, 1990; Sayavedra-Soto *et al.*, 1988; Fusseder *et al.*, 1988). Using this technique, we were able to determine the cytokinin status of *U. dioica* as depending on the nitrogen status and to quantify the cytokinin fluxes from the root to the shoot (Wagner, 1991). The nettles were grown at low (3 m*M*)- and optimal (15 m*M*)-nitrate supply and under the same environmental conditions as described above. Plants 5–6 months old, with 12 to 14 leaf pairs, were analyzed. Figure 6 shows the cytokinin patterns of roots, stems, and young and fully expanded leaves as depending on the nitrogen status of the nettles. In all organs zeatin-type cytokinins were by far the dominant compounds, representing more than 90% of the total cytokinin content. Dihydrozeatin and its derivatives were almost absent and the isopentenyl types were found only in minor amounts. Except the fully expanded source leaves, where zeatin and its O-glucoside were the major cytokinins, zeatinriboside was in large excess over the other zeatin conjugates in roots, stalks, and the young leaves. Again apart from the fully expanded leaves, the O-glucosides, which are considered as storage conjugates (McGaw, 1988; Fusseder *et al.*, 1989), did not accumulate to a remarkable degree. However, zeatin-7-glucoside, which is interpreted as a stabilized but inactive form (Letham and Palni, 1983), was detected in substantial amounts.

Expanding (sink) leaves contained a high concentration of zeatinnucleotide and a relatively high concentration of isopentenyladenosine monophosphate. According to the current theory on the biosynthesis and metabolism of cytokinins (Fig. 7), the nucleotides represent the first cytokinin-type compounds in the biosynthetic pathway from 5′-adenosine monophosphate to the great variety of species of this phytohormone group.

Ribosylation and phosphorylation of cytokinin bases have been reported for a number of plant *in vivo* systems (for review, see Koshimizu and Iwamura, 1986) and therefore the high content of the nucleotides

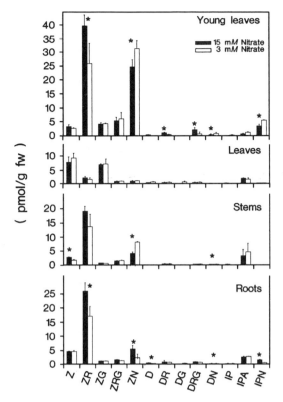

Figure 6 Cytokinin patterns of roots, stems, and young and adult leaves of *Urtica dioica* as influenced by the nitrogen supply to the plants (3 and 15 mM NO$_3^-$, respectively). The columns are statistical means (plus standard error) of four samples, each sample consisting of the respective organs of four plants. *Statistically significant difference between the corresponding data of the low- and high-nitrogen plants. Abbreviations: Z, zeatin; ZR, zeatinriboside; ZG, zeatin O-glucoside; ZRG, zeatinriboside O-glucoside; ZN, zeatinnucleotide; D, dihydrozeatin; DN, dihydrozeatinnucleotide; DR, dihydrozeatinriboside; DRG, dihydrozeatinriboside O-glucoside; IP, isopentenyladenine; IPA, isopentenyladenosine; IPN, isopentenyl-nukleotid.

in the meristematic parts of the shoot does not necessarily indicate cytokinin *de novo* synthesis. However, the root pressure exudate contained mainly zeatinriboside and small amounts of zeatin and only negligible amounts of nucleotides which may be considered as unavoidable impurities (Fig. 8). The relatively high amounts of zeatinnucleotide found in growing leaves must have been produced *in situ*, either by *de novo* synthesis or by phosphorylation of the imported zeatinriboside. Due to the fact that the cytokinin (zeatin and its derivatives) content of the young leaves is about four times as high as that of the adult leaves, both routes are presumably operating in the sink leaves. A controlled interaction of im-

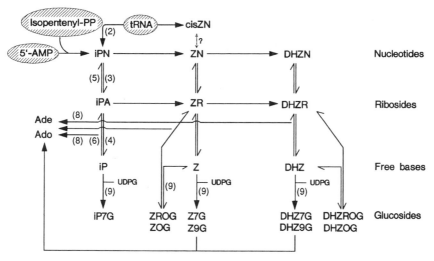

Figure 7 Biosynthetic relations and metabolic interconversions of natural cytokinins (from Koshimizu and Iwamura, 1986, greatly changed). Abbreviations: iP, isopentenyladenine; Z, zeatin; DHZ, dihydrozeatin; G, glucoside; O-G, O-glucoside; Ade, adenin; Ado, adenosin. Enzymes: 1, isopentenyl transferase; 2, ribonuclease; 3, 5'-nucleotidase; 4, adenosine-nucleosidase; 5, adenosine kinase; 6 + 5, adenine phosphoribosyltransferase; 7, zeatin reductase; 8, cytokinin oxidase; 9, glucosyl transferase.

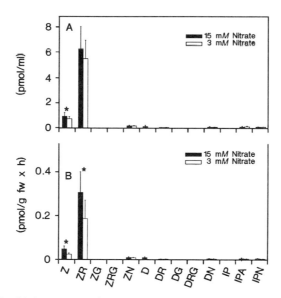

Figure 8 Cytokinin patterns of root pressure exudates of *Urtica dioica* as influenced by the nitrogen supply to the plants. (A) Concentration in the exudate; (B) rate of cytokinin export from the roots. Symbols and abbreviations as in Fig. 6.

ported and *in situ* synthesized cytokinins is suggested by the finding that under low nitrate the nucleotides apparently accumulate in the growing tissues of the shoot (but not in the root) and in the case of young leaves even represent the dominating fraction. A similar interaction has been reported recently for ABA produced upon soil water shortage in the root which subsequent to its transmission into the shoot induces ABA synthesis in the leaves (Neales and McLeod, 1991; see also Schulze, this volume, Chapter 7).

Adding the concentrations of the various cytokinins determined in the individual plant parts reveals the root to be the only organ whose overall cytokinin content clearly responds to the nitrogen status of the plant (Fig. 6). In source leaves, the contents of the individual cytokinins were identical in low- and optimal-nitrate plants. In developing leaves as well as in the stems of low-nitrate plants the substantially lower contents of zeatinriboside were completely balanced by larger pools of the nucleotides of zeatin and isopentenyladenosine. In the roots of low-nitrate plants, however, both the ribosides and the nucleotides were present at significantly lower concentrations. A plot of the concentrations of the isopentenyl-, the zeatin-, and the dihydrozeatin-type cytokinins in the roots vs the R/S ratio revealed a nearly linear relationship with the zeatin-type cytokinins but no dependence on the others (Fig. 9). This correlation suggests that the zeatin-type cytokinins, especially zeatinriboside, act as

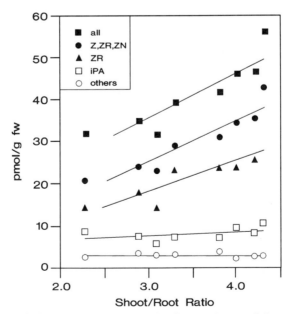

Figure 9 Correlation of the shoot/root ratio of *Urtica dioica* and the cytokinin content of the roots.

an internal signal which mediates the growth response of *Urtica* in accordance with its nitrogen status (see also Komor, this volume, Chapter 6). Inhibition of root growth by application of several native and synthetic cytokinins has been reported (Stenlid, 1982). Inhibition of root growth by 50% was recorded after treatment of wheat and flax roots with 3×10^{-9} and $2 \times 10^{-7} M$ zeatin solutions, respectively. Zeatinriboside was less inhibitory on root growth in wheat and flax by 2 and 1 order of magnitude, respectively. Roots of *U. dioica* that were grown at optimal nitrate supply contained $5 \times 10^{-9} M$ zeatin and $2.5 \times 10^{-8} M$ zeatinriboside. The zeatinriboside concentration of the roots of the corresponding low-nitrogen nettles was lower by one-third (Fig. 6) and thus may indeed be less inhibitory to root growth.

Lower cytokinin content under nitrogen deficiency has been reported for a number of plants (for literature references, see above); however, it has not yet been reported for roots alone. Our findings lead to the question of how the other plant parts of the nettle maintain their overall cytokinin content. Undoubtedly, the root represents the major site of cytokinin production (Letham and Palni, 1983) and irrespective of cytokinin production in other meristematic tissues, the root-produced cytokinins are interpreted as master signals in the control of shoot growth (Sitton *et al.*, 1967; see also Komor, this volume, Chapter 6). Although the cytokinin concentration in root pressure exudates has been determined in a few cases (Upadhyaya *et al.*, 1991; Cahill *et al.*, 1986), a realistic estimation of the daily cytokinin gain of the shoots is still missing. The correlation between the volume of root pressure exudate and the volume of actual xylem sap is unknown and there is no information on the root's capacity for cytokinin synthesis. Information concerning both questions was obtained for nettles which were grown in the root pressure chamber (Passioura and Munns, 1984) which allowed the flow of the root pressure exudate into the xylem stream to be increased. As shown in Fig. 8 the xylem sap of the nettles contained only zeatinriboside and small amounts of the free base zeatin. Expectedly, the concentrations in the xylem fluid of these compounds decreased with increasing exudation rates. The total amount of exported cytokinin approached a maximum of 90 pmol h^{-1} at flow rates of 15 ml h^{-1} and higher (Fig. 10). This rate, which was found in plants grown at optimal nitrate supply, was interpreted to reflect the maximal rate of cytokinin biosynthesis. Due to the nocturnal decrease of the transpiration rate the cytokinin concentration of the xylem sap collected from a source leaf petiole was high during the early morning hours when transpiration started (3.5–4 nM) but continuously decreased until noon to a constant value of about 1.5–2 nM. This concentration was then constant for the rest of the light period. During the night cytokinins accumulated in the root and maximal concentrations were obtained at the end of the dark period.

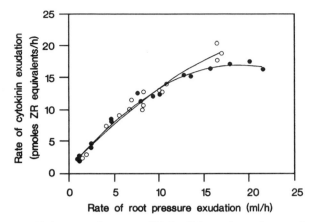

Figure 10 Cytokinin export by roots of *Urtica dioica* as dependent on the volume and production rate of the root pressure exudate. The experiments were performed with nettles that were grown in the root pressure chamber system.

This type of a diurnal course was measured in low-nitrate plants as well, however, at a substantial lower level (1.5–2.5 nM in the morning and 0.5 nM in the afternoon). From the diurnal courses, the daily cytokinin gain by the shoots of optimally N-supplied plants was calculated as 1.2 nmol and that of the low-nitrogen plants as 0.83 nmol which is approximately 30–40% of the cytokinin content of the shoots and about 20% of that of the roots (Table II). Plotting the maximal rates of cytokinin production by the roots versus the S/R ratio revealed a significant relation (Fig. 11), which again suggests the regulatory function of the cytokinins in the biomass partitioning in the plant.

The controlling function of cytokinins was finally established by a study of the carbon distribution in *U. dioica* at an artificially elevated zeatinriboside concentration of the xylem sap (Fetene and Beck, 1993). ^{14}C pulse chase experiments were performed, as described above, with

Table II Cytokinin Relations of *Urtica dioica*

	15 mM	3mM
Nitrate concentration of the nutrient solution		
Total cytokinin content of the shoot (pmol)	3115.6	2695.8
Total cytokinin content of the root (pmol)	2096.8	1718.4
Daily cytokinin export from the root to the shoot (pmol day $^{-1}$)	1204.4	829.4
Ratio of the amount of cytokinin in the shoot to the daily gain	2.6	3.2
Ratio of the amount of cytokinin in the root to the daily export	1.75	2.1

Four plants and xylem fluid collections were combined to form a sample and the data are statistical means of four of such samples.

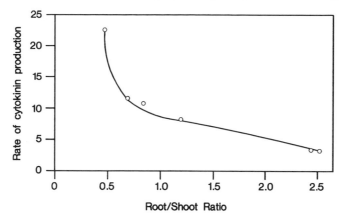

Figure 11 Relation between the root/shoot ratio of *Urtica dioica* plants and the rate of cytokinin production in the roots. The rate of cytokinin synthesis was determined at a rate of root exudation which was adjusted to the rate of transpiration of the intact plants.

nettles grown at 15 mM nitrate, of which a minor portion of the root system was dug out and excised in order to allow unrestricted uptake of exogenously supplied zeatinriboside solutions into the xylem stream. The hormone treatment started 6 h prior to the ^{14}C pulses. Table III shows that the additional zeatinriboside, even when taken up in minute amounts, dramatically enhanced the sink strengths of the shoot tips and more or less completely abolished those of the roots.

Table III Influence of zeatin riboside administered at different concentrations via cut roots to three-month-old *Urtica dioica* plants on photosynthesis, ^{14}C-export from the 6th leaves and relative sink strength (of the root).

	Concentration (mol l^{-1}) of applied ZR solution					
	0 (control)	10^{-7}	10^{-6}	10^{-5}	10^{-4}	10^{-3}
ZR (nmol) taken up by the plant	—	0.24	5.0	56	40	400
Net CO_2 uptake rate (μmol m^{-2}s^{-1})	2.0	1.9	2.8	2.2	2.3	2.1
Exported carbon by the ^{14}C-fed leaf as percent of the total fixed ^{14}C	35.1	19.2	28.7	19.0	19.0	26.2
Relative sink strength (root sink/shoot sink)	54.3	0.05	0.25	0.05	0.004	0.004
Respiration during pulse-chase experiment (% of total fixed ^{14}C)	0.7	5.3	2.7	2.7	5.2	1.9

Under various stress conditions, such as drought, flooding, heat, and high salt concentrations, lower cytokinin concentrations have been measured in the plant tissues as well as in the root pressure exudates (Itai and Vaadia, 1971; Burrows and Carr, 1969; Itai *et al.,* 1977; Torrey, 1976). Because of the above-mentioned restrictions such data are not very conclusive. Nevertheless the obvious question is whether unfavorable conditions cause a decrease of the cytokinin production in the root or whether there is an antagonism counteracting the sink-strengthening effect of cytokinins. An antagonistic interplay has been reported for ABA and cytokinins in the regulation of stomatal aperture. Interruption of

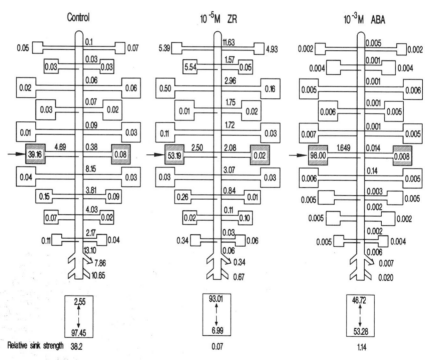

Figure 12 Distribution of ^{14}C in plants of *Urtica dioica* after assimilation of ^{14}CO$_2$ by the sixth leaves (arrows) from 3 to 6 PM and subsequent export of ^{14}C during the dark phase from 6 PM to 8 AM. The amount of ^{14}C which was incorporated into the whole plants was set 100% and the numbers represent percentage thereof. The size of the squares gives an estimate of the average dry weight of the leaves. The arrowheaded parts of the root system indicate the cut roots. The special pretreatment (6 h prior to ^{14}CO$_2$ application) of the plants is indicated on the respective tops of the schemes. The numbers in the squares at the bottom of the schemes show the distribution of ^{14}C between the root and the shoot sink. Relative sink strength was calculated as ^{14}C transported to the root divided by ^{14}C transported to the shoot apex.

the cytokinin supply to the shoot was assumed in explaining restricted stomatal opening under drought (Blackmann and Davies, 1985). As mentioned in the introduction, root etiolement is typical also of drought-stressed plants. Using again the ^{14}C pulse chase technique and the split-root system, the effects of ABA alone and of mixtures of ABA and zeatinriboside on photosynthate distribution in optimally nitrogen-supplied *Urtica* plants were studied. Slightly higher $^{14}CO_2$ concentrations than ambient were required to get radiocarbon photosynthetically fixed when the application of ABA had resulted in an increase of the stomatal resistence. A clear-cut relief of that resistance by application of zeatinriboside together with ABA could not be established. Treatment with ABA strongly inhibited export of labeled photosynthates from the source leaves (Fig. 12). Inhibition of carbon export from source leaves upon soil water depletion has been reported repeatedly (Munns and Pearson, 1974; Robinson *et al.,* 1983; Schurr, 1991) and this effect was traced to a root signal, such as ABA (Williams *et al.,* 1991; Schurr, 1991). However, the small, almost negligible amounts of photosynthate exported from source leaves of ABA-treated plants were distributed more evenly within the nettles (Fig. 12) as compared to the unipolar transport in the controls or the zeatinriboside-treated plants. Schurr's interpretation (1991) of the ABA effect on assimilate translocation in *Ricinus,* namely a decrease of the sink strength of the shoot apex and the concomitant feedback on the pressure flow in the phloem, is not quite in line with our observations. ABA treatment of the shoot alone, as in our experiments, should not have influenced the original sink strength of the intact part of the root system, as it is demonstrated in the controls (Fig. 12).

VII. Studies on the Mechanism of Cytokinin Action on Carbon Distribution

Interpreting the cytokinins as root signals for the distribution of the biomass between the two sinks (root and shoot) leaves us with two major black boxes: (i) The biochemical connection between the plant's N and cytokinin status, and (ii) the mode of stimulation of the sink strength by cytokinin.

While there is very little knowledge about the step(s) in cytokinin metabolism which respond to the plant's nitrogen status or about the type of metabolite controlling such a step, an experiment was made toward revealing the contents of the second black box. The question was whether cytokinins directly stimulate phloem unloading into a sink or whether they enhance phloem unloading by increasing the carbohydrate demand of the sink tissue.

Phloem unloading can occur symplastically, via plasmodesmata, or apoplastically (see also Komor, this volume, Chapter 6). The latter involves transmembrane transport across the plasmalemma of the phloem and the sink cells and passage through the interfacial cell wall (Ho, 1988). A model has been proposed (Eschrich, 1980; Eschrich and Eschrich, 1987), for the control of phloem unloading by phytohormones, which is based on the spatial activation of the cell wall-bound acid invertase. This enzyme is active in the narrow range between pH 4.0 and 6.5 (Fahrendorf and Beck, 1990) and thus sucrose unloaded from the sieve elements into the apoplast will not be cleaved at the natural pH of this compartment. Uncleaved sucrose may reenter the phloem, whereas its constituents glucose and fructose, at least at physiological concentrations, cannot (Kallarackal and Komor, 1989). Effective phloem unloading via the apoplastic route was therefore presumed to involve sucrose cleavage by the cell wall invertase and consequently to require acidification of the apoplast. Hypothetically, phytohormones could activate apoplastic phloem unloading by providing for an acid pH by stimulating a plasmalemma proton pump. However, such a mechanism has not yet been demonstrated (Ho *et al.*, 1991).

In *U. dioica*, a difference in apoplastic pH between sink and source leaves could not be detected even when several methods of investigation were used. The apoplastic pH measured in both types of leaves ranged between 7.0 and 7.2 and therefore is unfavorable to the acid cell wall invertase (Möller and Beck, 1992). Even artificial activation of that invertase by infiltration of the apoplast with a buffer solution of pH 5 was unsuccessful because the cells rapidly adjusted the pH to the normal value. Both types of leaves showed a similar capacity for cleavage of apoplastically administered sucrose at the normal pH, which probably takes place concomitantly with uptake into the cells. All these results are incompatible with the idea of an active cell wall invertase in growing leaves and consequently are contradictory to the proposed control of phloem unloading by phytohormones (see also Komor, this volume, Chapter 6). Expectedly, feeding a detached growing leaf with zeatinriboside did not increase its capacity of sucrose cleavage. More likely, cytokinins, by stimulating cell division activity of a sink (Bernier *et al.*, 1977), may indirectly increase a tissue's demand for photosynthates and nutrients and thus cause enhanced sink strength and, in turn, phloem unloading. Rapid production of mRNAs from nuclear-encoded genes was observed after treatment of cotyledons with benzyladenine (Ohya and Suzuki, 1991), an observation which is in line with our interpretation of the mode of sink strength regulation by cytokinins.

VIII. A Molecular Biological Approach to the Action of Cytokinins

Investigating the action of cytokinins at the molecular level (meristematic activity), due to its complexity, is presently very difficult. When examining at the molecular level the phytohormone-mediated effect of the nitrogen status of a plant on its growth strategy it was obvious that a nitrogen metabolism enzyme is chosen. In addition to the transmembrane carriers for nitrogenous compounds, nitrate reductase represents the bottleneck of the metabolic pathway from nitrate to the amino acids. This has been shown for *Urtica* (Ludwig, 1981; Renner, 1982) as well as for suspension-cultured cells of *C. rubrum* (Renner and Beck, 1988; Beck and Renner, 1989). Nitrate reductase is controlled by various environmental factors, such as light (Bakshi *et al.*, 1979; Kakefuda *et al.*, 1983; Renner and Beck, 1988), the effect of which is presumably mediated by phytochrome (Rajasekar *et al.*, 1988; Melzer *et al.*, 1989), and nitrate concentration (Beevers and Hageman, 1983; Srivastava, 1980; Curtis and Smarelli, 1986; Renner and Beck, 1988). Both factors stimulate *de novo* synthesis of the enzyme. On the other hand, ammonium in many cases represses nitrate reductase synthesis (Syrett and Morris, 1963; Orebanjo and Stewart, 1975; Flores *et al.*, 1980), but not in cell suspension cultures (Heimer and Filner, 1971; Mohanty and Fletcher, 1976; Beck and Renner, 1989), where it stimulates the production of that enzyme. Molecular biology of nitrate reductase from a variety of plant species is under intense investigation (Remmler and Campbell, 1986; Caboche and Rouze, 1990).

Nitrate reductase is also one of the classic examples of an enzyme that is under cytokinin control (Borriss, 1967). Likewise classical is the plant material for studying cytokinin effects on this enzyme, namely the embryo of *Agrostemma githago* (Kende and Shen, 1972; Kende *et al.*, 1971; Hirschberg *et al.*, 1972; Schmerder and Borriss, 1986; Dilworth and Kende, 1974). Phytohormonal effects on nitrate reductase in leaves have been addressed to a much lesser extent (Rao *et al.*, 1984; Roth-Bejerano and Lips, 1970; Parkash, 1982; Lu *et al.*, 1990). Many studies suffer from the fact that induction of nitrate reductase was achieved with synthetic cytokinins such as benzyladenine and that the effect of cytokinin treatment was pleiotropic. Notwithstanding these flaws, there is no doubt that cytokinins stimulate transcription of inducible nitrate reductase genes. However, a more selective system is required (i) to trace the special cytokinin that acts *in vivo* on that process and (ii) to investigate the mechanisms by which the stimulation is accomplished.

Suspension-cultured photoautotropic cells of *C. rubrum,* grown under long-day conditions can be maintained in the stationary phase for several weeks (Fig. 13). During this period, two forms of nitrate reductase are

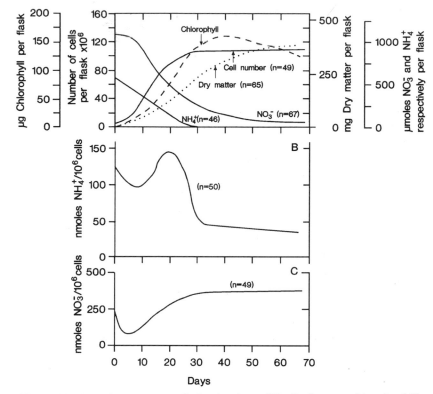

Figure 13 Growth parameters of a batch culture (30 ml) of autotrophic cells of *Chenopodium rubrum* as described by Beck and Renner (1990). (B and C) Changes of the cellular pools of NH_4^+ and NO_3^-, respectively.

suggested by the diurnal course of the enzyme activity: (1) an inducible form, the production of which depends on illumination, and (2) a basic form, for which the level of activity becomes obvious during the dark period (Renner and Beck, 1988; Füchtbauer, 1992). Only the inducible nitrate reductase can be readily suppressed with cycloheximide. At the beginning of the stationary phase of the cell culture the maximal activity of the inducible enzyme is about twice as high as that of the stable form. Induction of that form requires ammonium (Beck and Renner, 1989). The levels of endogenous cytokinins are low during the stationary phase, except those of the O-glucosides (Peters and Beck, 1991), which are deposited in the vacuole (Fusseder and Ziegler, 1988) and may serve as transitory storage conjugates (Fusseder *et al.*, 1989). Among the free bases and ribosides, zeatinriboside reaches the highest concentrations (200 pmol g^{-1} protein) during the course of the day (Füchtbauer, 1992).

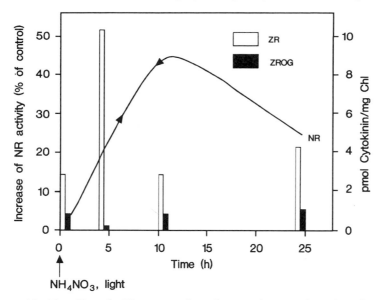

Figure 14   The effect of adding ammonium nitrate at time zero (arrow) on the zeatin-type cytokinins and the nitrate reductase activity of suspension-cultured autotrophic cells of *Chenopodium rubrum* during the stationary phase.

When the formation of inducible nitrate reductase was triggered with ammonium, the cells responded 4 h later with a transitory threefold increase of zeatinriboside (Fig. 14). After 6 h, nitrate reductase reached its maximal activity. Formation of nitrate reductase could also be triggered with cytokinins in the presence of NO_3^-. A comparison of the efficacy of the individual cytokinins showed zeatinriboside as the most effective form. Induction of nitrate reductase activity with this cytokinin did not result in cell number augmentation and hence appeared to be rather specific. A more detailed time kinetics of the induction showed a lag phase of approximately 2 h, followed by a 40% increase of nitrate reductase over the control, a 2-h stable level, and a second substantial increase 6 h after the induction (Fig. 15). The greater part of the first increase could be abolished by pretreatment with cycloheximide but was not impaired by actinomycin D which completely eliminated the second stimulation. The first enhancement of nitrate reductase activity by zeatinriboside was therefore interpreted as stimulation of post-transcriptional processes while the second increase was attributed to *de novo* synthesis, starting with transcription.

The small initial stimulation which was insensitive to both cyclohexi-mide and actinomycin D could have originated from activation of pre-formed nitrate reductase protein, as has been observed also with maize

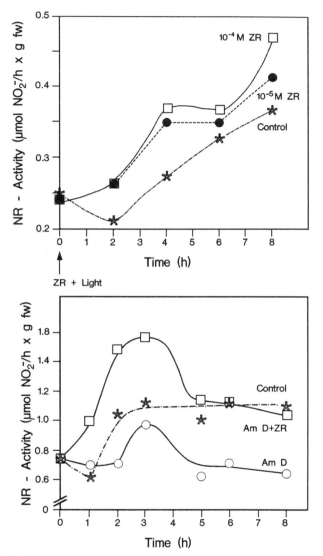

Figure 15 Time course of nitrate reductase activity of suspension cultured cells of *Chenopodium rubrum* subsequent to the addition of zeatinriboside (arrow) simultaneously with the onset of illumination (top). In the lower panel the effect of supplying actinomycin D alone and together with $10^{-5}M$ zeatinriboside is shown.

leaves (Lillo, 1991). It has been shown recently that nitrate reductase can exist in an inactive phosphorylated form and an active dephosphorylated form (Kaiser and Brendle-Behnisch, 1991; Kaiser and Spill, 1991). Presumably, the initial, cycloheximide-insensitive increase in activity is due to dephosphorylation of inactive nitrate reductase.

A modulating, mostly promoting effect of cytokinins on transcription has been reported for several nuclear-encoded proteins (Crowell *et al.*, 1990; Abdelghani *et al.*, 1991) and recently also for nitrogen reductase (Lu *et al.*, 1992). By contrast, an effect on translation has not yet been described. Therefore that effect was examined in more detail.

The structures of 10 nitrate reductase genes (*nia*-genes; Caboche and Rouze, 1990) were aligned and sense and antisense oligonucleotides (18 b) were synthesized using the phosphoramitide method in an automatic synthesizer. These were used as primers for PCR with a RNA extract of *C. rubrum* suspension-cultured cells as template. With the resulting cDNA probe of 147 base pairs, Northern blots were produced which showed an increased nitrate reductase mRNA level in actinomycin-treated cells too.

Elevation of the level of mRNA that precedes accumulation by *de novo* transcription could have resulted from an increased stability against the attack of nucleases or from the liberation of mRNA from a storage form (Shapiro *et al.*, 1987). Increase of the level of mRNA that is due to a greater stability requires ongoing mRNA production. Stimulation of mRNA production in the presence of actinomycin D would imply activation of untranslatable precursor RNA.

The system is under further investigation to elucidate the interaction of zeatinriboside with the respective mRNA. By efficiency and selectivity, our *C. rubrum* system appears to be superior to the classical *A. githago* embryo for the examination of the action of cytokinins on the molecular level.

IX. Conclusion

Regulation of intracellular short-term fluxes of metabolites is presumably independent of phytohormones as messengers (see this volume, Chapter 1). By contrast, the textbooks of plant physiology leave no doubt that long-term phenomena such a growth patterns clearly respond to treatment of plants with phytohormones. However, all such examples result from external application of artificial or natural hormones and even in the split-root experiment reported in this chapter (Table III), a natural though exogenous cytokinin was administered to the plants. On the other hand, endogenous production of hormones by plants, predominantly by their meristematic tissues is textbook knowledge, too. And there is a quite obvious connection between both sets of information resulting in the conclusion that phytohormones control and regulate plant growth, in particular biomass distribution between the individual plant parts. Several correlations corroborate this idea: The nitrogen status of a plant, more accurately the plant's root (Fig. 1), correlates with the cytokinin status which is expressed as cytokinin production and export by the root (Table II). On the other hand, growth patterns as quantified by the root/shoot ratio correlate with the nitrogen status of the root and the cytokinin gain by the shoot. Do we really need more information to conclude that cytokinins are phytohormones which control the biomass distribution within a plant? We do! Not only with respect to the black boxes of reaction mechanisms mentioned in Section VIII but also with respect to the question of whether signals, in particular root signals, are consistently involved in biomass distribution or whether they merely change a given metabolically controlled pattern under suboptimal or adverse environmental conditions. This is not a pure theoretical question because flux control is a principle of regulation quite different from control of assimilate consumption in the sinks (see this volume, this chapter and Chapter 7).

Despite several missing links in the sequence of reactions from nitrogen supply to a certain pattern of assimilate distribution in herbs our data suggest that cytokinins are indispensable root signals, irrespective of the conditions under which the plants are growing. Of course, cytokinins are not the only phytohormones involved in biomass distribution, but presumably they are the most important ones because of their capability of controlling the activity of sinks.

The main results which indicate an obligatory role of cytokinins as root signals for assimilate distribution are the following:

- The root signal could be traced to one individual cytokinin, namely zeatinriboside, whose production and export rates are in agreement with those postulated for a pertinent root signal. It should be emphasized that zeatinriboside is produced and exported into the shoot also under optimal growth conditions.
- Biomass distribution following flux control should comprise control of phloem unloading in sinks. With respect to a utilization sink, such as a developing leaf, unloading passively followed the rate of consumption and thus cannot be considered as being directly controlled (see Section VIII).
- Apparently, there is a hierarchy of control mechanisms in the nitrogen–carbon interaction. Irrespective of where nitrogen assimilation and concomitant carbon utilization took place in the plant, the distribution of assimilates responded to the level of the nitrogen source rather than to the localization of the nitrogenous carbon sinks (Fig. 5).
- Finally, involvement of zeatinriboside in the induction of enzyme (nitrogen reductase) activity was demonstrated, which evidences the efficacy of cytokinins as signals to an addressed plant organ.

To allow clear, though not exhaustive, answers to clear questions, the plant system chosen in this work was as simple as possible. The system involved only vegetative growth of the roots and the shoots, comparable biomass production in the different experiments, a constant supply of nutrients, and the nitrogen concentration being the only variable. The cytokinin fluxes determined so far may represent only part of the total cytokinin balance. They are solely based on cytokinin production in the roots and on distribution of the phytohormones via the transpiration stream. Actually, this means control of the cytokinin gain of the shoot by the production rate only. However, cytokinins have been identified in the phloem sap, at least of trees (Weiler and Ziegler, 1981), and thus may be redistributed via the phloem. Phloem sap collection was impossible with stinging nettles and therefore an eventual involvement of that transport system could not be established. Although the results of this work indicate an indirect control of assimilate fluxes by cytokinins, the importance of the compartments involved in that kind of flux control is as of yet unknown.

Acknowledgments

The data discussed in this chapter have been gathered by a number of guest scientists and co-workers to whom I express my sincere gratitude; these are W. H. Campbell, O. Dauberschmidt, T. Fahrendorf, M. Fetene, B. Füchtbauer, M. Ludwig, I. Marques, I. Möller, W. Peters, U. Renner, I. Rosnitschek-Schimmel, R. Stumm, and B. Wagner. In addition thanks for skillful technical assistance are due to C. Benker, A. Fischbach, H. Simper, and D. Wiesner.

References

Abdelghani, M. O., Suty, L., Chen, J. N., Renaudin, J.-P., and Teyssendier de la Serve, B. (1991). Cytokinins modulate the steady-state levels of light-dependent and light-independent proteins and mRNAs in tobacco cell suspensions. *Plant Sci.* **77,** 29–40.

Ågren, G. I., and Ingestad, I. (1987). Root:shoot ratio as a balance between nitrogen productivity and photosynthesis. *Plant, Cell Environ.* **10,** 579–586.

Bakshi, I. S., Faroogi, A. H. A., and Maheshwari, S. C. (1979). Control of circadian rhythm in nitrate reductase activity in *Wolffia microscopia* Griff. *Plant Cell Physiol.* **20,** 957–963.

Barta, A. L. (1975). Effect of nitrogen nutrition on distribution of photosynthetically incorporated $^{14}CO_2$ in *Lolium perenne. Can. J. Bot.* **53,** 237–242.

Beck, E., and Renner, U. (1989). Ammonium triggers uptake of NO_3^- by Chenopodium rubrum suspension culture cells and remobilization of their vacuolar nitrate pool. *Plant Cell Physiol.* **30,** 487–495.

Beck, E., and Renner, U. (1990). Net fluxes and pools of nitrogenous compounds during suspension culture of photoautotrophic Chenopodium rubrum cells. *Plant, Cell Environ.* **13,** 111–122.

Beevers, L., and Hageman, R. H. (1983). Uptake and reduction of nitrate: Bacteria and higher plants. *Encycl. Plant Physiol., New Ser.* **15A,** 351–375.

Bernier, G., Kinet, J. M., Jacqmard, A., Havelange, A., and Bodson, M. (1977). Cytokinin as a possible component of the floral stimulus in *Sinapis alba. Plant Physiol.* **60,** 282–285.

Blackmann, P. G., and Davies, W. J. (1985). Root to shoot communication in maize plants of the effects of soil drying. *J. Exp. Bot.* **36,** 39–48.

Bloom, A. J., and Chapin, F. S., III (1981). Differences in steady-state net ammonium and nitrate influx by cold- and warm adapted barley varieties. *Plant Physiol.* **68,** 1064–1067.

Bloom, A. J., Chapin, F. S., III, and Mooney, H. A. (1985). Resource limitation in plants—an economic analogy. *Annu. Rev. Ecol. Syst.* **16,** 363–392.

Borriss, H. (1967). Untersuchungen über die Steuerung der Enzymaktivität in pflanzlichen Embryonen durch Cytokinine. *Wiss. Z. Univ. Rostock, Math.-Naturwiss. Reihe* **16,** 629–639.

Brouwer, R. (1962). Nutritive influences on the distribution of dry matter in the plant. *Neth. J. Agric. Sci.* **10,** 361–375, 399–408.

Burrows, W. J., and Carr, J. D. (1969). Effects of flooding the root system of sunflower plants on the cytokinin content in the xylem sap. *Physiol. Plant.* **22,** 1105–1112.

Caboche, M., and Rouze, P. (1990). Nitrate reductase: A target for molecular and cellular studies in higher plants. *Trends Genet* **6,** 187–192.

Cahill, D. M., Weste, G. M., and Grant, B. R. (1986). Changes in cytokinin concentrations in xylem extrudate following infection of *Eucalyptus marginata* Donn. ex Sm. with *Phytophthora cinnamomi* Rands. *Plant Physiol.* **81,** 1103–1109.

Chung, G. C., Rowe, R. N., and Field, R. J. (1982). Relationship between shoot and root of cucumber plants under nutritional stress. *Ann. Bot. (London)* [N. S.] **50,** 859–861.

Crowell, D. N., Kadlecek, A. T., John, M. C., and Amasino, R. M. (1990). Cytokinin-induced mRNAs in cultured soybean cells. *Proc. Natl. Acad. Sci. U.S.A.* **87,** 8815–8819.

Curtis, L. T., and Smarelli, J., Jr. (1986). Metabolite control of nitrate reductase activity in soybean seedlings. *J. Plant Physiol.* **126,** 135–143.

Darral, N. M., and Wareing, P. F. (1981). The effect of nitrogen nutrition on cytokinin activity and free amino acids in *Betula pendula* Roth. and *Acer pseudoplatanus* L. *J. Exp. Bot.* **32,** 369–379.

Dauberschmidt, O. (1992). Anzuchtoptimierung der Grossen Brennessel, *Urtica dioica* L. Diploma Thesis, University of Bayreuth.

Dilworth, M. F., and Kende, H. (1974). Comparative studies on nitrate reductase in *Agrostemma githago* induced by nitrate and benzyladenine. *Plant Physiol.* **54,** 821–825.

Doddema, H., and Telkamp, G. P. (1979). Uptake of nitrate by mutants of Arabidopsis thaliana, disturbed in uptake or reduction of nitrate. II. Kinetics. *Physiol. Plant.* **45,** 332–338.

Ericsson, R. (1981). Effects of varied nitrogen stress on growth and nutrition in three *Salix* clones. *Physiol. Plant.* **51,** 423–429.

Eschrich, W. (1980). Free space invertase, its possible role in phloem unloading. *Ber. Dtsch. Bot. Ges.* **931,** 363–378.

Eschrich, W., and Eschrich, B. (1987). Control of phloem unloading by source activities and light. *Plant Physiol. Biochem. (Paris)* **25,** 625–634.

Fahrendorf, T., and Beck, E. (1990). Cytosolic and cell wall-bound acid invertases from leaves of *Urtica dioica* L.: A comparison. *Planta* **180,** 237–244.

Fetene, M., and Beck, E. (1993). Reversal of sink-source relations in *Urtica dioica* L. plants by increasing cytokinin import into the shoot. *Bot. Acta* **106,** 235–240.

Fetene, M., Möller, I., and Beck, E. (1993). The effect of nitrogen supply to *Urtica dioica* plants on the distribution of assimilate between shoot and roots. *Bot. Acta* **106,** 228–234.

Flores, E., Guerrero, M. G., and Losada, M. (1980). Short-term ammonium inhibition of nitrate utilization by *Anacystis nidulans* and other cyanobacteria. *Arch. Microbiol.* **128,** 137–144.

Füchtbauer, B. (1992). Das Cytokininmuster stationärer Zellkulturen von *Chenopodium rubrum* L. im Verlauf eines Tages und seine Beziehung zur Aktivität der Nitratreduktase. Diploma Thesis, University of Bayreuth.

Fusseder, A., and Ziegler, P. (1988). Metabolism and compartmentation of dihydrozeatin exogenously supplied to photoautotrophic suspension cultures of *Chenopodium rubrum*. *Planta* **173**, 104–109.

Fusseder, A., Wagner, B., and Beck, E. (1988). Quantification by ELISA of cytokinins in root-pressure exudates of *Urtica dioica* plants grown under different nitrogen levels. *Bot. Acta* **101**, 214–219.

Fusseder, A., Ziegler, P., Peters, W., and Beck, E. (1989). Turnover of O-glucosides of dihydrozeatin and dihydrozeatin-9-riboside during the cell growth cycle of photoautotrophic cell suspension cultures of *Chenopodium rubrum*. *Bot. Acta* **102**, 335–340.

Geiger, D. R., and Fondy, B. R. (1991). Regulation of carbon allocation and partitioning: Status and research agenda. *In* "Recent Advances in Phloem Transport and Assimilate Compartmentation" (J. L. Bonnemain, S. Delrot, W. J. Lucas, and J. Dainty, eds.), pp. 1–9. Quest Editions, Presses Academiques, Nantes.

Goyal, S. S., and Huffaker, R. C. (1986). The uptake of NO_3^-, NO_2^-, and NH_4^+ by intact wheat (*Triticum aestivum*) seedlings. *Plant Physiol.* **82**, 1051–1056.

Heimer, Y. M., and Filner, P. (1971). Regulation of the nitrate assimilation pathway in cultured tobacco cells. *Biochim. Biophys. Acta* **230**, 362–372.

Hirschberg, K., Hübner, G., and Borriss, H. (1972). Cytokinin-induzierte de novo-Synthese der Nitratreduktase in Embryonen von *Agrostemma githago*. *Planta* **108**, 333–337.

Ho, L. C. (1988). Metabolism and compartmentation of imported sugars in sink organs in relation to sink strength. *Annu. Rev. Plant Physiol. Plant Mol. Biol.* **39**, 355–378.

Ho, L. C., Lecharny, A., and Willenbrink, J. (1991). Sucrose cleavage in realtion to import and metabolism of sugars in sink organs. *In* "Recent Advances in Phloem Transport and Assimilate Compartmentation" (J. L. Bonnemain, S. Delrot, W. J. Lucas, and J. Dainty, eds.), pp. 178–186. Quest Editions, Presses Academiques, Nantes.

Hofstra, R., Lanting, L., and de Visser, R. (1985). Metabolism of *Urtica dioica* as dependent on the supply of mineral nutrients. *Physiol. Plant.* **63**, 13–18.

Hüsemann, W., and Barz, W. (1977). Photoautotrophic growth and photosynthesis in cell suspension cultures of *Chenopodium rubrum*. *Physiol. Plant.* **40**, 77–81.

Itai, C., and Vaadia, Y. (1971). Cytokinin activity in water-stressed shoots. *Plant Physiol.* **47**, 87–90.

Itai, C., Ben-Zioni, A., and Ordin, L. (1977). Correlative changes in endogenous hormone levels and shoot growth induced by short heat treatments to the root. *Physiol. Plant.* **29**, 355–360.

Johnson, I. R. (1985). A model of partitioning of growth between the shoots and roots of vegetative plants. *Ann. Bot. (London)* **55**, 421–431.

Kaiser, W. M., and Brendle-Behnisch, E. (1991). Rapid modulation of spinach leaf nitrate reductase activity by photosynthesis. I. Modulation *in vivo* by CO_2 availability. *Plant Physiol.* **96**, 363–367.

Kaiser, W. M., and Spill, D. (1991). Rapid modulation of spinach leaf nitrate reductase by photosynthesis. II. *In vitro* modulation by ATP and AMP. *Plant Physiol.* **96**, 368–375.

Kakefuda, G., Duke, S. H., and Duke, S. O. (1983). Differential light induction of nitrate reductases in greening and photobleached soybean seedlings. *Plant Physiol.* **73**, 56–60.

Kallarackal, J., and Komor, E. (1989). Transport of hexoses by the phloem of *Ricinus communis* L. seedlings. *Planta* **177**, 336–341.

Kende, H., and Shen, T. C. (1972). Nitrate reductase in *Agrostemma githago*. Comparison of the inductive effects of nitrate and cytokinin. *Biochim. Biophys. Acta* **286**, 118–125.

Kende, H., Hahn, H., and Kays, S. E. (1971). Enhancement of nitrate reductase activity by benzyl-adenine in *Agrostemma githago*. *Plant Physiol.* **48**, 702–706.

Koshimizu, K., and Iwamura, H. (1986). Cytokinins. *In* "Chemistry of Plant Hormones" (N. Takahashi, ed.), pp. 153–199. CRC Press, Boca Raton, FL.

Kuchenbuch, R., Weigelt, W., and Jungk, J. (1988). Modification of root-shoot-ratio of sunflower (*Helianthus annuus* L.) by nitrogen supply and a triazole-type plant growth regulator. *Z. Pflanzenernaehr. Bodenkd.* **151**, 391–394.

Kuiper, D., Schuit J., and Kuiper, P. J. C. (1988). Effects of internal and external cytokinin concentrations on root growth and shoot to root ratio of *Plantago major* ssp. *pleiosperma* at different nutrient conditions. *Plant Soil* **111**, 231–236.

Kuiper, D., Kuiper, P. J. C., Lambers, H., Schuit, J., and Staal, M. (1989). Cytokinin concentration in relation to mineral nutrition and benzyladenin treatment in *Plantago major* ssp. *pleiosperma*. *Physiol. Plant.* **75**, 511–517.

Kursanov, A. L. (1984). "Assimilate Transport in Plants." Elsevier, Amsterdam.

Letham, D. S., and Palni, L. M. S. (1983). The biosynthesis and metabolism of cytokinins. *Annu. Rev. Plant Physiol.* **34**, 163–197.

Lillo, C. (1991). Diurnal variations of corn leaf nitrate reductase: An experimental distinction between transcriptional and post-transcriptional control. *Plant Sci.* **73**, 149–154.

Lu, J.-L., Ertl, J. R., and Chen, C. M. (1990). Cytokinin enhancement of the light induction of nitrate reductase transcript levels in etiolated barley leaves. *Plant Mol. Biol.* **14**, 585–594.

Lu, J.-L., Ertl, J. R., and Chen, C. M. (1992). Transcriptional regulation of nitrate reductase mRNA levels by cytokinin-abscisic acid interactions in etiolated barley leaves. *Plant Physiol.* **98**, 1255–1260.

Ludwig, M. (1981). Untersuchungen zum Stickstoffmetabolismus der Brennessel (Urtica dioica L.): Aktivitätsdynamik wichtiger Enzyme während der vegetativen und reproduktiven Entwicklung der Pflanze. Diploma Thesis, University of Bayreuth.

Lundegårdh, H. (1954). "Klima und Boden." G. Fischer, Fourth edit. Jena.

McGaw, B. A. (1988). Cytokinin biosynthesis and metabolism. *In* "Plant Hormones and Their Role in Plant Growth and Development" (P. D. Davies, ed.), pp. 76–93. Kluwer Acad. Publ., Dordrecht, The Netherlands.

Melzer, J. M., Kleinhofs, A., and Warner, R. L. (1989). Nitrate reductase regulation: Effects of nitrate and light on nitrate reductase mRNA accumulation. *Mol. Gen. Genet.* **217**, 341–346.

Mohanty, B., and Fletcher, J. S. (1976). Ammonium influence on the growth and nitrate reductase activity of Paul's Scarlet Rose Suspension cultures. *Plant Physiol.* **58**, 152–155.

Möller, I., and Beck, E. (1992). The fate of apoplastic sucrose in sink and source leaves of *Urtica dioica*. *Physiol. Plant.* **85**, 618–624.

Munns, R., and Pearson, C. J. (1974). Effect of water deficit on translocation of carbohydrate in *Solanum tuberosum*. *Aust. J. Plant Physiol.* **1**, 529–537.

Neales, T. F., and McLeod, A. L. (1991). Do leaves contribute to the abscisic acid present in the xylem sap of 'droughted' sunflower plants. *Plant, Cell Environ.* **14**, 979–986.

Neumann, D. S., Rood, S. B., and Smit, B. A. (1990). Does cytokinin transport from root-to-shoot in the xylem sap regulate leaf response to root hypoxia? *J. Exp. Bot.* **41**, 1325–1333.

Nissen, P. (1987). Multiple or multiphasic uptake mechanisms in plants? *Plant, Cell Environ.* **10**, 475–485.

Ohya, T., and Suzuki, H. (1991). The effect of BA on the accumulation of messenger RNAs that encode the large and small subunits of Rubisco and light-harvesting chlorophyll a/b protein in excised cucumber cotyledons. *Plant Cell Physiol.* **32**, 577–580.

Olsen, C. (1921). The ecology of *Urtica dioica*. *J. Ecol.* **9**, 1–19.

Orebanjo, T. O., and Stewart, G. R. (1975). Ammonium repression of nitrate reductase formation in *Lemna minor*. *Planta* **122**, 27–36.

Oscarson, P., Ingemarsson, B., and Larsson, C.-M. (1989). Growth and nitrate uptake properties of plants grown at different relative rates of nitrogen supply. II. Activity and affinity of the nitrate uptake system in Pisum and Lemna in relation to nitrogen availability and nitrogen demand. *Plant, Cell Environ.* **12**, 787–794.

Parkash, V. (1982). Involvement of cytokinins and other growth regulating substances in nitrate assimilation. *Plant Physiol. Biochem. (Paris)* **9,** 48–53.

Passioura, J. B., and Munns, R. (1984). Hydraulic resistance of plants. II. Effects of rooting medium and time of day, in barley and lupin. *Aust. J. Plant Physiol.* **11,** 341–350.

Pate, J. S., Layzell, D. B., and Atkins, C. A. (1979). Economy of carbon and nitrogen in nodulated and non-nodulated (NO_3^--grown) Legume. *Plant Physiol.* **64,** 1083–1088.

Patrick, J. W. (1991). Control of phloem transport to and short-distance transfer in sink regions: An overview. *In* "Recent Advances in Phloem Transport and Assimilate Compartmentation" (J. L. Bonnemain, S. Delrot, W. J. Lucas, and J. Dainty, eds.), pp. 167–177. Quest Editions, Presses Académiques, Nantes.

Peters, W., and Beck, E. (1991). Endogenous cytokinins in suspension cultured cells of *Chenopodium rubrum* at different growth stages. *In* "Physiology and Biochemistry of Cytokinins in Plants" (M. Kamínek, D. W. S. Mok, and E. Zazímalová, eds.), pp. 71–73. SPB Academic Publishing, The Hague, The Netherlands.

Rajasekar, V. K., Gowri, G., and Campell, W. H. (1988). Phytochrome-mediated light regulation of nitrate reductase expression in squash cotyledons. *Plant Physiol.* **88,** 242–244.

Rao, L. V. M., Datta, N., Mahadevan, M., Guha-Mukherjee, S., and Sopory, S. K. (1984). Influence of cytokinins and phytochrome on nitrate reductase activity in etiolated leaves of maize. *Phytochemistry* **23,** 1875–1879.

Reif, A., Teckelmann, M., and Schulze, E.-D. (1985). Standortansprüche von *Urtica dioica. Flora (Jena)* **176,** 365–383.

Remmler, J. L., and Campbell, W. H. (1986). Regulation of corn leaf nitrate reductase. *Plant Physiol.* **80,** 442–447.

Renner, U. (1982). Untersuchungen an zwei Enyzmen aus dem Stickstoff-Metabolismus von *Urtica dioica* L. Diploma Thesis, University of Bayreuth.

Renner, U., and Beck, E. (1988). Nitrate reductase activity of photoautotrophic suspension culture cells of *Chenopodium rubrum* is under the hierarchical regime of NO_3^-, NH_4^+ and Light. *Plant Cell Physiol.* **29,** 1123–1131.

Robinson, M., Havevy, A. H., Galili, D., and Plaut, Z. (1983). Distribution of assimilates in *Gladiolus grandiflorus* as affected by water deficit. *Ann. Bot. (London)* [N. S.] **51,** 461–468.

Romanov, G. A. (1990). Cytokinins and tRNAs: A hypothesis on their competitive interaction via specific receptor proteins. *Plant, Cell Environ.* **13,** 751–754.

Rosnitschek-Schimmel, I. (1982). Effect of ammonium and nitrate supply on dry matter production and nitrogen distribution in *Urtica dioica. Z. Pflanzenphysiol.* **108,** 329–341.

Rosnitschek-Schimmel, I. (1983). Biomass and nitrogen partitioning in a perennial and an annual nitrophilic species of *Urtica. Z. Pflanzenphysiol.* **109,** 215–225.

Rosnitschek-Schimmel, I. (1985). The influence of nitrogen nutrition on the accumulation of free amino acids in root tissue of *Urtica dioica* and their apical transport in xylem sap. *Plant Cell Physiol.* **26,** 215–219.

Roth-Bejerano, N., and Lips, S. H. (1970). Hormonal regulation of nitrate reductase activity in leaves. *New Phytol.* **69,** 165–169.

Rufty, T. W., Raper, C. D., and Huber, S. C. (1984). Alterations in internal partitioning of carbon in soybean plants in response to nitrogen stress. *Can. J. Bot.* **62,** 501–508.

Rufty, T. W., Huber, S. C., and Volk, R. J. (1988). Alterations in leaf carboydrate metabolism in response to nitrogen stress. *Plant Physiol.* **88,** 725–730.

Salama, A. M. S. E. D. A., and Wareing, P. F. (1979). Effects of mineral nutrition on endogenous cytokinins in plants of sunflower (*Helianthus annuus* L.). *J. Exp. Bot.* **30,** 971–981.

Sattelmacher, B., and Marschner, H. (1978). Nitrogen nutrition and cytokinin activity in *Solanum tuberosum. Physiol. Plant.* **42,** 185–189.

Sayavedra-Soto, L. A., Durley, R. C., Trione, E. J., and Morris, R. O. (1988). Identification of cytokinins in young wheat spikes (*Triticum aestivum* cv. Chinese Spring). *J. Plant Growth Regul.* **7,** 169–178.

Schmerder, B., and Borriss, H. (1986). Induction of nitrate reductase by cytokinin and ethylene in *Agrostemma githago* L. embryos. *Planta* **169,** 589–593.

Schulze, E.-D., and Bloom, A. J. (1984). Relationship between mineral nitrogen influx and transpiration in radish and tomato. *Plant Physiol.* **76,** 827–829.

Schulze, E.-D., and Chapin, F. S., III (1987). Plant specialization to environments of different resource availability. *Ecol. Stud.* **61,** 120–148.

Schulze, E.-D., Koch, G., Percival, F., Mooney, H. A., and Chu, C. (1986). The nitrogen balance of *Raphanus sativus raphanistrum* plants. I. Daily nitrogen use under high nitrate supply. *Plant, Cell Environ.* **8,** 713–720.

Schurr, U. (1991). Die Wirkung von Bodentrockenheit auf den Xylem und Phloemtransport von *Ricinus communis* und deren Bedeutung für die Interaktion zwischen Wurzel und Spross. Ph.D. Thesis, University of Bayreuth.

Shapiro, D. J., Blume J. E., and Nielsen, D. A. (1987). Regulation of messenger RNA stability in eukaryotic cells. *BioEssays* **6,** 221–226.

Siddiqi, M. Y., Glass, A. D. M., Ruth, T. J., and Rufty, T. W., Jr. (1990). Studies of the uptake of nitrate in barley. *Plant Physiol.* **93,** 1426–1432.

Sitton, D., Itai, C., and Kende, H. (1967). Decreased cytokinin production in the roots as a factor in shoot senescence. *Planta* **73,** 296–300.

Spek, L., and Van Oijen, M. (1988). A simulation model of root and shoot growth at different levels of nitrogen availability. *Plant Soil* **111,** 191–197.

Srivastava, H. S. (1980). Regulation of nitrate reductase activity in higher plants. *Phytochemistry* **19,** 725–733.

Stenlid, G. (1982). Cytokinins as inhibitors of root growth. *Physiol. Plant.* **56,** 500–506.

Strebel, O., and Duynisveld, W. H. M. (1989). Nitrogen supply to cereals and sugar beet by mass flow and diffusion on a silty loam soil. *Z. Planzenernaehr. Bodenkd.* **52,** 135–142.

Strnad, M., Vanek, T., Binarová, P., Kaminek, M., and Hanuš, J. (1990). Enzyme immunoassays for cytokinins and their use for immunodetection of cytokinins in Alfalfa cell culture. *In* "Molecular Aspects of Hormonal Regulation of Plant Development" (M. Kutácek, M. C. Elliott, and I. Macháčková, eds.), pp. 41–54. SPB Academie Publishing, The Hague, The Netherlands.

Syrett, P. J., and Morris, I. (1963). The inhibition of nitrate assimilation by ammonium in *Chlorella. Biochim. Biophys. Acta* **67,** 566–575.

Teckelmann, M. (1987). Kohlenstoff-, Wasser- und Stickstoffhaushalt von *Urtica dioica* L. an natürlichen Standorten. Ph.D. Thesis, University of Bayreuth.

Thornley, J. H. M. (1972). A balanced quantitative model for root: Shoot ratios in vegetative plants. *Ann. Bot. (London)* [N. S.] **36,** 431–441.

Torrey, J. G. (1976). Root hormones and plant growth. *Annu. Rev. Plant Physiol.* **27,** 435–459.

Trewavas, A. (1991). How do plant growth substances work? *Plant, Cell Environ.* **14,** 1–12.

Upadhyaya, N. M., Parker, C. W., Letham, D. S., Scott, K. F., and Dart, P. J. (1991). Evidence for cytokinin involvement in Rhizobium (IC3342)-induced leaf curl syndrome of pigeonpea (*Cajanus cajan* Millsp.). *Plant Physiol.* **95,** 1019–1025.

Van Staden, J., and Davey, J. E. (1979). The synthesis, transport and metabolism of endogenous cytokinins. *Plant, Cell Environ.* **2,** 93–106.

Wagner, B. M. (1991). Der Cytokininhaushalt der Brennessel. Ph.D. Thesis, University of Bayreuth.

Wagner, H., and Michael, G. (1971). Der Einfluss unterschiedlicher Stickstoffversorgung auf die Cytokininbildung in Wurzeln von Sonnenblumenpflanzen. *Biochem. Physiol. Pflanz.* **162,** 147–158.

Wardlaw, I. F. (1990). The control of carbon partitioning in plants. (Tansley rev. #27). *New Phytol.* **116,** 341–381.

Wareing, P. F., Khalifa, M. M., and Treharne, K. J. (1968). Rate-limiting processes in photosynthesis at saturating ligth intensities. *Nature (London)* **220,** 453–457.

Warner, R. L., and Huffaker, R. C. (1989). Nitrate transport is independent of NADH and NAD(P)H nitrate reductases in barley seedlings. *Plant Physiol.* **91,** 947–953.

Weiler, E. W., and Ziegler, H. (1981). Determination of phytohormone in phloem exsudate from tree species by radioimmunoassay. *Planta* **152,** 168–170.

Williams, J. H. H., Minchin, P. E. H., and Farrar, J. F. (1991). Carbon partitioning in split root systems of barley: The effect of osmotica. *J. Exp. Bot.* **42,** 453–460.

6

Regulation by Futile Cycles: The Transport of Carbon and Nitrogen in Plants

E. Komor

I. Introduction

By a series of specific transport processes, carbon and nitrogen compounds produced by photosynthetic assimilation in leaves or absorbed by roots from the soil are distributed throughout the plant. The transport processes are governed by membrane proteins which facilitate the movement of hydrophilic solutes across the lipophilic barriers of membranes. In that respect, the transport of organic compounds is fundamentally different from that of water which follows thermodynamic gradients largely unhindered by cellular membranes. Gene analysis of procaryotic and eucaryotic microorganisms has led to the molecular identification of several transport proteins which show a conspicuous structural homology of 12 (in a few cases 6) transmembrane spans (Maiden *et al.*, 1987). The two transport proteins isolated from plants follow the same pattern (Sauer, 1991). In some cases passage of solutes mediated by transport proteins may be passive and the solute will be distributed at the same concentration on both sides of the membrane. Mostly, however, solutes are accumulated on one side of the membrane because the transmembrane voltage (internally negative) maintained by ion pumps attracts the flow of positively charged compounds. Because the flow of neutral substances (such as sugar) or of anions is stoichiometrically coupled with the flow of proton(s), their accumulation is also powered by the protonmotive force (Komor, 1982). In the plant kingdom all these possibilities are present. For example, there is an uptake of hexoses by proton symport (Komor, 1973) and an uptake of (positively charged) arginine by uniport (Komor *et al.*, 1981b). However, there are also apparently passive flows of sugars at high concentrations (Opekarova and Kotyk, 1973).

Membrane-integrated transport systems catalyze both the uptake of nutrients into cells and the transfer from cell to cell. They are equally important for the long-distance transport in the phloem and xylem because nutrients also have to reach the long-distance transport path (Geiger, 1975; Pate, 1975). This requires transmembrane steps to load the nutrients into the vascular systems where they are accessible to a convective water flow and are eventually distributed between plant organs.

Since active transport has been identified as an enzyme-catalyzed, genetically regulated process, transport studies should no longer be confined to phenomenological descriptions to prove that *there is* catalyzed transport of particular solutes (although the latter approach had to be the first step of research in this field). The integration of transport processes, plant metabolism and plant communication are considered in this chapter to evaluate the contribution of uptake systems for the nutrition of plants, their responses to environmental factors, and their role in the regulation and coordination of plant growth and performance. The chapter focuses on sugars and amino acids which are probably the most important assimilation products conveying the macronutrients carbon and nitrogen in the plant.

II. Export from the Source: Phloem Loading

Assimilation of carbon proceeds to the largest extent in fully developed leaves. The same is partly true for nitrogen, as some assimilation also occurs in the root in some species (Beevers and Hageman, 1983). In both cases the sites of assimilation (sources) are not identical with the sites of the main requirement for the nutrient (sinks) which are the apex of growing leaves and root tips. Since the latter are not the targets of transpirational water flow, the delivery of carbon and nitrogen to these places must be via the phloem whose solute and water flow is driven by the active process of phloem loading. The pathway and the possible mechanisms of phloem loading have been a matter of extreme controversy (e.g., Lucas, 1985), which, in part, originated from the lack of simple experimental systems for an unambiguous testing of hypotheses. Phloem loading has been tested by circumstantial means such as radioactive labeling of leaf discs or analysis of sieve tube exudate from severed aphid stylets. In the first case, no clear proof for the solute identity (and location) could be given. In the second case, no controlled substrate feeding was possible. To resolve the dilemma, an experimental system was established using

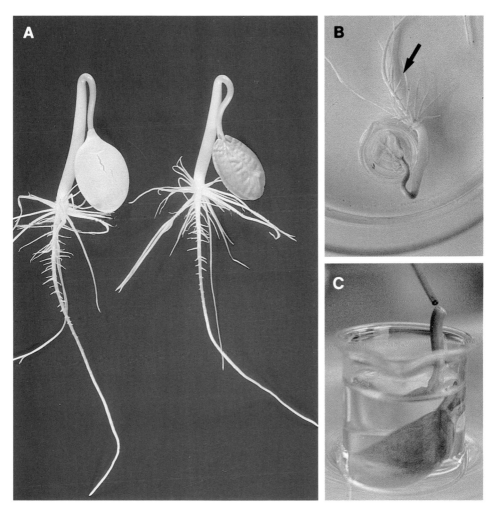

Figure 1 The castor bean seedling: (A) 6 days after soaking in water, with and without endosperm; (B) an excised seedling, the arrow shows the place of the cut; (C) the cotyledons incubating in a medium and collection of sieve tube sap with a microcapillary. *(Figure continues on next page)*

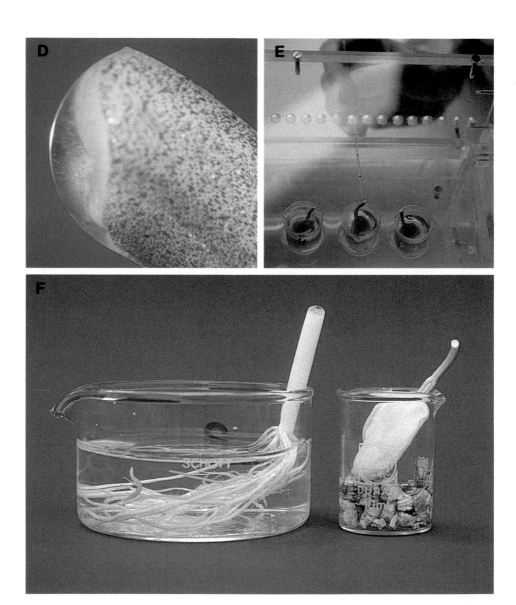

Figure 1 *(continued)* (D) the cut hypocotyl stump with a droplet of exuding sieve tube sap; (E) temperature-controlled water bath with Plexiglas hood and several beakers containing cotyledon pairs; (F) the cut root system of the seedling (left) used to collect xylem exudate and an endosperm-containing cut seedling (right) for collection of phloem sap under quasi-natural conditions.

the *Ricinus* seedling. This system allowed both the analysis of sieve tube sap and the controlled feeding of the loading site (Kallarackal *et al.*, 1989).

A. The Experimental System: The Seedling of Castor Bean (*Ricinus communis*)

The cotyledons of the castor bean seedling take up nutrients which are mobilized in the endosperm during germination (Kriedemann and Beevers, 1967). The lower epidermis of the cotyledons is closely attached to the endosperm and lacks a cuticle so that hydrophilic solutes originating from the endosperm can be easily taken up by the mesophyll cells. Part of the nutrients are used by the cotyledons for their own growth. However, the larger part is loaded into the phloem and translocated to the axis of the seedling to sustain the growth of hypocotyl and roots. When the endosperm is carefully removed, the exposed cotyledons can be incubated in a medium with test solutes, which may be loaded into the phloem. Castor bean is one of the very few plants whose sieve tubes do not instantaneously plug after cutting (Milburn, 1974). Sieve tube sap will exude at a rate of 5–40 $\mu l \cdot h^{-1}$ for several hours when the hypocotyl is cut. Figure 1 shows the standard experimental setup, i.e., a 6-day-old seedling with and without endosperm. The cut hypocotyl stump exudes sieve tube sap, which is collected with a microcapillary. Excised cotyledons are incubated in beakers and placed in a temperature and humidity-controlled Plexiglas hood. The excised root system may also be used to collect xylem exudate driven up by root pressure.

Sucrose is the major compound in sieve tube sap. In addition there is a series of amino acids present (Table I). Sucrose plus amino acids constitute ca. 90% of solutes. Potassium is the dominant cation followed by much lower amounts of magnesium and sodium. The anions phosphate, chloride, sulfate, and sugar phosphates are all present at roughly similar but low concentrations.

In general, the composition of sieve tube sap from *Ricinus* seedlings with the endosperm attached is similar to that from adult *Ricinus* plants (Komor *et al.*, 1987) and it is also similar to that of other dicots. However, it is very dissimilar to that of cucurbitaceae (Ziegler, 1975). The high concentration of sucrose, amino acids, and potassium ions and the low concentration of hexoses, nitrate, calcium, and glycolytic enzymes indicate that the collected exudate represents pure sieve tube sap, not contaminated by solutes from the cut parenchyma cells of the hypocotyl.

The sucrose concentration of the exudate from the cut hypocotyl will remain constant at 300–400 mM for several hours (Fig. 2) if the cotyledons are embedded in the endosperm. If the endosperm is removed the sucrose concentration falls continuously until it reaches a level of

Table I Major Compounds of Sieve Tube Sap from
Ricinus Seedlings (with the Endosperm Attached, First
row) and from Adult *Ricinus* Plants (Second row)

Compound	Concentration (mM)	
	Seedlings	Mature plant
Sucrose	270	259^a
Glucose	1.8	trace^a
Fructose	0.6	trace^a
Amino acids	158	113^a
Neutral	119	86.5^a
Basic	32.4	2.7^a
Acidic	5.9	23.8^a
K$^+$	25	68.1^a–86.6^b
Na$^+$	3	1.0^b–6.7^a
Mg^{2+}	4	3.9^a–7.1^b
Ca^{2+}	0.1	1.3^a–5.6^b
Chloride	6	3.9^b–$8.9^{a,c}$
Phosphate	5	7.6^c–27.9^b
Sulfate	2.5	25.8^b
Nitrate	0.1	$3.0^{a,c}$–4.1^b
Malate	0.5	5.2^c–8.8^b
Other organic anions	nm	10.9^b–15.5^a
Sugar phosphates	3	nm

^{a} Smith and Milburn, 1980.
^{b} Van Beusichem *et al.*, 1985.
^{c} Allen and Smith, 1986.
nm, not measured.
   The exudate from the seedlings was collected from the cut hypocotyl. The first 5 μl exudate was discarded because of possible contami-
nation by the wounded hypocotyl parenchyma. The data for adult
plants were taken from the literature (see also Schulze, this volume,
Chapter 7, Table 1).

ca. 100 mM after 3 h (Fig. 2). At that level it stays constant for many
hours. The results indicate that phloem loading of sucrose depends on
the delivery of sucrose from outside (in this case, the endosperm), but
that there are also carbohydrate sources inside the cotyledon, which
supply sucrose to the phloem though at a lower rate.

   The sieve tube sap composition changed rapidly in response to sucrose
in the medium surrounding the cotyledons as is shown by addition of
radioactively labeled sugar (Fig. 3). Within 15 min radioactivity appeared
in the sieve tube sap at the cut hypocotyl, although it took 2 h to reach
a steady state. Therefore, it became routine practice to incubate the
excised cotyledons for 3 h under any new experimental condition to
allow steady state to be established.

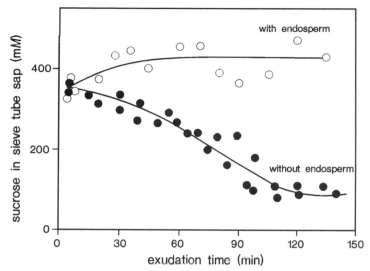

Figure 2 Sucrose concentration in phloem exudate collected from the cut hypcotyl of a seedling which still contains the endosperm and from a seedling without endosperm. In the first case the endosperm was covered in wet vermiculate (see Fig. 1F), in the second case the excised cotyledons were incubated in 5 mM sodium phosphate buffer, pH 6.0. Sucrose was determined by enzymatic test.

The decrease of solute concentration in sieve tube sap after removal of the endosperm was different for the different solutes, since it depended on the size of the storage pools in the cotyledons. In the case of amino acids, the pools seemed to be exhausted quite quickly and the amino acid concentration in the sieve tube sap fell to 10% of the initial value. In

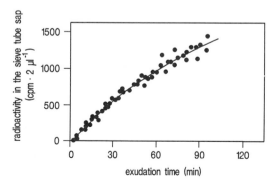

Figure 3 Labeling of sieve tube sap collected from the cut hypocotyl. Radioactive [^{14}C]sucrose was added to the medium of the excised cotyledons at Time 0. Exudate was collected in 2-μl samples and the radioactivity was measured by scintillation spectrometry.

contrast, it was impossible to decrease the magnesium ion concentration of the sieve tube sap by starvation.

In conclusion, the similarity of the sieve tube sap composition with that of other plants, the relative ease of sieve tube sap collection, and the possibility of feeding substrates to the phloem justifies the use of the *Ricinus* seedling as a model plant for the study of mechanism and regulation of phloem loading.

B. Concentration Dependence and Substrate Specificity of Phloem Loading

The composition of sieve tube sap obtained from various plants showed certain common features such as high content of sucrose and K^+ and low content of hexoses and nitrate. The question is open, which process(es) control(s) solute specificity? Is it the phloem, which only takes up particu-

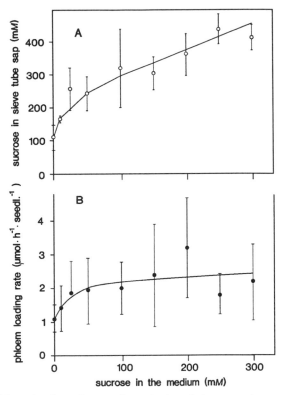

Figure 4   Phloem loading of sucrose by excised cotyledons at various sucrose concentrations in the medium. (A) Sucrose concentrations in sieve tube sap. (B) Loading rate for sucrose (i.e., volume flow rate × sucrose concentration in sieve tube sap).

lar solutes from a series of offered compounds, or is it the mesophyll tissue of the leaf, which delivers only particular solutes to the phloem? With the *Ricinus* seedling the specificity of the phloem loading system can be directly tested, since substrates can be offered to the phloem at will, irrespective of whether a given substrate will occur at the loading site naturally.

Sucrose loading was first investigated to check the concentration dependence for a definitely phloem mobile solute. Incubation of cotyledons in media of increasing sucrose concentrations resulted in increasing concentrations of sucrose in the phloem sap (Fig. 4A). The sucrose level in the sieve tube was always higher than that in the medium. When the rate of sucrose loading was determined (concentration × volume flow rate), a saturation-like behavior became evident (Fig. 4B). Obviously, phloem loading is stimulated by up to 100 mM sucrose in the medium. Above that level, the rate of loading is saturated, though the sucrose concentration in the sieve tube sap increases. This means that even though sucrose is always accumulated in the phloem, an increase of sucrose production above the level corresponding to 100 mM sucrose in the apoplast will not result in a higher rate of delivery of carbon to a sink. If there is a shift to higher sucrose concentrations in the leaf due to, e.g., an adaption to low water potentials (Zrenner and Stitt, 1991), this shift will not be counteracted by an increased sucrose export via the phloem (provided that the concentration was already saturating for phloem loading).

When glucose or fructose were applied to the cotyledons, each of these hexoses was found in the phloem sap at concentrations much higher than those found in the intact system (Fig. 5). This indicates that the phloem is able to load hexoses and that in the intact system hexoses usually do not reach the phloem-loading site, i.e., the control on hexoses is exerted in the mesophyll. Glucose and especially fructose applied to the cotyledons result in an increase of sucrose concentration in the sieve tube sap (Fig. 5). Labeling experiments have shown that sucrose in the sieve tube sap originated from the added hexoses (Kallarackal and Komor, 1989). The larger sucrose production from added fructose compared to that from added glucose is explained by the dominant role of sucrose synthase (Geigenberger and Stitt, 1991) in *Ricinus* cotyledons. The faster metabolism of fructose may be the reason for the lower rate of fructose appearance in the phloem. Thus phloem loading may be controlled by the mesophyll by delivery of solutes (e.g., sucrose) and by rapid removal of solutes (e.g., fructose).

The specificity of phloem loading for amino acids is another example of the various aspects of control exerted by mesophyll and by phloem on substrate specificity. Addition of glutamine to the cotyledons strongly increased the glutamine level in sieve tube sap (Fig. 6). Also, alanine was

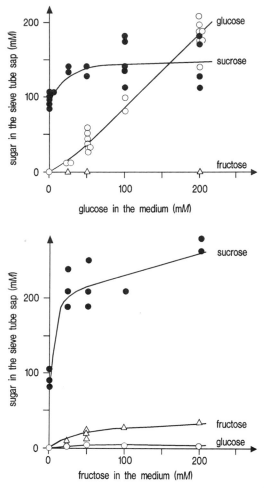

Figure 5 Phloem loading of hexoses. Glucose and fructose were added for 3 h to the medium at the indicated concentrations and then the concentration of sucrose (closed circles), glucose (open circles), and fructose (triangles) in the sieve tube sap was determined. The sugars were measured by enzymatic test.

loaded, though to a lower level; on the other hand, arginine and especially glutamate were hardly loaded even when supplied to the cotyledons at the same concentration as glutamine or alanine (Schobert and Komor, 1989). All these amino acids also led to a small increase of other amino acids in the phloem sap, apparently derived from metabolic conversions. The efficiency of amino acid species loaded into the phloem is not paralleled by their ability to be taken up into the cotyledon (Robinson and Beevers, 1981). This may be indicative of different substrate specificities of amino acid transport systems at the mesophyll and at the phloem.

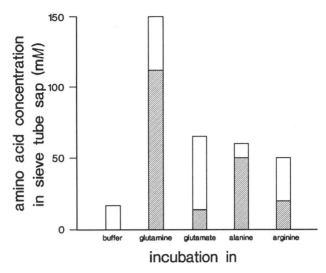

Figure 6 Phloem loading of various amino acids. The cotyledons were incubated for 3–4 h in a 10 m*M* solution of one of the indicated amino acids (glutamine, glutamate, arginine, or alanine). The sieve tube sap was collected and analyzed by an automated amino acid analyzer. The columns show the total amino acid concentration in sieve tube sap after the particular treatment, and the shaded areas indicate the concentration of that amino acid species which was added to the medium.

C. The Route of Phloem Loading: Apoplastic vs Symplastic

In recent years, there has been a passionate debate on the route of phloem loading, whether it is through the symplast or through the apoplast. In case of a purely symplastic route, sugars and amino acids are taken up by mesophyll cells and guided from cell to cell across plasmodesmata to the sieve tube. A necessary prerequisite for that route is the abundant presence of plasmodesmata between all cells from the mesophyll to the sieve tubes. Since plasmodesmata are not known to be capable of selective active uptake, the accumulation of sugars and amino acids must occur primarily by mesophyll cells to achieve the high concentrations in the sieve tube sap. Furthermore, this route must entail a similar solute composition in the mesophyll cytoplasm and in the sieve tube sap, otherwise the plasmodesmata would have to be selectively permeable for solutes.

In a purely apoplastic route the assimilates which are translocated in the sieve tubes are taken up from the cell wall space (apoplast) by plasmalemma-bound transport systems in the phloem cells. This model does not rule out the possibility that mesophyll cells are capable of active uptake. However, solutes in the mesophyll would have to leave the cell symplast first before they can be loaded into the phloem. According to this model several predictions can be made, namely, (i) assimilates can be loaded into the phloem without having ever gone through the mesophyll, (ii) the assimilates coming from the mesophyll have the "choice" between uptake into the phloem and efflux into the medium, and (iii) there is no strict, if any, correlation between solute composition of mesophyll cells and that of phloem sap.

The two different routes of phloem loading have different implications for the regulation of assimilate allocation by the source. In the symplastic case, only the mesophyll cells would determine the nature and the quantity of solutes to be loaded. In the apoplastic case both mesophyll and phloem could equally and independently participate in control of phloem loading, since only those solutes will be loaded which are set free by the mesophyll and taken up by the phloem.

Owing to the possibility that it feeds solutes for phloem loading from the medium (apoplast) and from internal storage pools (symplast) the *Ricinus* seedling offered a chance for tests of the different models of phloem loading. Anatomical studies showed that the *Ricinus* cotyledon has the same organization as a "normal" dicotyledonary foliage leaf (Fig. 7A) with upper and lower epidermis (together 20% of cotyledon's volume), one palisade layer (25% of volume), five to seven layers of spongy parenchyma (including a layer of "extended bundle sheath," making up 42% of volume), and interspersed large and small bundles (11% of volume). The minor veins intrude into the intercostal fields and end blindly in the mesophyll (Fig. 7B). Usually, the individual minor veins are separated by four to six mesophyll cells. The intracellular organization of the cotyledon cells is very different from a foliage leaf, even a juvenile one, because the degree of vacuolization is very low and the cytoplasmic compartment dominates. The spongy mesophyll and the bundle sheath contain many starch grains, a few lipid bodies, and some phytin-Mg crystals. In electron micrographs, some plasmodesmata can be detected between mesophyll cells as well as between sieve tubes and companion cells. However, there is, in general, no indication for an intense symplastic continuum from mesophyll to sieve tubes. Injection of fluorescein into sieve tubes showed a rapid passage of fluorescence to neighboring sieve tubes and to companion cells, but not to bundle cells (Fig. 8A). Fluorescein injected into mesophyll passed to neighboring mesophyll cells, but not to the bundle (Fig. 8B). In conclusion, there is, from an anatomical point of view, no likelihood for a purely symplastic route of phloem loading.

The comparison of solutes in phloem sap and cotyledon under different conditions such as with endosperm or without endosperm after 3 h in buffer (Table II) showed that there was no obvious correlation between the solute composition in cotyledons and in sieve tube sap. However, there was a tendency for a decrease in solutes in the cotyledons by starvation that

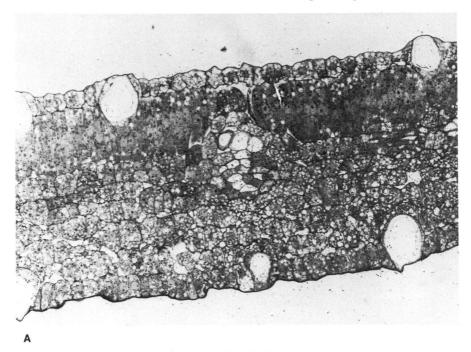

A

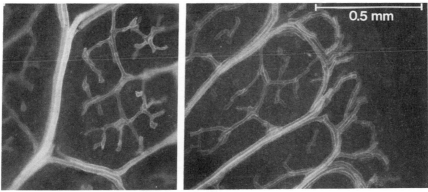

B

Figure 7   Cross-section (1-mm thick) of a cotyledon of a 6-day-old *Ricinus* seedling fixed in Epon–Araldite and stained with toluidine blue (A) and surface view of a chloralhydrate-treated cotyledon with the xylem vessels showing (B).

was paralleled by a decrease in sieve tube sap, but the amount of change shows no pattern whatsoever.

Feeding radioactive sucrose to cotyledons in relatively short time intervals and locating the label by microautoradiography of the tissues allowed the quantification of radioactivity in the different tissues during sucrose

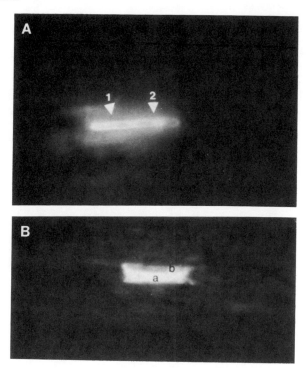

Figure 8 Passage of fluorescein which was injected by pressure: (A) Injection into a sieve tube (1) with fluorescein passing into the next sieve tube (2). (B) Injection into a companion cell (a) and passage of fluorescein into the adjacent sieve tube (b). The cotyledons had been longitudinally sliced along the main vein to allow a lateral microscopic view on bundles, so that directing the micropipette into the sieve tube was possible.

Table II Concentration of Some Solutes in Cotyledons and Sieve Tube Sap under Natural Conditions (Endosperm Attached) and without Endosperm (after 3 h Incubation in Buffer)

		+ Endosperm (mM)	− Endosperm (mM)
Sucrose	In cotyledon	160	80
	In sieve tube sap	270	100
Glucose	In cotyledon	5	3
	In sieve tube sap	1.8	1.0
Glutamine	In cotyledon	17	5.1
	In sieve tube sap	50	1.5
Arginine	In cotyledon	20	12
	In sieve tube sap	11	0

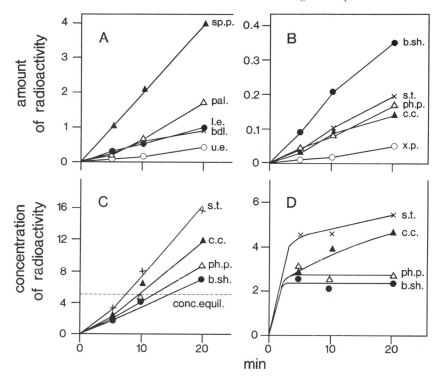

Figure 9   Labeling of cotyledonary cells by radioactive sucrose, revealed by microautoradiography. Cotyledons were incubated in [^{14}C]sucrose for indicated periods (3, 10, and 20 min) and then lyophilized, embedded in Epon–Araldit, and sliced in 1-μm sections. The fixed sections were overlayed with photoemulsion and developed. The radioactivity, indicated by silver grains, was counted for each cell type under the microscope. The counts were related to cell volume, which had been determined by planimetric evaluation of many cross sections. (A,B,C) A continuous labeling experiment, (D) A pulse chase (2 min pulse) experiment. Amount of radioactivity and concentration are given in arbitrary units. (A) Radioactivity in leaf tissues and (B) in bundle tissues. (C) Concentration in bundle tissues in continuous labeling and (D) in pulse chase labeling. Abbreviations: sp. p., spongy parenchyma; pal., palisade parenchyma; l.e., lower epidermis; u.e., upper epidermis; bdl., bundle; b.sh., bundle sheath; s.t., sieve tube; ph.p., phloem parenchyma; x.p., xylem parenchyma; c.c., companion cell. The dashed line indicates the concentration equilibrium with the label in the medium.

uptake. Figure 9A shows that there is a more or less linear uptake of sucrose into all tissues of the cotyledon. The majority of label is found in the mesophyll but the concentration is highest in sieve tubes and companion cells from the beginning. Replacing labeled sucrose after a short incubation time by unlabeled sucrose arrested labeling in all tissues (Fig. 9B). Thus there was no indication of a precursor-like behavior (emptying of one pool and filling up of another pool) of label in any of the tissues, as would have been expected if there was an obligate symplastic movement of sucrose from mesophyll via bundle sheath to phloem cells. Rather it appears that all cell types in the cotyledon take up sucrose independently from each other (Köhler *et al.*, 1991).

This conclusion was supported by electrophysiological experiments. Since active uptake of sucrose (and of amino acids) proceeds by proton symport, i.e., by charge movement, those cells which are capable of active sucrose (or amino acid) transport should respond to addition of sucrose (or amino acids) by depolarization of the membrane. This was found. The cells responded by membrane depolarization to sucrose (Fig. 10), but not to glucose, indicating an active uptake of sucrose, but not of glucose. Electrophysiological tests of the various cell types of the cotyledon revealed that all cell types except the palisade parenchyma are capable of active sucrose uptake (Table III). (Regrettably the sieve tubes could not be tested because the membrane potential did not stabilize for a sufficient amount of time.)

The low frequency of plasmodesmata, the fact that solute concentrations in mesophyll and sieve tube sap are different, and the capacity of nearly all the cells in the cotyledons to take up sucrose from the apoplast independently of neighboring cells are good evidence against a predominant symplastic route of phloem loading.

Direct evidence for an apoplastic route was sought in experiments which studied the competition between intracellular (mesophyll-based) and extracellular (apoplastic) sucrose. Cotyledons of intact seedlings were loaded with radioactive sucrose to label the internal carbohydrate pool. Then the cotyledons were transferred either to buffer or to medium containing unlabeled sucrose, and the flow of labeled sucrose into the

Table III   Capacity for Active Sucrose Uptake by Various Cell Types of the Cotyledon

Cell type	Resting potential (mV)	Change after addition of sucrose (mV)
Lower epidermis	−122	42
Spongy parenchyma	−116	41
Palisade parenchyma	− 95	0
Upper epidermis	− 92	27
Bundle sheath	−102	34
Phloem parenchyma	−116	32
Xylem parenchyma	−102	34

The uptake capacity for sucrose was tested by monitoring the membrane depolarization after addition of 20 mM sucrose.

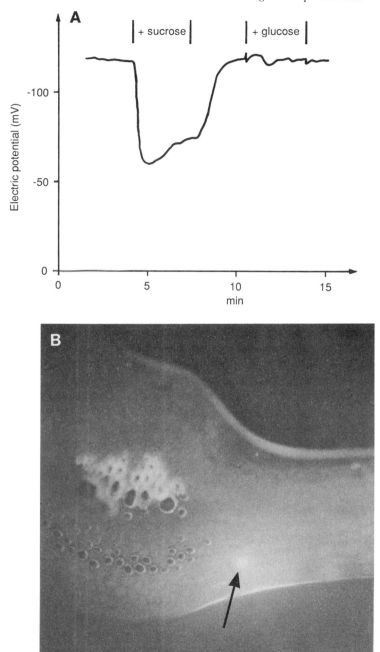

Figure 10 (A) Membrane potential of a parenchyma cell during application of sucrose or glucose to the cotyledon and (B) fluorescent labeling of the cell by electrophoretic infusion of Lucifer yellow after membrane potential recording.

Table IV Partitioning of Intracellular, Labeled Sucrose between Phloem Loading and Efflux into the Medium

| | % Partitioning of labeled sucrose after incubation in | |
	Buffer	Sucrose
Cotyledons	37	31
Phloem exudate	62	27
Medium	1.6	43

The cotyledons of intact seedlings had been preloaded with [^{14}C]glucose overnight, then the cotyledons were transferred to either a medium containing unlabeled 100 mM sucrose or buffer. The hypocotyl was cut to collect phloem sap. The radioactivity in phloem sap, bathing medium, and the cotyledons was measured for a 3-h incubation period.

sieve tubes and into the medium was measured. It was found that in the presence of unlabeled sucrose in the medium (i.e., apoplast) a certain amount of labeled sucrose left the mesophyll and did not enter the phloem, instead escaped into the medium (Table IV). This result indicates a competition between apoplastic (unlabeled) sucrose and mesophyll-derived (labeled) sucrose for phloem loading. The place of competition must have been the apoplast, where the labeled sucrose could exchange with that offered from the medium.

Comparison of the specific radioactivity of sucrose in phloem sap, cotyledons, and bathing medium revealed that sucrose from the phloem sap was labeled less by a factor of 2 than that of the cotyledon (Table V). Therefore, approximately 50% of sucrose in the sieve tube sap must have been directly loaded from the apoplast without having ever intermingled with the sucrose of the mesophyll cells (Orlich and Komor, 1992).

The other 50% of loaded sucrose might have passed through the mesophyll, but most likely via an apoplasmic rather than symplasmic route since a close relationship between release of label into the medium and into the phloem was observed. For example, slicing the hypocotyl stump during an exudation experiment to open plugged sieve tubes immediately slowed down the efflux of labeled sucrose into the medium and stimulated sucrose flow via the phloem (Orlich and Komor, 1992). In conclusion, there is a lot of evidence in favor of an apoplastic route. Therefore there are at least two feasible sites of independent regulation of phloem loading, namely the plasmalemma of mesophyll cells as the

Table V Specific Radioactivity of Sucrose
from Sieve Tube Sap, Cotyledons, and
Bathing Medium

| | Specific radioactivity of sucrose (cpm · nmol^{-1}); incubation | |
	In buffer	In sucrose
In cotyledons	8.2	4.5
In phloem sap	9.7	2.4

Cotyledons of intact seedlings had been preloaded with [^{14}C]sucrose overnight and were then placed into 100 m*M* unlabeled sucrose or into buffer. The hypocotyl was cut to collect phloem sap. After 3 h exudation the specific radioactivity of sucrose from phloem sap and cotyledon extract was determined by enzymatic test and scintillation counting.

site of assimilate release (and retrieval) and the plasmalemma of the sieve tubes as the site of uptake into the long-distance transport route. The apoplast of the leaf is the meeting place for solutes coming from the mesophyll (assimilation) and from the xylem stream (nutrient absorption). Phloem loading will be simultaneously affected by these solutes.

D. Regulation by Substrate Interaction

Sucrose loading is an active, saturable process, accompanied by pH changes and changes in membrane potential. Since transport of other compounds such as amino acids or ions also proceeds by either proton symport or charge uniport, a complex interaction between transport of various substrates via pH and membrane potential changes is expected. Indeed, feeding of salts such as KCl or MgCl$_2$ led not only to an increased phloem loading of the respective salts, but also to a strong decrease of sucrose loading (Table VI). Most likely this effect is caused by the membrane potential change which is strongly depolarized by these salts. NaCl which does not depolarize is less inhibitory for sucrose loading. Phosphate loading decreases the pH value in phloem sap from 7.6 to 5.9 and probably thereby inhibits sucrose loading. But there is also a stimulating interaction; for example, phosphate stimulates the transport of Mg ions (data not shown).

Loading of amino acids, the major solute in phloem sap after sucrose, decreased the concentration of sucrose in phloem sap, and sucrose loading decreased the amino acid concentration in phloem sap (Fig. 11), an effect which also points to an inhibition via the membrane potential. But that interpretation is preliminary, because the membrane depolarization

Table VI   Effect of Salts on Phloem Loading of Sucrose and
on the Membrane Potential of Cotyledon Cells

Salt	Sucrose loading (μmol · h^{-1} · seedling^{-1})	Membrane potential (mV)
Control	2.7	− 120
KCl	0.3	− 64
NaCl	2.6	− 101
MgCl$_2$	0.7	− 86
CaCl$_2$	0.7	− 92
Na-phosphate	1.6	− 135

The cotyledons of seedlings were incubated with 20 m*M* sucrose and
30 m*M* of the respective salts for 3 h, then the hypocotyl was cut and phloem
sap was collected for analysis. The membrane potential was measured with
microelectrodes.

by sucrose and amino acids is transient and recovers partially after several
minutes, in contrast to the depolarization caused by KCl or MgCl$_2$. Therefore, another reason for the apparently mutual inhibition between sucrose and amino acid should be considered, namely, the osmotic effect
caused by solute uptake which will change the concentration of all other
solutes in phloem sap (see also Steudle, this volume, Chapter 8).

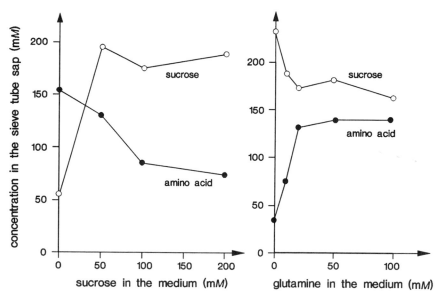

Figure 11   Mutual interaction between sucrose loading and amino acid loading. Cotyledons were (A) incubated in 10 m*M* glutamine and different concentrations of sucrose or
(B) in 100 m*M* sucrose and different concentrations of glutamine.

E. Phloem Loading and Water Flow

The active transport of the major solutes into the sieve tubes is coupled to an osmotic flow of water which will tend to dilute the sieve tube sap and to accelerate the speed of solution flow in the sieve tubes (see also Steudle, this volume, Chapter 8). Water "extraction," e.g., by application of an apoplastic osmoticum, will have the inverse effects. For example, it has been shown previously that the concentration of sucrose in sieve tube sap rises at high sucrose concentrations in the medium, even when the actual phloem loading rate does not change (Fig. 4; (as a result the flow rate will fall, but the mass transfer remains unaltered).

When cotyledons were transferred into a medium with ratioactively labeled sucrose either from a solution with the same concentration of unlabeled sucrose (i.e., no osmotic change) or from buffer (i.e., increase of osmoticum) the exudation of phloem sap from the hypocotyl was very different. With no osmotic change there was a steady volume flow rate and an increasing appearance of label in sieve tube sap (as already described above). With an osmotic change exudation stopped for a while and then resumed (Fig. 12). The first droplet of exudate from the cotyledons with osmotic change contained the same amount of labeled sucrose as had been exuded until that time from the cotyledon with no osmotic change (Fig. 12). This shows that phloem loading itself, i.e., transport of labeled sucrose into sieve tubes, was *not* affected by the osmotic change. Only the speed of the solution flow was affected. Thus, it appears that at least in the short-term water potential exerts its influence more on long-distance volume flow and not on the membrane-located transport processes.

In conclusion, there are two basically different interactions between solutes at the phloem loading site: one is caused by changes of the loading rate, exerted via membrane potential and pH changes, and the other is caused by osmotic effects, which influence the concentration of solutes in sieve tube sap, but not the amount of solutes transported per unit time.

III. Long-Distance Solute Transport

A. Effect of Water Potential Gradients

Solutes loaded into the sieve tubes at the source are translocated along the phloem strands by flow of water to the sink. Since water movement is relatively unhindered by membranes and cell walls, water potential gradients in the plant tissues outside of the phloem along the bundles are expected to modulate the characteristics of solution flow (see Steudle, this volume, Chapter 8). Since no enzyme-catalyzed reactions are directly

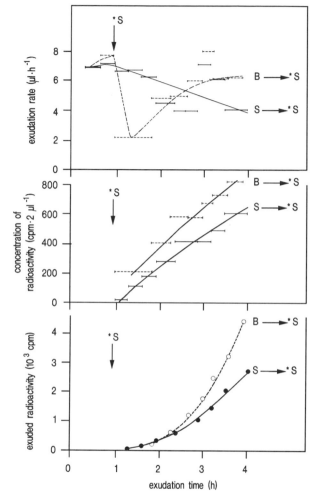

Figure 12   Phloem loading of radioactive sucrose and volume flow either under constant osmotic condition (S→*S) or under osmotic change (B→*S). Cotyledons were incubated either in 100 mM sucrose (S) or in 5 mM buffer (B) and then, at 1 h, transferred to 100 mM radioactive sucrose (*S). Phloem sap was continuously collected in 2-μl portions and the radioactivity was determined by scintillation spectrometry.

involved in long-distance solution flow, its properties must obey simple physical laws (Christy and Ferrier, 1973), which may be modeled. We developed a detailed physical model to give an idea what might happen under different situations which cannot be checked experimentally. In general, water potential gradients in the plant will have, depending on the direction of the gradient, some steepening or flattening effect on the distribution of solute concentration along the sieve tube from source to sink (Fig. 13), but the effects are relatively small. For a water potential gradient of 1.6 MPa · m⁻¹ (i.e., 0.48 MPa more negative in a 0.3-m distant source) the solute concentration would increase only 30% at the source compared to the situation with no water potential gradient and the speed of volume flow will be 30% less (Fig. 13). Thus, less water availability in the source due to transpiration will cause a small concentrating effect on phloem sap in the source and slow down the volume flow slightly. A reverse effect is expected if the gradient is reversed.

B. Phloem Unloading and Solute Circulation

Two models concerning the control of phloem unloading are possible (see also Beck this volume, Chapter 5). One considers the sieve tube plasmalemma as the crucial control point in the sense that the properties of the membrane will be changed according to the sink requirements. That way the phloem in the sink will determine which compounds are unloaded and at which rate.

The other model postulates that all solutes, which arrive in the sink, are unloaded in proportion to their concentration, irrespective of whether they are needed or not. Sink tissues would consume those nutrients which they need. The other solutes are recirculated by the xylem stream. In that case the control is placed on the plasmalemma and the metabolism of the sink cells.

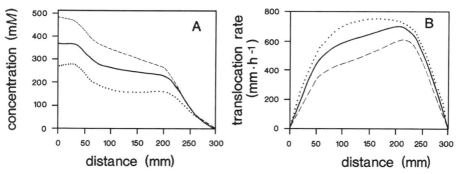

Figure 13 Calculated distribution of (A) solute concentration and (B) flow speed along a (closed) sieve tube in absence and presence of a water potential gradient. (solid line) No gradient; (dotted line) a gradient of -1.6 mPA·m⁻¹, i.e., the sink is more negative than the source; (dashed line) a gradient of 1.6 MPa·m⁻¹ (the source is more negative). The model was based on realistic values (for *Ricinus* seedling) of hydraulic conductivities, phloem loading rate by a saturable process, and efflux of solutes from the sieve tube proportional to the solute concentration. In the model the first 5 cm were defined as source and the last 5 cm as sink.

In a model the phloem unloading rate was modulated and the solute concentration in the source region was calculated. It was found that even a 100-fold increase of the unloading rate in the sink would have no substantial effect on solute concentration and pressure in the source, if it was 0.3 m apart from the sink (Fig. 14), because the longitudinal resistance to water flow dampens the events in the distant sink to such an extent that the source is hardly affected.

In case some solutes are unloaded at a lower rate (because they are needed less), they would accumulate in the sieve tube sap of the sink region (Fig. 15). This accumulation, however, would hardly change the concentration of the particular solute in the sieve tubes at the source, because hydraulic resistance and the continuous flow of the major solute sucrose would lead to the piling up of the solutes at the sink. The increase of pressure is a small signal too unspecific to regulate phloem loading of a particular solute. An exception is sucrose which, if unloaded at a 100-fold lower rate in the sink, would rise by 50% at the source. Thus, a modulation of phloem loading of specific solutes by a variation of phloem unloading is not feasible if sink and source are separated by a sufficient distance. Consequently, the only suitable mechanism of a communication (and interaction) between source and sink is by a circulation flow of solutes. In this way, solutes which are not used by the sink will recirculate via the xylem to the source, where they may compete in the apoplast with source-derived solutes for phloem loading.

Collection of root pressure exudate from *Ricinus* seedling showed that sugar and amino acids may indeed be present at a high concentration (50 and 40 mM, respectively) in the xylem fluid, if the first exudate samples after cutting are considered (Schobert and Komor, 1990). Thus there is some experimental evidence for the existence of appreciable carbon and nitrogen flow circulating between phloem and xylem.

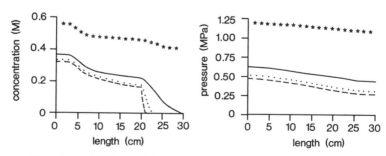

Figure 14   Effect of different phloem unloading rates on the profile of solute concentration and pressure along the sieve tubes. (solid line) Control; (dotted line) 10-fold higher unloading rate; (dashed line) 100-fold higher unloading rate; (* line) 10-fold lower unloading rate. The curves are calculated from the same model as used in Fig. 13.

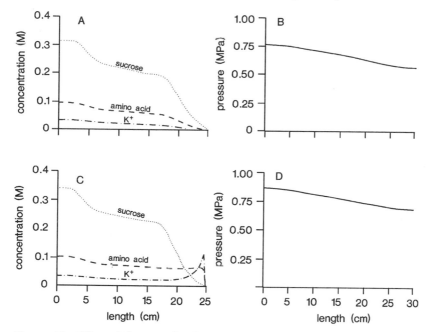

Figure 15 Effect of disproportional phloem unloading rates of different solutes on the profile of the solutes sucrose, amino acid, and K^+ and on the pressure along the sieve tubes. (A) Concentration under control conditions where the unloading of the three solutes is governed by the same diffusion constant; (B) pressure under this condition; (C) concentration if the diffusion constant of phloem unloading is decreased by a factor of 30 for amino acid and 10 for K^+; (D) pressure under this condition. The graphs were calculated with the same model as used in Fig. 13.

C. The Control of Apoplastic Solute Concentration by Solute Circulation and Growth

Measurement of the turgor of cortex cells in the expanding zone of the *Ricinus* hypocotyl revealed a radial turgor gradient with lower turgor in the cells close to the surface (Fig. 16; see also Steudle, this volume, Chapter 8). Turgor pressure was balanced by the osmotic pressure of the cell sap. Interrupting water flow through the xylem resulting in an immediate decrease of turgor of the innermost cell layers (close to the bundles), whereas the osmotic pressure in the cell increased (Fig. 16). Obviously turgor decreased because solutes accumulated in the apoplast. Flooding of the apoplast by water perfusion via the xylem immediately restored turgor pressure. From the difference between turgor and osmotic pressure the concentration of apoplastic solutes could be roughly estimated, namely <20 mM under control conditions and 200–300 mM

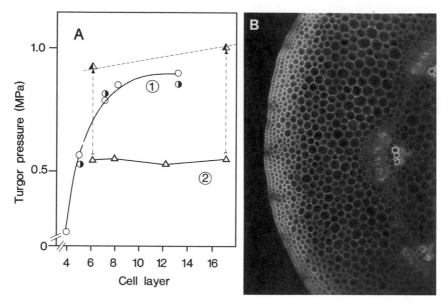

Figure 16 (A) Turgor (open symbols) and osmotic pressure (half-closed symbols) of cells from the hypocotyl of a well-watered seedling (1) and a seedling whose root system had been cut off 15 min prior to measurement (2). The turgor was measured by microcapillary pressure probe. The cell sap was withdrawn with capillary and the osmotic value was measured by a microosmometer. (B) Part of a cross section of the *Ricinus* hypocotyl, stained with calcofluor.

shortly after stoppage of xylem flow. The result may be interpreted in terms of a phloem unloading, which proceeds irrespective of the solute concentration in the apoplast, and of the sink demand and that the unloaded and unconsumed solutes (which are usually swept away by xylem flow) pile up if water flow is reduced. On the other hand, radial water flow may influence the availability of phloem solutes. This may occur during the expansion growth of the hypocotyl just below the hook. It was found that expansion was accompanied by a degradation of starch and an accumulation of soluble sugars (especially sucrose). The sugar accumulation was larger than could be accounted for by starch degrada-

Figure 17 (A) Distribution of growth (shaded area), starch, and sugar content, and (B) the incorporation of phloem-derived radioactive sucrose in the *Ricinus* seedling. Growth measured between Day 5 and Day 6 by determination of the dry weight increase of hypocotyl segments; sugar and starch in the individual segments were determined enzymatically. The export was measured as radioactivity in hypocotyl segments of seedlings, after 5 h incubation of the cotyledons in [^{14}C]sucrose.

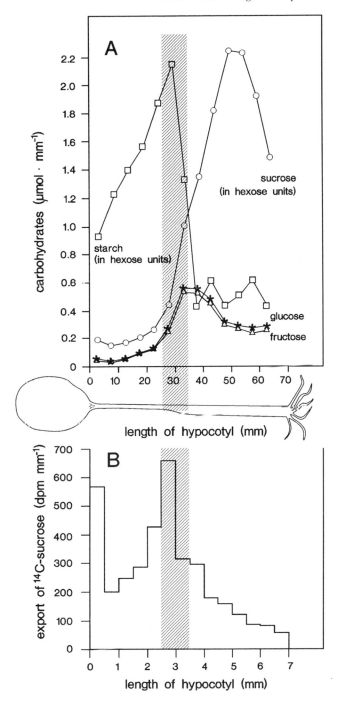

tion; therefore, phloem-born sucrose should have contributed to it. There was an increased incorporation of labeled sucrose from phloem (Fig. 17).

The sequence of events was not known, but since it was unlikely that increased sucrose import itself caused starch degradation, it was likely that starch degradation was the first event. The following sequence could be envisaged: Due to an (unknown) signal starch was degraded to soluble sugar, which (together with alteration of cell wall properties) may have caused osmotic water flow from the xylem in the radial direction away from the bundle to the cortex. In this way phloem-born unloaded sucrose which otherwise is circulated upward by the xylem stream was transported to the expanding cortex cells and taken up there. Water flow created by sink metabolism may have led locally to an increase of allocation of phloem-born solutes, though the mechanism of the unloading process itself was not influenced. The key role for nutrient allocation would therefore be played by the metabolism and the transport properties of the sink tissues, not the plasmalemma of sieve tubes of the sink region (see also Stitt and Schulze, this volume, Chapter 4).

IV. Uptake by the Sink

A. The Experimental System: Suspension Culture and Plants of Sugarcane (*Saccharum officinarum*)

Since assimilates are set free by the phloem irrespective of demand, the sink tissues take up the assimilates according to their requirements and thereby control the net dry matter allocation. The transport properties of cells in the sink and the integration of nutrient uptake and metabolism was studied in storage parenchyma of sugarcane which is known for high rates of growth and storage (Figs. 18A–18D). For simple experimental handling, a cell suspension culture derived from this tissue was often used. The cell suspension culture (Figs. 18E–18J) was grown in 100 ml nutrient medium containing salts, vitamins, amino acids, and sucrose. The incubation of cells with cellulases and hemicellulases released protoplasts, from which vacuoles could be prepared by osmotic shock or by rapid centrifugation. When the influence of a particular nutrient in growth and sugar storage was investigated, a continuous culture in a 2-liter chemostat was used.

Sugarcane is a tropical grass with C_4 photosynthesis, which stores sucrose in the stalk (Van Dillewijn, 1952). The cells in the internodes of the apical region are small and rich in proteins, and amino acids and contain small vacuoles. Between internode numbers 1–3 (counted from the top), the cells expand; between internodes 5–7, they start to load

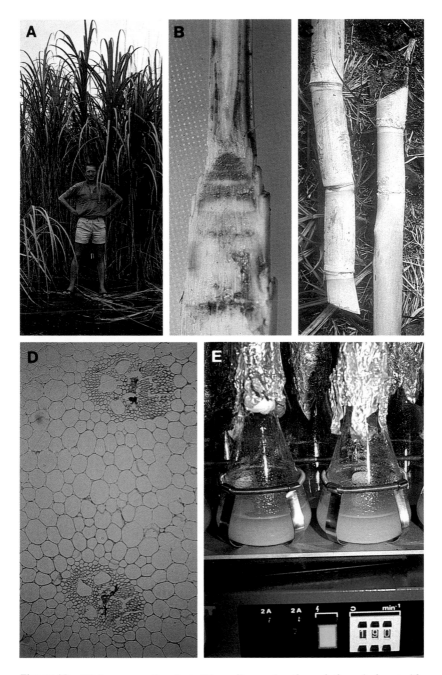

Figure 18 (A) Sugarcane, the plant; (B) median section through the apical part with several short immature internodes; (C) ripe internodes;(D) cross section of storage parchenchyma from the eighth internode; (E) sugarcane suspension culture in 100-ml batch volume cultures on a shaker. *(Figure continues on next page)*

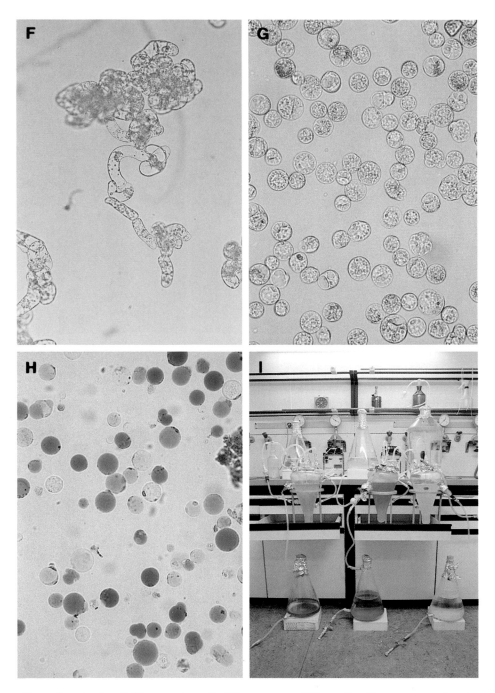

Figure 18 *(continued)* (F) suspension cells; (G) protoplasts; (H) vacuoles; (I) suspension culture in a chemostat.

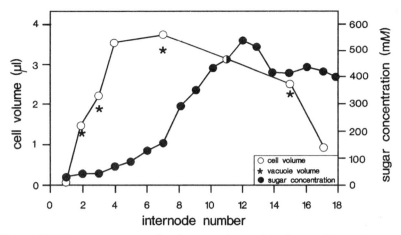

Figure 19 Sucrose content, cell volume, and vacuole volume of parenchyma from internodes of different ages. The internodes were counted from top to bottom, i.e., the youngest internodes are those with the lowest numbers. The internode with the first visible, partly unrolled leaf is counted as number one.

sucrose; and internodes 12 and older have finally accumulated about 500 mM sucrose (Fig. 19).

In batch culture, the time course of events resembled that during the internode development of the plant. A phase of rapid nitrogen acquisition is followed by a phase of net sucrose storage and finally a leveling off to a stable, high-sucrose content (Fig. 20, Komor *et al.*, 1987).

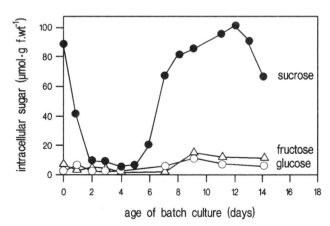

Figure 20 Time course of sucrose and hexose content of a batch culture of suspension cells. The cells had been inoculated into fresh medium at Day 0 (6 ml of a 14-day-old culture was added to 100 ml fresh medium).

B. Transport Systems for Sugar and Amino Acids: Kinetics and Substrate Specificity

A typical feature of solute transport in plant cells is the so-called biphasic kinetics. There is a saturable phase exhibiting a relatively low K_m value (below 1 mM) and a linear diffusion-like phase which becomes obvious only at high concentrations (Stanzel *et al.*, 1988a,b). Uptake of glucose, fructose, and sucrose by tissue slices of sugarcane (Fig. 21A) and by suspension cells (Fig. 21B) more or less follow this pattern. Detailed kinetic measurements were performed using suspension cells because

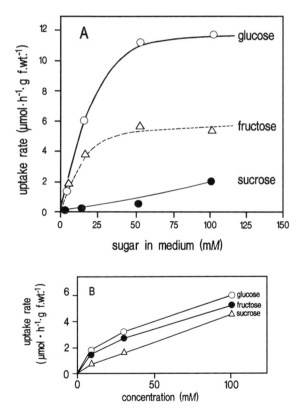

Figure 21 Concentration dependence of glucose, fructose, and sucrose uptake by sugar-cane tissue slices (A) and sugarcane suspension cells (B). Tissue slices of 1 mm thickness had been prepared from the sixth internode; the suspension cells were from the 7th day of batch culture. The uptake was measured by adding radioactive sugar to the medium and taking samples in 15-min intervals in the case of tissue slices or 1 min intervals in the case of suspension cells. The slices were quickly washed in cold buffer and extracted and the extract was measured in the scintillation counter. The suspension cells were filtered and washed with cold buffer and then measured by scintillation spectrometry.

Table VII K_m Values for Glucose and Fructose in
Sugarcane Suspension Cells

	Glucose (μM)	Fructose (μM)
K_m, measured by uptake	50	550
K_m, measured by Competitive inhibition	500	6000

The K_m values were either measured directly by uptake experiments of the respective hexose (e.g., glucose) at different substrate concentrations or determined by competition experiments, where the uptake of the other hexose (e.g., fructose) at variable concentrations was inhibited by a constant concentration of, e.g., glucose. From the "pseudo-K_m" (K_m*) the real affinity constant for glucose was calculated according the equation (K_m* = k $(1 + I \cdot k_i^{-1})$, where k is the K_m value of the varied hexose (e.g., fructose), I the concentration of the competing hexose (e.g., glucose); and k_i the K_m value of the competing hexose (e.g., glucose).

access of substrates is less hindered by diffusion barriers. The question was whether only one or several transport systems are responsible for sugar uptake. Competition between glucose and fructose uptake was observed, but the K_m value calculated by competition experiments was by a factor of 10 higher than that measured by uptake experiments (Table VII). If the same transporter were responsible for uptake of glucose and fructose, then the same affinity value should have been obtained whether measured by direct uptake studies or determined by competition. Thus it appears that there are at least two separate hexose transporters at the plasmalemma of sink cells, one more specific for glucose and the other more specific for fructose.

The uptake of sucrose seemed to be characterized by a relatively low contribution of the saturable uptake phase (Fig. 21). Since there is a cell wall-bound, extracellular acid invertase in sugarcane cells, some hydrolysis of sucrose and uptake of hexoses was suspected to be the reason for the uptake of labeled sucrose. This suspicion was strengthened by the observation that proton symport after sucrose addition occurred after a short lag phase (Komor *et al.*, 1981c), in contrast to hexose addition where proton uptake commenced immediately (Fig. 22). Also, cells without a cell wall (protoplasts) did not take up sucrose by saturable kinetics, proving that invertase was involved in what seemed to be sucrose uptake. Thus carbon from apoplastic sucrose cannot be taken up by an active transport system in sugarcane parenchyma unless sucrose is first hydrolyzed by an invertase.

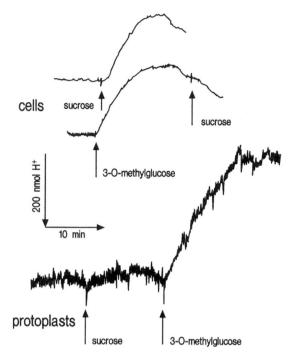

Figure 22 Transient proton uptake by cells and protoplasts initiated after addition of sucrose or 3-*O*-methylglucose. Cells and protoplasts were incubated in a sugar-free medium of which the pH was continuously monitored with an electrode, then 1 m*M* sucrose or 3-*O*-metylglucose was added.

The uptake of nitrogen in sugarcane parenchyma takes place exclusively via amino acid transport systems. There are at least three different uptake systems, each specific for one amino acid group according the net charge of the respective amino acid at neutral pH (Wyse and Komor, 1984). There is a system for neutral amino acids, which works by symport with one proton, a system for basic amino acids (without proton symport), and a system for acidic amino acids, which are transported together with two protons. All three uptake systems work at relatively low K_m values (10–100 μM). Nitrate is hardly used by sugarcane cells, whereas ammonium can be efficiently assimilated if overacidification (due to proton antiport) of the extracellular space is prevented (Veith and Komor, 1991).

C. Regulation of Uptake and Interaction between Uptake of Different Nutrients

In microorganisms the regulation of sugar and amino acid uptake usually proceeds on the genetic level, i.e., by induction or derepression of genetic information. For cells of higher plants there is no evidence for strong induction or repression of transport. Incubation of sugarcane cells with sugar for a short time interval led to a decrease of sugar uptake, tested by the addition of labeled sugar which could be relieved if internal sugar was removed by metabolism (Fig. 23); which was not possible in case of 3-*O*-methylglucose). A similar result was observed in other plant tissues, e.g., *Ricinus* cotyledons (Komor, 1977). This decrease in transport rate was most likely caused by kinetic inhibition of the transport reaction from inside the cell (transinhibition), since inhibition and relief of inhibition are unaffected by inhibitors of RNA or protein synthesis (Komor *et al.*, 1981c). The effect can be explained by a slower recycling of the loaded transporter relative to the empty transporter. Together with the increase in intracellular sugar the percentage of loaded transporter increased and thereby the transporter cycling (in kinetic terms) through the membrane slowed down. This effect was especially strong in the case of nonmetabolizable sugar analogues (3-*O*-methylglucose). Amino acid transporters seem to be regulated the same way. The purpose of that type of regulation may be envisaged as a method to "save" energy, since each transporter cycle of an active uptake system consumes part of the proton gradient which needs ATP to be restored. It presumes that there are also natural situations in which plant cells contain appreciable concentrations of sugars or amino acids in the cytosol. Compartmentation analysis of sugarcane cells proved that to be true (Preisser *et al.*, 1992).

Another type of regulation or interaction of transport systems works through the effect of each individual transport process on the energetic

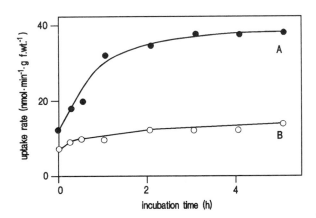

Figure 23 Uptake rates for hexose by sugarcane cells which had been preloaded with glucose (A) or with 3-*O* methylglucose (B) and which were then transferred to sugar-free medium. The uptake rates were tested at each time point by addition of labeled 3-*O*-methylglucose and measurement of the uptake kinetic for 5 min.

status of the membrane (similar to that already explained for phloem loading). The transfer of charge through the membrane, either by proton symport or by uniport, causes a depolarization of the membrane which is compensated by an increase of proton and potassium export. How complete this compensation is and which of the two compensating reactions prevails depends on the concentration of protons and potassium and their availability in the cytosol and on the stoichiometry of proton flow together with the transported substrate (Table VIII). In the case of uptake of sugar or neutral amino acids, there is at steady state a balanced export of protons and potassium and a partly depolarized membrane potential (Komor *et al.*, 1989). In the case of an uptake of a positively charged compound (arginine, cations) the compensation mostly proceeds by K^+ efflux and is accompanied by a strongly depolarized membrane. In the case of uptake of a compound, which carries more protons than net charge (e.g., glutamate anion plus $2 H^+$) the compensation is incomplete and occurs predominantly by proton export together with a hyperpolarization of the membrane.

The membrane potential being part of the driving force for transport may have a direct effect on the rate of charged compounds moved across the plasmalemma; therefore, the change of the membrane potential caused by the transport of one charged compound inevitably influences the transport of others including those which do not share the same transport system. Therefore, antagonistic effects between uptake of solutes are expected even if there are separate transport proteins for these compounds. In a similar manner, the proton flow caused by uptake of one compound would influence the uptake of another unrelated compound depending on whether there is a pH change near the plasmalemma of the cells (i.e., in the apoplast).

D. Charge and Acidity Balance for Uptake and Assimilation

As described in the previous section, the uptake of charged compounds has to be balanced in a way that avoids the buildup of a high membrane voltage. The transport systems determine which ions are selected and

Table VIII   Effect of Sugar and Amino Acid Uptake on Proton and Potassium Flow and on Membrane Potential

Transported substrate	H^+ taken up per substrate	K^+ lost per substrate	Membrane potential
Glucose	0.87	0.94	Depolarized
Glutamine	0.98	1.0	Depolarized
Arginine	−0.27	0.73	Depolarized
Glutamate	1.9	1.1	Hyperpolarized

taken up from a complex growth medium. For sugarcane cells a charge balance sheet has been elaborated for two growth phases, one where nutrition is mostly on organic nitrogen and the other where nitrate is the major nitrogen source (Komor *et al.*, 1981a). Table IX shows that net charge balance is mostly achieved by ion flow and not by protons especially in the phase where nitrate is assimilated. Nitrate influx seemed to stimulate Na^+ uptake in particular. However, the cyclic proton flow which proceeds inward by proton symport with hexoses and outward by the action of the plasmalemma-bound proton ATPase is far larger than the sum of all net ion flows.

The major net demand for protons originates from intracellular assimilation reactions, especially from the reduction of nitrate (Table X). However, the rate of metabolic proton production is small compared to that of cyclic proton fluxes during nutrient uptake. Hence, in normal circumstances the transport processes themselves adjust the levels of acidity in the cytoplasma and in the wall space.

Table IX   Charge Balance of Nutrient Uptake and Proton Fluxes by Sugarcane Cells in Two Growth Phases

		$mmol \cdot liter^{-1}$ suspension	
		Phase I	Phase II
Amino acids:	neutral	2.15	0.28
	basic	1.09	0.22
	acidic	1.07	0.30
Inorganic cations:	K^+	1.11	1.61
	Na^+	0.29	3.52
	Ca^{2+}	0.05	0.15
	Mg^{2+}	0.45	0.33
	$NH_4{}^+$	0	0
Inorganic anions:	Chloride	0.14	0.33
	Phosphate	0.57	0.65
	Sulfate	0.19	1.14
	Nitrate	0.43	2.14
Extruded H^+		0.55	0.05
Positive Charges taken up		3.49	6.31
Negative Charges taken up		2.74	5.87
Positive Charges Extruded		0.55	0.05
Hexoses taken up		20.4	54.9

Phase I is where organic nitrogen is the major N source and phase II, where nitrate is the only N source. The data were obtained from elemental analysis at the start and at the end of each growth phase. The proton fluxes were estimated from the measured stoichiometries of nutrient and proton symport.

Table X Acidity Balance of the Cell for Transport and Metabolic Reactions

	mequiv. protons · liter^{-1} suspension	
	Phase I	Phase II
Metabolic reactions consuming protons		
Reduction of nitrate to ammonia and amino nitrogen	0.58	3.62
Reduction of sulfate to sulphydryl	0.08	0.1
Conversion of acidic amino acids to neutral amino acids	0.19	0
Protein ejection for charge balance	0.55	0.05
Total	1.40	3.77
Metabolic reactions producing protons		
Production of carboxylic acids for amination	0.22	1.66
Conversion of basic amino acids to neutral amino acids	0.36	0
Total	0.58	1.66
Difference (consumption - production)	0.72	2.11
Measured acidity production of cells	0.70	2.07

The acidity changes were calculated from the analytical data of sugarcane cells of growth phases I and II (see Table IX). The metabolic acidity production was measured as so-called proton deficit after combustion of cell homogenates (i.e., the homogenate of a particular pH was combusted; the ash was taken up in the previous volume with water and titrated back to the original pH; the necessary acid is equivalent to the number of combusted ionized carboxyl groups from the homogenate).

E. Role of Transport Processes for Growth and Sugar Storage

The active, high-affinity uptake systems are also used to perform the transport of amino acids in parallel with diffusion-like systems. The latter are especially important at high substrate concentrations. To evaluate the role of these transport systems for sugarcane, cell suspensions were grown in continuous culture (chemostat) at different low levels of nutrient concentrations. For sugar as well as for amino acids the high-affinity systems may fully support maximal growth rates which are saturated at 0.5 mM of hexose far below the concentration ranges where the diffusionlike uptake proceeds (Fig. 24). This means that the parenchyma cells in the sink are equipped to support fast growth even at low concentrations of the main nutrients N and C, i.e., at concentrations much lower than those occurring in the apoplast.

What then is the function of the uptake phase working only at high concentrations? Sugarcane suspension cells store sucrose, but the maximal concentration reached is still only one-third of that in a mature

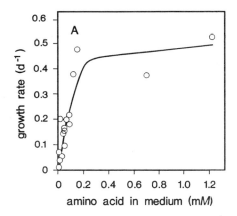

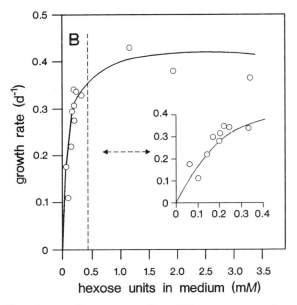

Figure 24 Growth rate of sugarcane cells at different concentrations of amino acids (A) or sugar (B). The cells were grown in a chemostat at constant cell density and nutrient composition. All nutrients were in excess except amino acids (A) or sugar (B).

sugarcane stalk. Incubation of suspension culture cells at very high concentrations of sucrose (or hexose) such as 200 or 300 mM raised the intracellular sucrose to levels of 250–300 mM (Fig. 25). This effect was not an osmotic one (such as extraction of intracellular water), since equivalent concentrations of mannitol did not lead to an elevation of sucrose storage. The applied extracellular sugar concentrations seem unusually

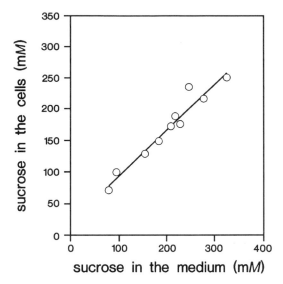

Figure 25   Intracellular sucrose concentration in suspension cells at very high apoplastic sucrose concentration.

high, but apoplastic sugar concentrations in sugarcane internodes are indeed in the order of 300 mM (Welbaum and Meinzer, 1990). It may be suggested that it is a regular natural situation during ripening for nutrients to pile up in the apoplast so that the diffusionlike systems will tend to equilibrate sucrose concentrations between apoplast and cell.

F. Regulation and Integration of Sucrose Storage with Sucrose Metabolism

Sugarcane cells growing at an optimal growth rate used half of the re-sorbed sugar for respiration and the other half for synthesis of cellular material. There was virtually no storage of sucrose (Fig. 26). It appeared that the use of sugars for growth had priority over that for storage. A reduction of growth rate by limitation of an essential nutrient seemed to be a way to induce sugar storage. This idea is supported by field experience which indicated that sugar yield in sugarcane plantations is improved if nitrogen fertilization is withheld before harvest (Das, 1936). Similarly, in sugarcane suspension cells the sucrose content rose when the growth rate was reduced by nitrogen limitation. At the same time a decrease of carbon input to respiration and structural cell material oc-curred (Fig. 26). However, a simple inverse relationship between storage and growth was not generally valid. Growth limitation by phosphate shortage did not result in increased storage of sucrose as could be shown

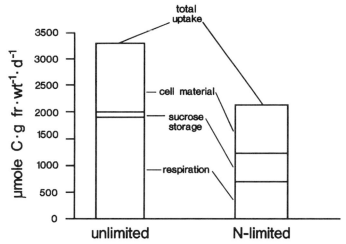

Figure 26 Partitioning of carbon between respiration, sucrose storage, and structural cell material in sugarcane cells grown under unlimited and under N-limited conditions. The cells had been grown at a constant growth rate in the chemostat. In the so-called unlimited condition all nutrients were at nonlimiting concentrations; for N limitation the amino acid content in the medium was reduced to 10% of the nonlimiting value. Respiration was measured with an oxygen electrode; the carbon uptake was determined from the disappearance of carbon from the medium; sucrose and cell material were determined by chemical methods.

most clearly with cells grown in a chemostat when shifting from N to P limitation. Growth rate was decreased even further, the content of amino acid increased, and intracellular sucrose declined (Fig. 27). Thus, it is not the growth rate per se which determines carbon partitioning between storage and dissimilation of sucrose. Rather, there must be a more specific regulation mechanism related perhaps to the amino acid pool or products thereof (see also Stitt and Schulze, this volume, Chapter 4).

The determination of the metabolites of the glycolytic and gluconeogenic pathways from cells performing the shift from rapid growth to sucrose storage gave no hint for a particularly stringent regulatory step, since there were only slight changes in metabolite pools. Measuring the rate of initial sucrose synthesis adding radioactive hexoses to the cells and taking samples after a short time to determine the labeling of intracellular sucrose showed that the rate of sucrose synthesis was usually much higher than the net rate of sucrose storage (Fig. 28). Thus there is a fast cycle of sucrose synthesis and sucrose hydrolysis and only the small difference between these two reactions is responsible for net sucrose storage (Wendler *et al.*, 1990). It was observed that the onset of sucrose storage was correlated with a transient increase of the pyruvate pool which may

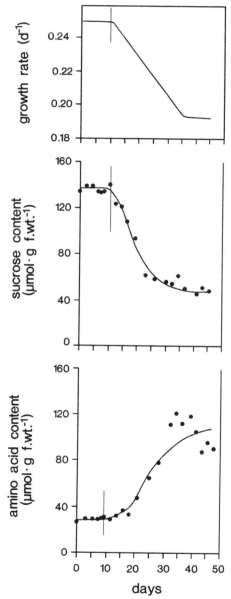

Figure 27 Growth rate and amino acid and sucrose content of sugarcane cells which were shifted from N limitation to P limitation at Day 10. The cells were grown in the chemostat with low, limiting amino acid concentration in the medium. At Day 10 (faint line) the medium was changed to a new medium with a higher amino acid concentration, but without phosphate.

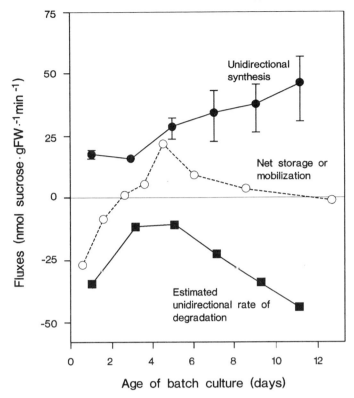

Figure 28 Measured rate of sucrose synthesis and sucrose storage and calculated rate of sucrose degradation. The rate of unidirectional sucrose synthesis was measured by pulse labeling of sucrose giving labeled hexoses to the medium. The rate of sucrose storage was determined from the change of sucrose concentration in the cells. The rate of sucrose degradation was calculated as the difference between the rate of sucrose synthesis and that of sucrose storage. All experiments were done with suspension cells in batch culture.

suggest that pyruvate kinase was involved in regulating storage. Furthermore, the substrates for sucrose synthesis, free hexoses, hexose phosphates, and UDP-glucose increased slightly. Triose-phosphate pools decreased indicating some regulation by phosphofructokinase. However, all these metabolite shifts were small.

The enzyme mostly responsible for sucrose synthesis in sugarcane suspension cells is sucrose-phosphate synthase. Labeling of intracellular sucrose was nearly uniform if labeled glucose plus nonlabeled fructose had been given (in case of sucrose synthase activity a strong incorporation of nonlabeled fructose into sucrose has to be expected). The activities of

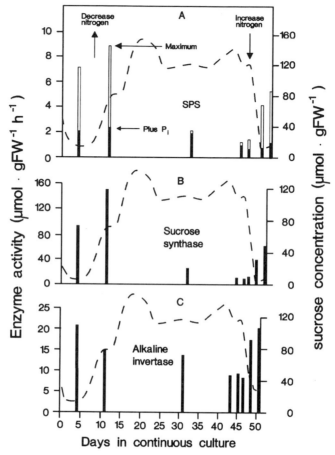

Figure 29 Activities, determined *in vitro,* of sucrose-phosphate synthase (SPS) measured with and without inhibitory phosphate concentration, sucrose synthase (SS), and soluble invertase in chemostat-grown suspension cells, shifted from unlimited growth to N limitation and back. The broken line indicates the intracellular sucrose concentration.

the individual enzymes involved in sucrose synthesis and sucrose hydrolysis did not show a clear correlation with the sucrose storage (Fig. 29), but rather were correlated with the growth rate of the cells.

In conclusion, there is a fast futile cycle of sucrose synthesis and sucrose degradation with the consequence that small shifts in metabolite pools and enzyme activities result in strong changes of net sucrose production, leading either to storage or to mobilization (scheme, Fig. 30). This futile cycle was found not only in suspension cells but also in internodes of the sugarcane plant with the modification that sucrose synthase was the major enzyme participating in sucrose synthesis (or shuttling of label) in imma-

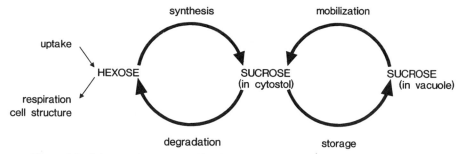

Figure 30   Scheme of sucrose cycling in sugarcane and input and output pathways.

ture internodes and sucrose-phosphate synthase in ripe internodes. In addition there was a steady decline of the rate of sucrose cycling as the internodes ripened.

G. Compartmentation of Sucrose in the Vacuoles

Since isolated vacuoles from sugarcane cells contain high concentrations of sucrose, the vacuoles were generally considered to be the storage compartment into which specific transport systems were accumulating the sugar (Willenbrink, 1982). Transport studies with isolated vacuoles from sugarcane suspension cells, however, showed that the permeation of sucrose across the vacuolar membrane was passive in the sense that it was independent of membrane energization, though it was definitely catalyzed by a transport system (Preisser and Komor, 1991). Compartmental analysis by differential efflux performed by specific permeabilization of the plasmalemma with $CuCl_2$ revealed that there was the same concentration of sucrose in vacuole and in cytosol (Preisser *et al.*, 1992). Therefore, it is probably not the sucrose transfer across the tonoplast itself which governs the sucrose content of the cell. Rather, the vacuole serves as a space which is inaccessible for some sucrose-degrading enzymes and thereby contributes to the regulation of the sucrose content.

In suspension culture cells the vacuole comprises about 50% of cell volume. In cells of sugarcane internodes it may be 90%. The strong degree of vacuolization of cells during differentiation from the meristematic to parenchymatic state may support sucrose storage. Determination of the vacuolar volume in internodes showed that the vacuolization definitely preceded sucrose storage (Fig. 19). In young internodal cells, vacuoles were most likely to be filled with potassium salts, since there was an inverse correlation between K^+ (together with sulfate, aconitate, and other anions) and sucrose content of internodes (Fig. 31). This feature held not only for different internodes of the same plant, but also for different sugarcane species of different sucrose storage capacity. The

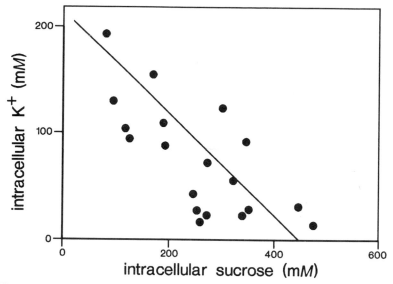

Figure 31 Relationship between K⁺ and sucrose content, measured in different internodes of sugarcane. The young internodes were those with low sucrose, the old ones those with high sucrose levels.

results indicated that vacuolization by itself did not govern sucrose storage. Therefore, the regulation of carbon partitioning between sucrose storage and growth was probably located in the metabolism and depended on the rate of delivery of nitrogenous compounds and the maintenance of high apoplastic sucrose concentrations.

V. Conclusions and Outlook

A. The Integration of Source and Sink by the Phloem and Xylem Streams

From the properties of source and sink cells described above and from the mechanisms of phloem loading and unloading a complete picture of the integration of plant organs by long-distance transport can be attempted (Fig. 32). In the apoplast of the source leaf there is a competition between the mesophyll and phloem for solutes. The mesophyll controls phloem loading either by withholding certain solutes or by their delivery. In addition, some specificity is exerted by the phloem so that certain solutes will be excluded from loading even if present at a high concentration (e.g., nitrate, calcium). On the other hand, the phloem is a strong sink in the source leaf because its capacity to accumulate sucrose and

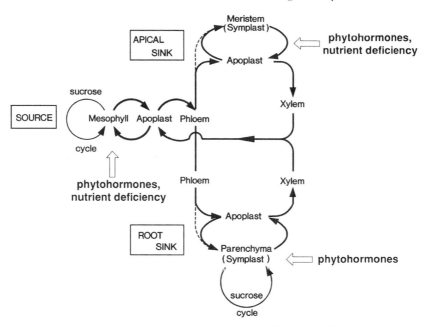

Figure 32 Scheme of nutrient cycling in the plant. The unloading of phloem might proceed into the apoplast or (directly) into the symplast (indicated by dashed line). In the latter case the superfluous nutrients must be excreted by the sink cells into the apoplast, otherwise no control on nutrient acquisition would be possible. The possible places of regulatory interaction by phytohormones or nutrient deficiency are indicated by faint arrows.

amino acids at high concentrations will create a water flow from the mesophyll cells to the phloem. Therefore, solutes which come up by the xylem or which are set free from the mesophyll will not be immediately flooded to the stomates, but rather be loaded into the phloem and translocated along the phloem (see also Schulze, this volume, Chapter 7). The unloading of solutes proceeds continuously irrespective of the demand of the sink cells. The nutrients not taken up by the sink cells will circulate back via the xylem stream and compete with mesophyll-born solutes for phloem loading. This cycling provides a well-balanced, constant supply of a nutrient mixture to all tissues irrespective of local and temporal demands, at least as long as the supply of a certain nutrient does not become limiting in the source. The cycling can definitely work between source leaves and basal sinks with the stalk in between. Whether it would also work for the path between source and apical sink would depend on the direction of water flow in the xylem. If there is water flux from the apex as occurs in growing fruits (Pate *et al.*, 1980), then cycling could also work in the apical regions. The demand of the sink cells determines the consumption of the delivered nutrients. The fact that the K_m-values for uptake of major nutrients such as sugars and amino acids are at one-thousandth of the concentration which is transferred in the phloem suggests that sink cells get their nutrients if they need them. On the other hand, directing the water flow away from the xylem, e.g., by the production of osmotic solutes in cells, may enhance the delivery of phloem-born nutrients to growing tissue. Interruption of xylem flow would lead to a rapid piling up of solutes in the apoplast. Such effects may be expected after leaf abscission, which cuts down transpirational water flow in certain bundles. If the vasculature still allowed some phloem transport into these bundles, which are not in the mainstream of transpirational water flow, the adjacent sink regions would be forced toward high levels of storage. Such an effect could also explain the sugar storage in ripening sugarcane internodes.

The role of the phloem in regulating nutrient allocation is not confined to the nutrients per se but also to the allocation of phytohormones which direct the differentiation and maturation of parenchyma cells. It must be considered that if there were a leaf-born hormone directly involved in regulation of phloem unloading and storage, then there must be a differential allocation of the respective hormone along the phloem path between the phloem sap leading out from a leaf to the apex and the phloem sap flowing to the base, since apex and base definitely behave differently. Root-born phytohormones translocated via the xylem can only play a role in the transpiring, i.e., mature, assimilating leaf, and not in the apex where shoot growth occurs because the transpiration of the latter is small. Therefore, if there is an effect of phytohormones on apical nutrient acquisition (see Beck, this volume, Chapter 5), then this effect must be mediated via the phloem. It seems unlikely that the root-born phytohormones directly affect the storage pattern of internodes (e.g., in sugarcane) along the xylem route, since the differentiation and storage status of internodes is not determined by their distance from the root but by their distance from the apex. Therefore, at our current stage of knowledge, the regulation of internode differentiation has to be either by an apically produced hormone or by a phloem-born hormone (whose quantity may be influenced by hormones coming up the xylem stream). Nevertheless, the roots may play a decisive regulatory role in assimilate allocation, because the root definitely has first access to two decisive nutrient factors, namely water and nitrogen. The case for water is self-evident. In addition, roots are able to signal water deficits to the leaves to stop transpiration and carbon assimilation (see Schulze, this volume, Chapter 7). In the case of nitrogen, even those roots which are poor in nitrate reductase activity can exploit ammonia and amino acids from the soil. Although the concentration of ammonia and amino nitrogen in the soil is often smaller than that of nitrate, the affinity of their specific uptake systems is by far higher. Also, those plants which are considered to reduce nitrate exclusively in the leaves export amino acids via the xylem even under "pure" nitrate nutrition, e.g., *Urtica* (Rosnitschek-Schimmel, 1985), so they probably can produce amino acids from nitrate in the roots. Once there is some growth in roots it will attract more sucrose (and with it amino acids) from the leaves. Such a small direct input of reduced N into the root would therefore trigger an additional flow via the phloem in which amino acids are also carried. The roots can always maintain a certain sink activity, even under extreme nitrogen shortage, so that their carbon dissimilation is superior to that of the nitrogen-starving apex.

B. Principles of Transport Regulation in Plants

The basic regulatory rules concerning transport processes in plants can be summarized into two seemingly contradictory terms, (i) the selfishness of each system or cell, and (ii) the complex interaction between systems controlled by different components.

The selfishness is expressed by transport systems in the membranes which are specific and take up "their" compounds, independent of other compounds and other transport systems; at best they sense the intracellular concentration and react by transinhibition. Examples are the following (Fig. 32):

- Mesophyll and phloem compete for solutes in the apoplast.
- Solutes, which are in the apoplast, are loaded into the phloem irrespective of whether they are needed in the sink.
- Solutes are unloaded irrespective of the needs of the sink and, if not used by sink cells, recirculated to the source.
- Sucrose is subject to hydrolysis and synthesis by similar and different enzymes (Fig. 30), which seem to act independently of each other.

The consequence of this apparent selfishness of reactions is a series of cycles, such as sugar moving in and out of the mesophyll, solutes moving down via the phloem and back via the xylem, and sucrose synthesis and splitting (Fig. 32). These cycles depend on the luxury supply of energy in the form of carbohydrates for dissimilation, a condition which probably is usual in plant life. Regarding the mechanisms of these futile cycles it is obvious that a shortage of energy would immediately slow them down; hence, they could be "regulated" by energy status. The selfishness might also be seen during growth under nutrient-limiting situations, where growth is favored for those plant organs which have first call on the limiting nutrient. This is the leaf in the case of carbon (or light) shortage, and the root in the case of nitrogen or water shortage.

These "selfish" responses are embedded in a network, which results in a complex, coordinated function of the plant as a whole. Examples of this network are the following:

- The individual, independent transport systems at the membrane level are subject to membrane energization, i.e., external and internal pH and membrane potential, with the consequence that charge balance is assured and acidity changes are minimized.
- The cyclic flow of nutrients through phloem and xylem assures a permanent supply of the necessary nutrients for plant cells with different requirements, also in cases of inappropriate phloem supply.
- Phloem transport is powered by the loading of the major solutes, especially of sucrose. This way carbon assimilation becomes a pacemaking reaction. But since sink cells direct the use of assimilates by their growth and storage activity, minor but indispensible nutrient elements can finally determine the net use of assimilates and the net allocation. Under situations of stress an "economic" shift of nutrient allocation is thereby achieved.
- The cycle of sucrose synthesis and hydrolysis in storage cells allows a fast response of net sucrose storage by only small changes of metabolite pools and enzyme activities. Such a system is extremely versatile in changing situations, where an irreversible decision (e.g., by a phytohormone) has not yet been made, for example, in a sugarcane internode in the phase between growth, expansion, and sucrose storage.

The work reported above concentrated on the flow of the major nutrients C and N; the role of minor compounds which may serve as signal substances for plants under stress was neglected in this chapter, because the role of phloem and xylem as pathways for signal substances is dealt with in Chapters 4 and 8 (this volume). Also not considered is the role of hormones in differentiation of phloem and xylem in sink and source tissues and in the transition from sink to source.

Acknowledgments

This chapter contains the assemblage of results from several Ph.D. studies from, namely, Dr. Bong-Heuy Cho, Dr. Christian Schobert, Dr. Renate Wendler, Dr. Robert Veith, and Dr. Jutta Köhler; from several Ph. D. students, Wen-Jun Zhong, Walter Köckenberger, and Oliver Zingsheim; and from several diploma students, Heidi Sprügel, Manfred Stanzel, and Martin Höfner and work from Dr. Gabriele Orlich. In addition several guest scientists participated: Dr. Richard Sjolund (Iowa City), Dr. Roger Wyse (New Brunswick), Dr. Jan Pavlovkin (Bratislava), Dr. Jose Kallarackal (Trichur, India), Dr. Anatoli Meshcheryakov (Moscow), and Margaret Thom and Dr. Andrew Maretzki, both from Aiea (Hawaii).

References

Allen, S., and Smith, J. A. C. (1986). Ammonium nutrition in *Ricinus communis:* Its effect on plant growth and the chemical composition of the whole plant, xylem and phloem saps. *J. Exp. Bot.* **37,** 1599–1610.

Beevers, L., and Hageman, R. H. (1983). Uptake and reduction of nitrate: Bacteria and higher plants. *Encycl. Plant Physiol., New Ser.* **15A,** 351–375.

Christy, L. A., and Ferrier, J. M. (1973). A mathematical treatment of Münch's pressure flow hypothesis of phloem translocation. *Plant Physiol.* **52,** 531–538.

Das, U. K. (1936). Nitrogen nutrition of sugarcane. *Plant Physiol.* **11,** 251–317.

Geigenberger, P., and Stitt, M. (1991). A futile cycle of sucrose synthesis and degradation is involved in regulating partitioning between sucrose, starch and respiration in cotyledons of germinating *Ricinus communis* L. seedlings when phloem transport is inhibited. *Planta* **185,** 81–90.

Geiger, D. R. (1975). Phloem loading. *Encycl. Plant Physiol., New Ser.* **1,** 395–431.

Kallarackal, J., and Komor, E. (1989). The transport of hexoses by the phloem of *Ricinus* seedlings. *Planta* **177,** 336–341.

Kallarackal, J., Orlich, G., Schobert, C., and Komor, E. (1989). Sucrose transport into the phloem of *Ricinus* seedlings measured by sieve tube sap analysis. *Planta* **177,** 327–335.

Köhler, J., Fritz, E., Orlich, G., and Komor, E. (1991). Microautoradiographic studies of the role of mesophyll and bundle tissues of the *Ricinus* cotyledon in sucrose uptake. *Planta* **183,** 251–257.

Komor, E. (1973). Proton-coupled hexose transport in *Chlorella vulgaris. FEBS Lett.* **38,** 16–18.

Komor, E. (1977). Sucrose uptake by cotyledons of *Ricinus communis* L.: Characteristics, mechanism and regulation. *Planta* **137,** 119–131.

Komor, E. (1982). Transport of sugar. *Encycl. Plant Physiol., New Ser.* **13A,** 635–676.

Komor, E., Thom, M., and Maretzki, A (1981a). Charge balance and acidity regulation during growth of sugar-cane cell suspensions. *Plant, Cell Environ.* **4,** 359–365.

Komor, E., Thom, M., and Maretzki, A. (1981b). Mechanism of uptake of L-arginine by sugarcane cells. *Eur. J. Biochem.* **116,** 527–533.

Komor, E., Thom, M., and Maretzki, A. (1981c). The mechanism of sugar uptake by sugarcane suspension cells. *Planta* **153,** 181–192.

Komor, E., Höfner, M., Wendler, R., Thom, M., and De La Cruz, A. (1987). Nitrogen nutrition of sugarcane cells: Regulation of amino acid uptake and the effect on cell growth and sucrose storage. *Plant Physiol. Biochem. (Paris)* **25,** 581–588.

Komor, E., Cho, B.-H., Schricker, S., and Schobert, C. (1989). Charge and acidity compensation during proton-sugar symport in Chlorella: The H^+-ATPase does not fully compensate for the sugar-coupled proton influx. *Planta* **177,** 9–17.

Kriedemann, P., and Beevers, H. (1967). Sugar uptake and translocation in the castor bean seedling. *Plant Physiol.* **42,** 161–173.

Lucas, W. J. (1985). Phloem loading: A metaphysical phenomenon? *In* "Regulation of Carbon Partitioning in Photosynthetic Tissue" (R. L. Heath, and J. Preiss, eds.), pp. 254–271. Am. Soc. Plant Chysiol., Rockville and Waverly, Baltimore, MD.

Maiden, M. C. Y., Davis, E. O., Baldwin, S. A., Moore, D. C. M., and Henderson, P. J. (1987). Mammalian and bacterial sugar transport proteins are homologous. *Nature (London)* **325**, 641–643.

Milburn, J. A. (1974). Phloem transport in *Ricinus:* Concentration gradients between source and sink. *Planta* **117**, 303–319.

Opekarova, M., and Kotyk, A. (1973). Uptake of sugars by tobacco callus tissue. *Biol. Plant.* **15**, 312–317.

Orlich, G., and Komor, E. (1992). Phloem loading in *Ricinus* cotyledons: Sucrose pathways via the mesophyll and the apoplasm. *Planta* **187**, 460–474.

Pate, J. S. (1975). Exchange of solutes between phloem and xylem and circulation in the whole plant. *Encycl. Plant Physiol.* **1**, 451–473.

Pate, J. S., Layzell, B., and Atkins, C. A. (1980). Transport exchange of carbon, nitrogen and water in the context of whole plant growth and functioning: Case history of a nodulated annual legume. *Ber. Dtsch. Bot. Ges.* **93**, 243–255.

Preisser, J., and Komor, E. (1991). Sucrose uptake into vacuoles of sugarcane suspension cells. *Planta* **186**, 109–114.

Preisser, J., Sprügel, H., and Komor, E. (1992). Solute distribution between vacuole and cytosol of sugarcane suspension cells: Sucrose is not accumulated in the vacuole. *Planta* **186**, 203–211.

Robinson, S. P., and Beevers, H. (1981). Amino acid transport in germinating castor bean seedlings. *Planta* **68**, 560–566.

Rosnitschek-Schimmel, I. (1985). The influence of nitrogen nutrition on the accumulation of free amino acids in root tissue of *Urtica dioica* and their apical transport in xylem sap. *Plant Cell Physiol.* **26**, 215–219.

Sauer, N. (1991). Cloning of a higher plant sugar transporter. *In* "Recent Advances in Phloem Transport and Assimilate Compartmentation" (J. L. Bonnemain, S. Delrot, W. J. Lucas, and J. Dainty, eds.), pp. 140–147. Ouest Editions, Presses Académiques, Nantes, France

Schobert, C., and Komor, E. (1989). The differential transport of amino acids into the phloem of *Ricinus* seedlings as revealed by sieve tube sap analysis. *Planta* **177**, 342–349.

Schobert, C., and Komor, E. (1990). Transfer of amino acids and nitrate from the roots into the xylem of *Ricinus communis* seedlings. *Planta* **181**, 85–90.

Smith, J. A. C., and Milburn, J. A. (1980). Osmoregulation and the control of phloem-sap composition in *Ricinus communis* L. *Planta* **148**, 28–34.

Stanzel, M., Sjolund, R. D., and Komor, E. (1988a) Transport of glucose, fructose and sucrose by *Streptanthus tortuosus* suspension cells. I. Uptake at low sugar concentration. *Planta* **174**, 201–209.

Stanzel, M., Sjolund, R. D., and Komor, E. (1988b). Transport of glucose, fructose and sucrose by *Streptanthus tortuosus* suspension cells. II. Uptake at high sugar concentration. *Planta* **174**, 210–216.

Van Beusichem, M. L., Baas, R., Kirkby, E. A., and Nelemans, J. A. (1985). Intracellular pH regulation during NO_3-assimilation in shoot and roots of *Ricinus communis. Plant Physiol.* **78**, 768–773.

Van Dillewijn, C. (1952). "Botany of sugarcane." Veenman, Wageningen.

Veith, R., and Komor, E. (1991). Nutrient requirement for optimal growth of sugarcane suspension cells: Nicotinic acid is an essential growth factor. *J. Plant Physiol.* **139**, 175–191.

Welbaum, G. E., and Meinzer, F. C. (1990). Compartmentation of solutes and water in developing sugarcane stalk tissue. *Plant Physiol.* **93**, 1147–1153.

Wendler, R., Veith, R., Dancer, J., Stitt, M., and Komor, E. (1990). Sucrose storage in cell suspension cultures of Saccharum sp. (sugarcane) is regulated by a cycle of synthesis and degradation. *Planta* **183**, 31–39.

Willenbrink, J. (1982). Storage of sugars in higher plants. *Encycl. Plant Physiol., New Ser.* **13A,** 689–699.

Wyse, R. E., and Komor, E. (1984). Mechanism of amino acid uptake by sugarcane suspension cells. *Plant Physiol.* **76,** 865–870.

Ziegler, H. (1975). Nature of transported substances. *Encycl. Plant Physiol., New. Ser.* **1,** 59–100.

Zrenner, R., and Stitt, M. (1991). Comparison of the effect of rapidly and gradually developing water-stress on carbohydrate metabolism in spinach leaves. *Plant, Cell, Environ.* **14,** 939–946.

7

The Regulation of Plant Transpiration: Interactions of Feedforward, Feedback, and Futile Cycles

E.-D. Schulze

I. Introduction

The daily water loss of a sunflower leaf may be equivalent to several times its total fresh weight under conditions of open stomata and high uptake of carbon dioxide. In contrast, a plant water deficit equivalent to only a small fraction of its total fresh weight would cause severe metabolic disorders due to water stress. Thus, the amount of water uptake necessary to balance loss can be very large, and plants do not have adequate structures which can change in volume sufficiently to store the amounts of water necessary to maintain transpiration if the water supply from roots becomes limited. The opportunity cost for storing water (i.e., the benefit achieved from the most favorable alternative pattern of allocation) is so high that it is rare to find higher plants capable of significant water storage, e.g., CAM plants (Chapin *et al.*, 1990). For this reason, the regulation of plant water relations directly interferes with growth and survival, even in CAM plants. During evolution, it has been mainly the refinements of regulating plant water status that has enabled plants to occupy terrestrial ecosystems ranging from sea level to mountain tops and from swamps to deserts.

In the following, I describe a number of control features which are involved in regulating water loss. Various aspects of this control have been reviewed in recent years (Schulze, 1986a,b; Farquhar and Sharkey, 1982; Davies and Zhang, 1991). An integration of the different controls at the stomatal and root level and their interaction with plant water status and growth are emphasized in this chapter.

II. The Regulation of Stomata in Response to Dry Air

In contrast to the view that plant bulk leaf water status regulates stomatal aperture and thus plant water loss, the observation of a "direct" response of stomata to air humidity (Raschke and Kühl, 1970; Lange *et al.*, 1971) changed conceptually the view of how plants interact with their environment (Cowan, 1977). Isolated epidermal strips of *Polypodium* and *Valerianella* were floated on water with a small air bubble below the guard cells representing the substomatal cavity (Lange *et al.*, 1971). Under these conditions, stomatal cells responded reversibly and individually to changes in humidity of the outside air. When exposed to moist air, stomata were open, but they closed in dry air despite a constant water potential at the inner side of the epidermis. The opening and closing process could be reproduced more than 20 times using the same pair of guard cells. It was even possible to independently open and close adjacent guard cells by exposing them to different vapor pressure deficits (VPD). However, when the entire epidermis was in contact with water, the responses to humidity disappeared and stomata remained wide open.

Cowan (1977) and Farquhar (1978) defined this response type as "feedforward response," i.e., the stomata respond with closure, if exposed to dry air before transpiration affected the water status in the mesophyll. Feedforward control may close stomata to the extent that the rate of transpiration would decrease in dry air, despite further increases in VPD (Schulze *et al.*, 1972; West and Gaff, 1976). Mechanistically, the response was explained by peristomatal transpiration (Seybold, 1961–1962; Maerker, 1965; Maier-Maerker, 1983), which is a transpirational water loss bypassing the stomatal pore through hydrophilic pores in the cuticle (Schulze, 1986a). The guard cells should lose water in relation to ambient vapor pressure deficit by cuticular transpiration independent of the water loss which passes through the aperture. This water would be recharged from surrounding cells and from the subepidermal air space. It was hypothesized that the resistance for these water flows would be so high that a water deficit arises in the guard cell as a function of peristomatal transpiration.

Considering this mechanism, the feedforward response of the stomata would be different from a feedforward in metabolic pathways, e.g., enzyme activation by the substrate. The signal was thought to be mediated by a bypass flux of water through the cuticle, which would cause a local cellular water deficit proportional to the gradient. Although mechanistic differences exist between the feedforward response of stomata and substrate-activated enzyme reaction, the overall effect is similar. Cowan (1977) illustrated the difference between feedforward and feedback control in a kybernetic diagram (Fig. 1). A feedback regulation will cause stomata to close as a consequence of a change in bulk leaf and xylem water potential (Fig. 1, transfer function G_1). In contrast, during feedforward regulation a change in VPD will directly affect stomata (Fig. 1, transfer function G_2) and close the aperture even before water loss has increased in proportion to VPD.

Numerous attempts have been made to demonstrate peristomatal water loss and its affect on guard cell turgor, because this is the key process of the feedforward hypothesis (Maier-Maerker, 1983). Due to cuticular transpiration, epidermal turgor should respond to changes in VPD independent of the stomatal opening, i.e., also if stomata were closed. There-

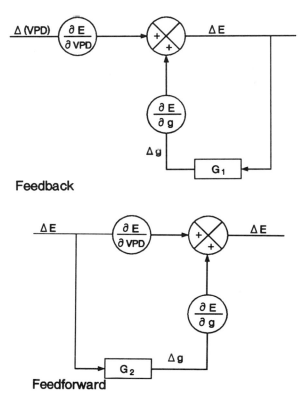

Figure 1   Schematic presentation of feedback (left) and feedforward (right) responses of stomata during changes in vapor pressure deficit (VPD) and the interaction with transpiration (E). G^1 and G_2 represent transfer functions of the response (after Cowan, 1977).

fore, epidermal turgor was measured (Fig. 2A) not only in the light but also in the dark, when stomata were closed, in order to avoid any interaction with the stomatal water flux (Fig. 3, right; Frensch and Schulze, 1988). Despite expectations, it was found that for *Tradescantia* epidermal turgor and transpiration did not change when vapor pressure deficit was increased. Apparently, peristomatal transpiration did not cause the humidity response in *Tradescantia*, although it may also be possible that the water supply of the epidermis with closed stomata was so high that water losses through the cuticle were balanced as was observed in water-flooded epidermal pieces (see Lange *et al.*, 1971). Only in the light (Fig. 3 left), when stomata were open, was epidermal turgor related to VPD. Obviously, the humidity response only functioned when stomata were at least partially open and stomatal transpiration was present. Only then did leaf conductance decrease with epidermal turgor at increasing VPD. This observation appears to support the suggestion by Sheriff and Meidner (1974) that evaporation from internal walls of guard cells and epidermal walls could cause stress-induced stomatal closure. This also appears to be supported by Mott and Parkhust (1991) who exposed leaves to water vapor in helox, a gas mixture in which N_2 is replaced by helium, which changes the diffusivity of water vapor. They found that stomatal response to humidity was based on the rate of water loss from the leaf rather than on cuticular transpiration.

One major criterion for identifying a feedforward response of stomata to changes in VPD was the observed decrease of transpiration despite increasing VPD (Schulze *et al.*, 1972; Schulze and Küppers, 1979). It was thought that a feedback response would only maintain constant transpiration but not decrease water vapor loss at increasing VPD. *Tradescantia virginiana* exhibits a response to VPD which meets the criterion for a feedforward-type response (Fig. 4). In well-watered plants, stomata did not respond very strongly to changes in VPD and transpiration increased with VPD. With increasing plant water stress stomata responded more strongly to changes in VPD and this resulted in a decrease of transpiration in dry air. This response was reversible. At severe drought stomata were closed and did not respond to humidity. On the basis of this result, I propose that this response is not restricted to *Tradescantia* but is a general feature of higher plants (Sheriff, 1979) if stomata respond over a sufficient range.

In order to clarify the mechanism of the humidity response, detailed measurements of cellular water relations are required. In *Tradescantia*, turgor pressure was always lower in epidermal cells than in mesophyll cells, while osmotic potentials were higher in the mesophyll than in the epidermis (Fig. 5). Thus, in contrast to our expectation, the water potentials were lower in mesophyll cells than in epidermal cells. This

Figure 2 (A) Experimental setup for measuring stomatal response of a plant in drying soil in a pressurized pot which can be seen at the bottom of the picture. When feeding ABA to the xylem sap of one leaf, the opposite leaf served as a control. Xylem sap coming from the root system was collected in a vial which can be seen in the left part of the picture. (B) Measuring cellular water relations in leaves of *Tradescantia virginiana*. The leaf is kept in position by a magnet while the glass capillary of the pressure probe is inserted into the cell and observed by a long-distance optic. (C) Measuring xylem sap in a pristine forest of Nothofagus in New Zealand using a compensating method (D) in which the electric current is measured in order to maintain a temperature gradient in the tree trunk constant. The tree is heated with electrodes and the temperatures are measured using thermocouples.

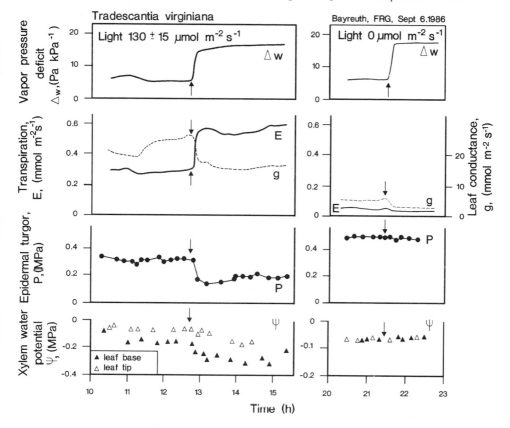

Figure 3　Effect of changing air humidity on stomatal aperture in the light (left) and in the dark (right). The arrow indicates the time at which VPD was changed (after Frensch and Schulze, 1988).

indicates that water is flowing from the epidermis to the mesophyll and not vice versa and that evaporation of water is mainly occurring from mesophyll cells. The differences in cellular water relations between epidermis and mesophyll are caused by a leaf internal cuticle which extends from the outside into the stomatal cavity covering the inner side of the guard cells and even part of the epidermal cells (Nonami *et al.*, 1990). During transpiration, water loss from the guard cells and the subsidiary cells is reduced significantly because of this cuticular cover. Therefore, the hypothesis of Sheriff and Meidner (1974) proposing water loss from the inner walls of the guard cells also is not supported by these measurements.

The effect of the leaf internal cuticle on epidermal water relations

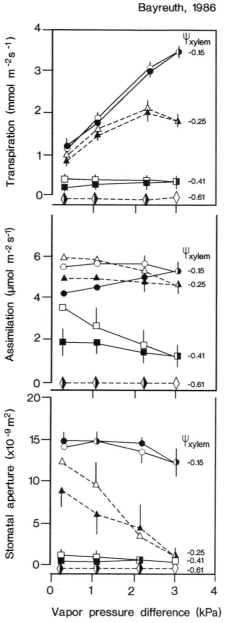

Figure 4 Transpiration, assimilation, and stomatal aperture of *Tradescantia* plants having various xylem water potentials in relation to vapor pressure differences between leaf
and air (after Nonami and Schulze, 1989).

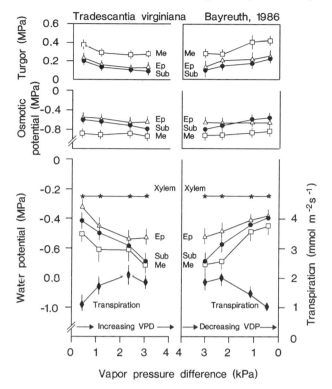

Figure 5 Cell turgor, cell osmotic potential, and cell water potential of mesophyll cells (Me), epidermal cells (Ep), and subsidiary cells (Sub) in relation to VPD when VPD was increased (left) and decreased (right) (after Nonami and Schulze, 1989)

becomes obvious when a decrease of transpiration takes place, i.e., at high VPD (Fig. 5). Turgor of all cell layers decreased only slightly with increasing vapor pressure deficit of the air. However, when water potentials decreased in the mesophyll at high VPD, water was drawn from the subsidiary cells to an extent that the water potential of subsidiary cells decreased and became similar to that in the mesophyll, while water potentials in other epidermal cells remained constant. It should be made clear (Nonami *et al.*, 1990) that the velocity of water movement through the decreasing pore aperture increases sharply at this point of inflection at increasing VPD. We suggest that the observed decrease of transpiration at increasing VPD is attributable to a withdrawal of water from the guard cell complex at high rates of evaporation from the mesophyll cell walls which exceed lateral transport of water in the epidermis. The leaf internal cuticle appears to play a special role in protecting guard cell water loss at low transpiration, but it cannot prevent stomatal closure when water is withdrawn from the guard cell complex as a result of high mesophyll water loss at high VPD. During these experiments xylem water potential, as measured by *in situ* psychrometry, was not affected. Cowan (1993) proposed a flip–flop mechanism of stomatal closure or opening at low turgor pressure. It is not clear if the "overshooting" response of stomata in dry air is similar to such a mechanism.

We may conclude that the stomatal response to air humidity is a typical feedback response, at least in *Tradescantia*. It is caused neither by peristomatal transpiration nor by water loss from the inner epidermal cell walls. However, water is drawn from the guard cell complex by high evaporation loss from the mesophyll. This induces a local water deficit in the guard cell complex. Since xylem water potential is not affected, it acts as if it is a feedforward response. We should also be aware that the humidity response is not a passive turgor-driven process but it is physiologically regulated (Lösch and Schenk, 1978), and the experiments by Schurr (1992) indicate that the daily course of stomatal conductance and the humidity response during the day may also be related to abscisic acid content (ABA, see below).

The hypothesis of VPD regulating stomata via a local water stress in mesophyll cells is supported by and has implications for transpiration of plant canopies. In a mature undisturbed *Nothofagus* forest xylem sap flux summed over a number of trees (Figs. 2C and 2D; Köstner *et al.*, 1992; Kelliher *et al.*, 1992) was equivalent to eddy-flux determinations of transpiration above the canopy. During the course of a day (Fig. 6), tree water loss increased in the morning with net radiation and VPD, but decreased in the afternoon despite further increases in VPD. More important than the average trend of water loss is the observation of high-frequency fluctuations of constant amplitude which are superimposed on the general pattern of canopy water loss. When the frequency of the xylem sap flux fluctuations were compared with meteorological parameters above the canopy, a great similarity became apparent. The frequency of short-term fluctuations in tree stem water flux was similar to those of vertical wind speed, temperature, and VPD in the atmosphere above the tree. More important, though, was that not only were tree transpiration and atmospheric fluxes characterized by fluctuations of similar length, but these fluctuations were also coherent (correlated over a range of frequencies; Fig. 7). The coherence between sap flow and atmospheric parameters reached a minimum at around 10^{-2} Hz and then increased again. This suggested that the maximum size of the coherent eddies may be about 100 m.

The size and distance between displacement events suggests that nearby trees should experience a similar evaporative environment. Indeed, fluctuations in sap flux were significantly correlated among neigh-

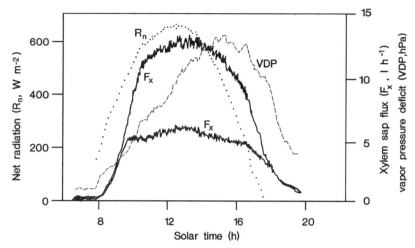

Figure 6 Typical clear-day pattern of net radiation (R_n), water vapor pressure deficit (VPD), and flux of water up two adjacent tree stems in a 30-m tall *Nothofagus* forest in New Zealand. Note the constant amplitude of high-frequency fluctuations in xylem sap flow during the day. These are analyzed in Figs. 7 and 8 (after Hollinger *et al.*, 1993)

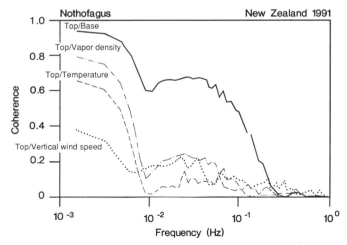

Figure 7 Coherence between sap flux measured at 15 m (top) and at 1.5 m (base), vapor density, air temperature, and vertical wind speed at a height above the canopy (after Hollinger *et al.*, 1993).

boring emergent canopy trees (Fig. 8). The intermittent nature of turbulence means that foliage experiences a non-steady-state evaporative environment. The saturation deficit above a forest may vary generally about 5 hPa during a period of 2 to 3 min. I suggest that these fluctuations of VPD in the turbulent flow of the atmosphere may act as the controlling mechanism that adjusts stomatal aperture to changes in humidity at the canopy level. Since stomata have a response time of about 1 to 10 min for closure and of 5 to 20 min for opening (Meidner and Mansfield, 1968), they would act as a low-pass filter to changes in atmospheric VPD, i.e., the period of dry air is longer than the period of moist air and therefore favors the closing response over the opening response of stomata, because the time needed for stomatal closing is shorter than that for opening. The coupling to the environment should lead to stomata that are relatively more closed under field conditions than they would be in a nonturbulent environment. The very strong response of conductance to VPD at the canopy level as described by Köstner *et al.* (1992) appears to be initiated by the turbulent changes in VPD which cause changes in transpiration and changes in cellular water relations which adjust the stomatal aperture such that the fluctuations of transpiration around a mean rate of water loss do not exceed plant internal water supply. Thus the band of fluctuations around a variable mean remains constant throughout the day.

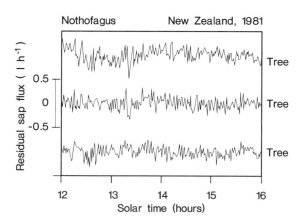

Figure 8 Synchronous changes of sap flux in neighboring dominant canopy trees of a *Nothofagus* forest. Only the residual sap flux is shown, which is the amplitude of the high-frequency fluctuations shown in Fig. 6.

III. Stomatal Response to Soil Water Deficits: Feedforward Control by Root Signals

There is very little evidence that stomata close in response to a decrease of whole plant xylem water potential (review by Schulze and Hall, 1982; Schulze, 1986b; Schulze *et al.*, 1988). Indeed, plants may reach the same level of water potential under well-watered conditions when transpiration is high or soil water stress when transpiration is low (Hall and Schulze, 1980). Also, during the course of a day, water potentials reached a daily minimum when the early morning maximum of leaf conductance and assimilation was achieved (Schulze *et al.*, 1974). *Vigna unguiculata* closed stomata in dry soil even though the plant internal water status did not change (Bates and Hall, 1981). This observation indicates that an interaction exists between root and shoot for regulating stomata. This is further supported by experiments with split roots (Blackman and Davies, 1984) or split pots (Coutts, 1981), in which one part of the root experienced drought while the other remained wet and stomata closed even if only part of the root system was dried while the whole-plant water status was not affected (review by Davies and Zhang, 1991). The response of stomata to a root signal may be regarded as a feedforward response, in which roots in dry soil produce a chemical signal to reduce water loss even before the plant experiences plant internal water stress.

The interactions between VPD and soil water in their effect on stomatal conductance were studied by changing VPD of an entire plant and an isolated leaf on the same plant independently, at different levels of extractable soil water (Turner *et al.*, 1984, 1985; Gollan *et al.*, 1985). As leaf water potential declined concomitant with a decrease in extractable soil water, leaf conductance of the isolated leaf also decreased but the curve for the relation of leaf conductance to leaf water potential for isolated leaves on plants in dry air was shifted about 0.75 MPa lower than the curve for those on plants in moist air (Fig. 9). This demonstrated that leaf water potential alone did not determine stomatal aperture when the whole plant was exposed to a constant humidity environment.

An experiment which demonstrated the existence of a root signal was carried out by Gollan *et al.* (1986). Plants were grown in a pressurized pot (Fig. 10), which compensated for any change in leaf water potential by applying a balancing pressure to the root system (Passioura, 1987). Thus, the leaf was fully turgescent at any time during the experiment. When soil dried, stomatal closure in wheat plants with fully turgescent leaves responded the same as in wheat plants in which leaves were not turgescent during soil drying (Fig. 10, left). Turgescent and nonturgescent leaves closed stomata when soil lost about 60% of its plant-available water. The experiment was repeated with sunflower (Fig. 10, middle)

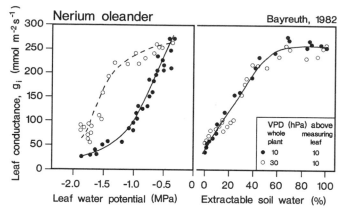

Figure 9　Changes in leaf conductance with soil drought of isolate leaves which were maintained at constant VPD while the rest of the plant experienced dry or moist air, respectively (after Gollan *et al.*, 1985)

and *Ricinus* (Fig. 10, right) exhibiting the same result (Schurr *et al.*, 1992; Schurr, 1991).

Very early in the discussion about root signals it was suggested that the stress hormone ABA may cause the observed stomatal closure (Raschke, 1979,), and it was proposed that the root tip is the actual stress sensor (Zhang *et al.*, 1987). Apparently, root tips produce ABA which is transported to the leaf via the xylem stream and cause stomatal closure. There is evidence that the root tip experiences loss in turgor earlier than the differentiated root cells (Spollen and Sharp, 1991) and that the hydraulic conductance of roots and stomatal aperature decreased concurrently due to soil water stress (Meinzer *et al.*, 1991). This suggests that the sensing mechanism for a feedforward control of stomata by roots would be a local water stress in the root tip which equilibrates with the water potential of the soil environment. The root tip would act as stress sensor and initiate a metabolic reaction resulting in a steady-state response of stomata. It is apparent that the response of roots to soil water stress is very similar to their response to compacted soils (Atwell and Newsome, 1990; Tardieu *et al.*, 1992), which may cast doubt on the hypothesis that root tip turgor is the only sensor for the ABA response.

Changes in xylem water ABA concentration in response to soil drought

Figure 10　Relationship between leaf conductance and soil water content of *Triticum aestivum* (left; after Gollan *et al.*, 1986), *Helianthus annuus* (middle; after Schurr *et al.*, 1992), and *Ricinus communis* (right; after Schurr, 1991). Closed symbols, leaf conductance of plants that were maintained fully turgescent by applying a balancing pressure to the root system; open symbols, leaf conductance of drying leaves not maintained turgescen.

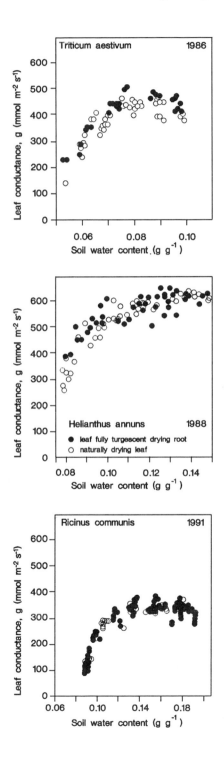

were observed in several studies (reviewed by Davies and Zhang, 1991), but there are only a few field experiments which demonstrate this effect. When *Prunus dulcis* was grown in large lysimeters (see Stitt and Schulze, this volume, Chapter 4) of different sizes (7, 14, and 21 m³ soil volume with 2, 4, or 6 m³ plant-available water), and was allowed to dry out, maximum stomatal conductance (g_{max}) was not correlated with xylem water potential in leaves (Fig. 11A), nor with mean soil water content (Fig. 11B). However, g_{max} decreased in a threshold manner with predawn water potentials (Fig. 11C) and whenever ABA concentration in the xylem sap increased beyond a certain level, independent of pot size (Wartinger *et al.*, 1990). This does not mean that the plant "measures" the soil water content only in the morning, but that local soil water potential gradients evolve during the day. This causes bulk soil water content to be an erroneous measure for the effective soil water status

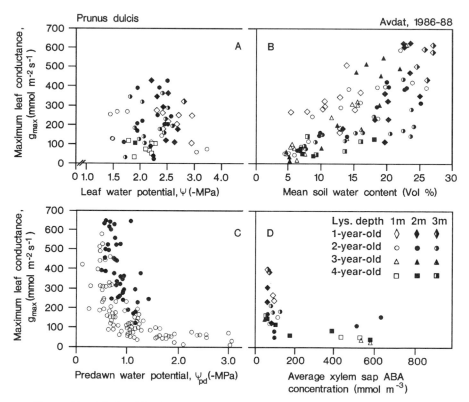

Figure 11   Relationships between maximum leaf conductance of *Prunus dulcis* and leaf water potential (A), mean soil water content (B), predawn water potential (C), and ABA concentration in the xylem sap (D) (from Wartinger *et al.*, 1990).

(Schurr, 1992; see also Horn, Part 3:B). At predawn only, the system is more or less equilibrated and can be used as a measure for correlations with xylem sap ABA. Stomatal ABA response to predawn water potentials was related to plant age which is in this case equivalent to soil age. Plants growing in newly filled pots with homogenized loess responded with stomatal closure at a lower plant water potential (-0.5 to -1 MPa) than plants growing in soils which had gone through a repeated dry-out cycle. These plants responded at -0.3 to -0.5 MPa. Repeated drying of soils caused local reorientation of particles which led to development of soil structure and the formation of soil crumbs (Semmel *et al.*, 1990; see Horn, this volume, Chapter 10). In the soils that developed structure the water transport from denser particles was decreased. Aggregation of soil resulted in more severe water stress. It appears that the plant responded to this cumulative stress with exactly the same sensitivity because the response of stomata to ABA was uniform.

In most investigations, concentrations of ABA in xylem water are related to stomatal response although Zhang *et al.* (1987) suggested that it is apparently the total amount of ABA which is important, not the concentration. In fact, it is the total amount of ABA that reaches the stomata which is physiologically important, because it is known that large proportions of ABA will also migrate into the mesophyll cells where ABA will be metabolized (Hartung and Davies, 1991) or it will be transported out of the leaf via the phloem (Schurr, 1992). When ABA was fed directly into the xylem (Fig. 2A) stomatal closure showed a dose–response to ABA in the xylem (Heckenberger, 1992). Following addition of ABA enriched xylem water (Fig. 12A), an immediate stomatal closure was observed. This immediate closure was followed by a slow recovery phase, which was due to ABA export via the phloem and due to ABA metabolization. When the phloem flow was reduced by cooling of the petiole, the recovery phase did not change significantly (Fig. 12B). However, when the rate of metabolization was changed by varying light intensity, the recovery phase was delayed by low light (Fig. 12C). This indicates that metabolization of ABA is a more important regulating process than ABA storage or export in controlling ABA concentration in the leaf and stomatal response.

IV. Coupled Fluxes and Futile Cycles Modulate the Root Signal

Cowan *et al.* (1982) pointed out that the flux of ABA in leaves is determined by the pH gradient between sources and sinks of ABA and by its state of protonation. At low pH in the xylem sap there is a steep gradient

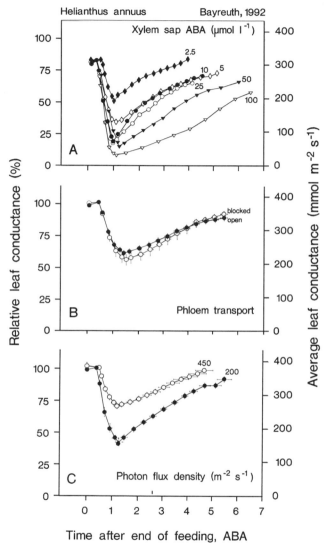

Figure 12 Responses of stomata following addition of ABA into the xylem vessels at known concentrations and for a period of 30 min. The data show leaf conductance before start of the experiment, the initial closing response, and the recovery phase. (A) Increasing concentrations of xylem ABA cause increasing initial closure and an increasing delay of recovery. (B) Phloem blockage by cold girdling causes no change in stomatal response following an application of 10 μM ABA. (C) Decreasing photonflux density, however, affects metabolization of ABA in the mesophyll and causes a significant change in the response of stomata to ABA following an application of 10 μM ABA (after Heckenberger, 1992).

to the chloroplast which supports uptake of uncharged ABA molecules from the xylem into the chloroplasts of the mesophyll in a form which could permeate mesophyll cell membranes. At higher pH of the xylem sap, ABA is charged and passage across membranes would be strongly reduced. However ABA anions could move through the cell walls and could reach the epidermis (Brinckmann *et al.*, 1990). There it could interfere with the potassium channels of the guard cell plasmalemma and thus inhibit stomatal responses (Hartung, 1983; Hornberg and Weiler, 1984). This migration scheme of ABA in the leaf, which was summarized in a complete ABA-flux model by Slovik and Hartung (1992), indicates that the action of ABA is modulated by the plant, especially through the coupled transport of other charged particles and the membrane proton pump, which affect the pH of the apoplast and thus of the xylem sap especially under water stress.

Evidence for an interaction of the pH of the xylem sap with stomatal sensitivity to ABA was presented by Schurr *et al.* (1992). In these experiments, stomatal conductance was not uniquely related to the ABA xylem sap concentration, but each plant showed a different response pattern (Fig. 13). It turned out that the sensitivity of stomata to changes in ABA was related to the pH, Ca, and nitrate in the xylem sap (Fig. 14). In dry soil, nitrate uptake of sunflower decreased but ABA concentration increased. However, plants with good nutrition responded less to soil drought. They maintained high levels of nitrate uptake and the pH did not rise as much as in those of low nutritional status.

The sunflower experiment of Gollan *et al.* (1992) and Schurr *et al.* (1992) was not designed to change xylem pH and nitrate uptake and the

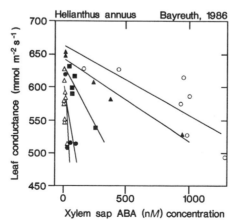

Figure 13   Relations between stomatal conductance and ABA in sunflower of plants which differed in their nutritional status (after Gollan *et al.*, 1992).

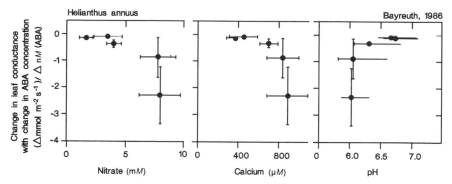

Figure 14 The sensitivity of stomata to ABA (g/ABA) as related to the xylem sap nitrate concentration (left), xylem sap calcium concentration (middle), and the xylem sap pH (after Schurr, 1992).

findings need further experimental confirmation. When *Ricinus* was used instead of sunflower (Schurr, 1992), this pattern did not emerge because xylem pH decreased with soil drought. In this case phloem unloading was affected in addition to changes in nutrient uptake. The balance between uptake of anions (mainly nitrate) and the circulation and uptake of cations (mainly potassium) caused xylem sap pH to decrease rather than to increase. In this case the "modulation" of stomatal response to ABA was not observed. This indicates that it is not only the coupled fluxes of nitrate and ABA which modulate the effect of ABA. Xylem pH depends also on the plant internal circulation of ions including ABA in the xylem and phloem. Schurr (1992) estimated that in well-watered *Ricinus* plants, the import of ABA into the leaf was equal to that of phloem export (Table I). Thus, in well-watered plants very little ABA did enter the mesophyll or the epidermis. However, at soil water stress not only phloem unloading was affected, but xylem import of ABA was also strongly increased. This enhanced the build up of ABA in the leaf. Neals and McLeod (1992) argued that the blockage of the phloem export might be an even more effective and rapid signal than the transport of new ABA from the root to the shoot. However, the xylem feeding experiment of Heckenberger (1992; see Fig. 12) demonstrated that the metabolization of ABA is more effective than ABA export. It is expected that during water stress not only ABA export but also ABA metabolization will decrease. The *Ricinus* experiment exemplifies that "futile cycles" could be highly effective in regulating or modulating plant processes (see also Komor, this volume, Chapter 6) and could also explain the modulation of the ABA response by plant water status in structured soil which was suggested by Tardieu and Davies (1992).

Table I   Xylem Import and Phloem Export of *Ricinus communis* Leaves at Two
Levels of Soil Water Content

		Soil water content			
	High, 21%			Low 9%	
Compound	Xylem import	Phloem export		Xylem import	Phloem export
			$(\text{mmol m}^{-2}\text{s}^{-1})$		
Saccharose	0	149.9		0	94.7
Nitrate	444.6	1.9		166.0	0.8
Sulfate	32.4	1.1		12.6	0.4
Chloride	20.7	0.5		11.5	0.3
Phosphate	16.2	0.6		3.6	0.5
Potassium	651.6	22.1		213.8	10.4
Calcium	92.7	0.5		30.6	0.2
Magnesium	35.1	1.8		15.1	0.8
Amino acids	62.1	34.8		30.6	16.3
Organic acids	0	11.4		0	8.7
			$(\text{pmol m}^{-2}\text{s}^{-1})$		
Abscisic acid	4.1	4.4		52.2	4.9

After Schurr, 1991.

There is another example where a futile cycle might be involved in stomatal regulation, namely, mistletoes. It is well known (review by Glatzel, 1983) that xylem-tapping autotrophic mistletoes accumulate large quantities of potassium because they are unable to return the potassium to the host via a phloem system connected to that of the host. At the same time, mistletoes maintain a higher leaf conductance than the host even under water stress. Thus, the mistletoe obtains large quantities of nitrogen and reduced carbon for growth from the xylem of the host (Schulze *et al.*, 1984, 1991; Ehleringer *et al.*, 1985). It is still unclear how the mistletoe is able to maintain a higher leaf conductance than the host and a high import rate of ABA from the host at the same time. It may well be that the modulation of stomatal sensitivity by futile cycles of organic acids and potassium causes this type of response. The pH of xylem sap of mistletoes is unknown.

The discussion about coupled ion fluxes and futile cycles suggests that although ABA may be a very important stress signal for stomatal response, it is not the only one. The response of stomata to a signal which is produced in the root and transported to the shoot may in fact be quite slow, especially in trees (Schulze, 1991). It may take weeks before the information of water shortage in the soil reaches the crown of a 100-m-high *Sequoia* tree. It is therefore interesting that Munns (1992) showed stomatal closure in wheat which was fed with xylem sap from which ABA was removed. Thus, other substances may have an effect similar to ABA or ABA may be released from the mesophyll (Slovik and Hartung, 1992).

Considering overall whole-plant functioning it appears that the maintenance of the plant internal circulation of phloem and xylem sap is a crucial process which has not been sufficiently investigated to date (Noble, 1974). One may hypothesize that plants maintain this internal circulation as much as possible even at increasing water stress. Noble (1974) estimated that diurnal variations of xylem water potential on the order of 1 to 2 MPa are consistent with the range of turgor and osmotic pressures which occur in the xylem and phloem and which still allow plant internal circulation. An example for this may be seen in *Hammada scoparia*. This plant maintains a diurnal variation in plant water potential of about 2 MPa at low and severe water stress (Fig. 15; Kappen *et al.*, 1975). In all seasons, it reached the level of osmotic potential at noon and the diurnal variation of water potential was constant while the absolute level decreased from -4 to -8 MPa. It is very likely that stomatal closure, loss of assimilatory organs, and osmotic adjustment maintained plant internal circulation even at extreme water stress. The present understanding and evidence for root signals still need to be connected with effects of changes of plant internal circulation (Schurr, 1992). It may well be that the ultimate "aim" of stomatal control and of root signals is a maintenance of this plant internal vital process (which is possible only within a limited range). Thus, the modulation of ABA by the futile cycle via the phloem may be the actual point of regulation of whole-plant function under water stress. This would also include protection against embolism (Lösch and Schulze, 1993).

V. Coupled Fluxes of Water Vapor and Carbon Dioxide Correlated with Maximum Leaf Conductance

Up to now I have discussed mechanisms by which stomatal conductance is reduced below a level of maximum opening. This, however, does not help define the level at which plants adjust their capacity to open stomata when conditions are favorable for transpiration. Large differences exist among plant life forms and leaf types in g_{max} (Körner *et al.*, 1979; Körner, 1993). Sun leaves of herbaceous species have a g_{max} about fivefold higher than needles of conifers (Fig. 16). Körner *et al.* (1979) already related g_{max} to photosynthetic capacity (A_{max}), because water vapor and carbon dioxide are strictly coupled flows. They diffuse across the same aperture

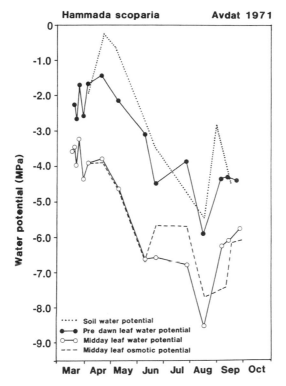

Figure 15 Seasonal changes in soil water potential, predawn water potential, midday water potential, and midday osmotic potential in the C4 plant *Hammada scoparia* (after Kappen *et al.*, 1975).

of the stomatal pore although in opposite directions. Both fluxes are quantitatively affected when stomata close (Farquhar and Sharkey, 1982). However, this may also depend on the rate at which these processes are affected by other environmental parameters.

Schulze and Hall (1982) applied the analysis of Körner *et al.* (1979) and showed that indeed g_{max} changed in relation to A_{max} under a broad range of conditions such as changes in light, climate, leaf age, nutrition, and genotype. The only exception where this relation did not hold was water stress. Stomata appeared to be more sensitive to water stress than was assimilation, although changes in assimilation capacity at saturated CO_2 and carboxylation efficiency appear to also affect leaf conductance (Tenhunen *et al.*, 1985).

Although the evidence for a correlation between g_{max} and A_{max} is very strong, experimental approaches to selectively disturb assimilation reveal that there is no mechanistic link between the two processes (Fig. 17).

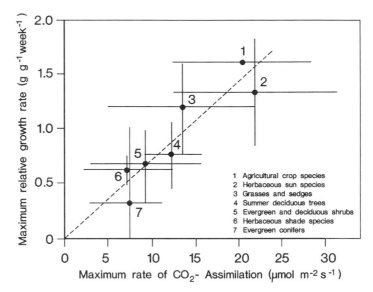

Figure 16 Relations between maximum leaf conductance and maximum rate of CO₂ assimilation of different life plant forms (after Körner *et al.*, 1979).

When assimilation capacity was decreased in isogenic mutants carrying antisense *rbcS* (see Stitt and Schulze, this volume, Chapter 4), leaf conductance was not related to this change in Rubisco, although the linear relation between conductance and photosynthesis was maintained (Lauerer, 1991). This indicated that the correlations of Schulze and Hall (1982) could describe the evolutionary result of long-term selection rather than a direct feedback of two coupled processes.

VI. The Regulation of Root-to-Shoot Ratios

Besides regulating water loss by adjustment of stomatal aperture, plants also control their water relations by adjustment of leaf area (leaf growth and abscision) and by root growth. However, the allocation pattern of plants under water stress is even less understood than the regulation of stomatal aperture (see also Stitt and Schulze, this volume, Chapter 4). Even at the cellular level, the mechanisms of cell expansion growth are not quite clear (Steudle, this volume, Chapter 8). Growth was described by an equation which equilibrated growth in relation to water supply and cell wall extensibility (Lockhart, 1965), and this equation was extended to include active solute transport also (Steudle, 1992). Growth-induced reduction in water potentials may limit extension growth (Boyer *et al.*,

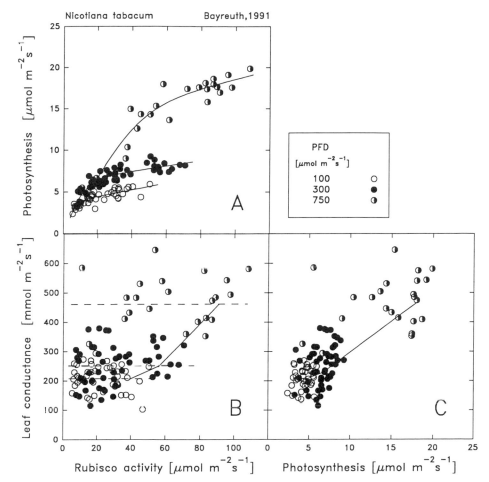

Figure 17 The effect of reduced Rubisco activity on CO_2 uptake (A) and leaf conductance (B) and relations between leaf conductance and photosynthesis (C). The experiments were carried out at three levels of light at which these plants were cultivated. The dashed line in (B) indicates the average leaf conductance which was reached at growing conditions. Variations in Rubisco activity were produced by using tobacco plants transformed with antisense rbcS (after Lauerer, 1991).

1985; Nonami and Boyer, 1990; Matyssck *et al.*, 1991) just as much as osmotic adjustment and cell wall extensibility may contribute to maintaining growth rate (Turner and Jones, 1980; Cleland, 1986). From very early on it was clear that extension growth was physiologically regulated in the cell wall by plant hormones (Trewavas, 1986). Fry (1989) suggested that auxins played a central role in loosening the primary wall by increasing the activity of extracellular cellulase and that xyloglucan-derived monosaccharides are incorporated into the cell wall in order to reconnect the loosened ends of cleaved cellulose (Passioura and Gardner, 1990; Smith and Fry, 1991; Jarvis, 1992). This mechanism would allow for expansion growth even at a constant low turgor. It also would explain the observation that cells resume growth at a constant rate after turgor disturbance (Cosgrove, 1986; Hsiao and Jing, 1987). But the compartment model of water flow (see Steudle, this volume, Chapter 8) indicates that the mechanistic basis for this observation may be even more complicated.

Dealing with whole plants makes the analysis of the underlying processes even more difficult, if not impossible, because of the feedback between root and shoot and because of their interaction with water uptake and water loss which, in turn, are linked to the external environment. When well-watered *V. unguiculata* was grown in humid and dry air (Nagarajah and Schulze, 1983), large differences were observed in total biomass (Fig. 18). Leaf growth was greatly reduced in well-watered plants growing in dry air. Also, the plants grown in dry air exhibited a transient higher root biomass than those grown in humid air. When these plants developed further, the reduction of leaf area had an overriding effect and, therefore, the whole biomass decreased, allowing for no further increase in root biomass. In dry soil, growth was further reduced. In dry soil, plants grown in dry air also produced a transient larger root biomass than those grown in humid air. However, plants in dry soil never produced a root biomass larger than that in those plants which were grown in moist soil even though the root/shoot ratio was increased.

Vigna unguiculata maintains a constant leaf water content even under extreme conditions of transpiration or soil water supply. If leaf water content is taken as a parameter for control, carbon allocation may be modeled such that the partitioning into leaves and supporting biomass may be determined by the transpiration rate of the leaf and the capacity of the root to take up water without changing leaf water content (Schulze *et al.*, 1983). At the whole-plant level, water uptake into roots is a curve of diminishing returns since more and more structure is necessary to support the root tip and to conduct water. In order to meet the additional requirement for water of the next growing leaf an increasing proportion of carbon must be invested into roots to meet the water demand. It can be shown (Schulze *et al.*, 1983) that during regulation of carbon allocation the plant water use is a highly sensitive system (Fig. 19). The balance between growth of new leaves and demand for water explains the differences in growth of *V. unguiculata*.

The problem of perennial biomass and its effect on growth is further complicated in woody species, which cannot adjust aboveground or be-

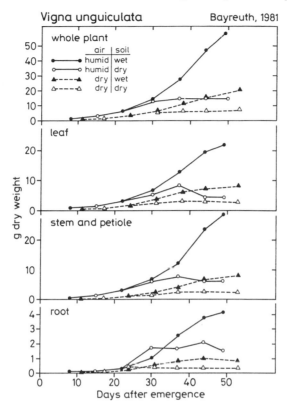

Figure 18 Whole-plant biomass development and its components (leaves, stems, roots) during growth in humid air (solid lines) and dry air (broken lines) with watered plants (closed symbols) and plants growing in progressively drying soil (open symbols) (after Nagarahja and Schulze, 1983).

lowground biomass as easily as the herbaceous *V. unguiculata*. The lysimeter experiment with *P. dulcis* (see above and Stitt and Schulze, this volume, Chapter 4) was designed to study root and shoot growth of trees over several years. In this case, trees were grown in pots or different soil volume watered once per season to field capacity. The plants had to adjust their performance to the amount of plant-available water in order to survive because the amount remained constant at increasing plant size. We wanted to know if shoot/root ratios would adjust to the amount of water available and if this ratio would change during plant development.

During the 4-year growing period, the trees maintained a constant shoot/root ratio and adjusted growth according to the amount of available water (Fig. 20). Water-use efficiency of biomass growth was constant. Obviously in addition to the strong root/shoot interaction during stomatal

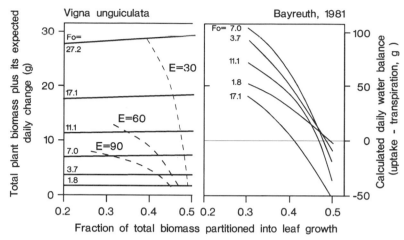

Figure 19   (Left) Biomass plus expected daily leaf growth as related to calculated variations in shoot/root carbon partitioning at a relative growth rate of 0.2 g g^{-1} day^{-1} (solid lines). Aboveground biomass increased with carbon allocation to leaves. The broken lines indicate the level of biomass which can be obtained for given root biomasses and different rates of transpiration without perturbing the plant water balance. (Right) Changes in the calculated water balance as related to hypothetical changes in carbon partitioning (after Schulze *et al.*, 1983).

regulation, there is also a root-to-shoot signal regulating plant growth under water stress. In this case, ABA may be a candidate also. Zhang and Davies (1990) and Bunce (1990) described ABA effects on leaf expansion. Hartung *et al.* (1990) found highest ABA concentrations in young leaves of *Anastatica hierochuntica* and a relation to reduced internode growth.

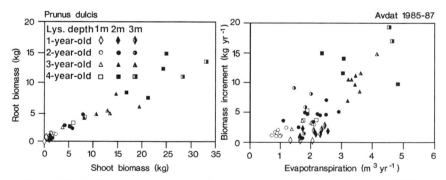

Figure 20   (A) Shoot biomass as related to root biomass in *Prunus dulcis* trees of different age and growth in different size lysimeters (see text); (B) Annual growth of wood as related to water consumption of trees growing in different size lysimeters (after Wartinger *et al.*, 1990).

In addition, Munns (1992) detected a new inhibitor for growth under water stress.

In the examples relating growth under stress to a root signal, we were dealing with whole plants having a closed phloem/xylem circulation. In the *P. dulcis* experiment, we were dealing with grafted plants, i.e., root and shoot were genetically different but were connected by xylem and phloem. Thus, these plants also represent a functional equilibrium between root and shoot as defined by Brouwer (1983) despite genetic differences. There is also an example of coordinated growth, in which this phloem/xylem connection does not exist. Xylem tapping mistletoes lack a phloem connection. They have only a xylem connection to the host. The mistletoe *Phoradendron juniperinum* germinates on 1-year old twigs of *Juniperus osteosperma* (Schulze and Ehleringer, 1984) and tends to kill the host shoot beyond the point of infection. Thus the mistletoe becomes the only green tissue to be supplied by that branch via the xylem. Nevertheless, a linear relation existed between the green biomass of the long-lived mistletoe and the xylem cross-sectional area in the subtending branch of the host (Fig. 21). The relation was similar to that found in trees where a linear relationship existed between foliage area and xylem cross-sectional area (Waring *et al.*, 1982; Oren *et al.*, 1986). The hostparasite relationship shows that stem enlargement is possible without a chemical signal from the leaf. Apparently in this case stem growth is proportional to the xylem sap flow through the host/parasite system.

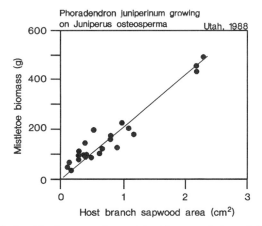

Figure 21   Relationship between dry weight biomass of the mistletoe *Phoradendron juniperinum* and the sapwood area of the host branch at the point of infection. The host, *Juniperus osteospermum*, grows proportional to the mistletoe biomass even though no phloem connection exists between the two organisms, and the host branch was dead beyond the point of infection. The mistletoe biomass represents the photosynthesizing organs, since *Phoradendron* assimilates CO_2 with its green stems (after Schulze and Ehleringer, 1984).

VII. Conclusions

The regulation of transpiration at the whole-plant level is based on a sequence of interacting control systems and the effects of decreasing soil water status are summarized in Fig. 22:

- The vapor pressure deficit causes a local cellular water stress in the epidermis which feeds back on stomatal aperture. The steady-state effect of this regulation behaves as if it were a feedforward response to air humidity, i.e., stomata close before the leaf experiences water stress.
- Root tips are an independent sensor for soil water status and soil compaction (Fig. 22). They may experience water stress while the whole plant is still well hydrated. In this sense, they have a function in sensing the soil environment similar to that of the guard cell complex of the leaf in sensing the aerial environment.
- During water stress, root tips produce abscisic acid as a hormonal signal which is transported through the xylem and effectively closes stomata independent of the water status of the leaf (Fig. 22).
- The sensitivity of stomata to changes in ABA is influenced by the chemical environment, which in turn depends on ion uptake (nitrate nutrition) and the phloem/xylem circulation of cations and organic acids (Fig. 22). Xylem sap composition over the long-term is correlated to changes in the leaf, and this could affect the sensitivity of stomata to ABA. The balance of ion uptake and ion circulation determines the pH in the xylem sap, which, in turn, appears to affect the movement of ABA through the leaf apoplast.
- A root signal also affects growth of the aboveground parts (Fig. 22). It is likely that ABA is involved as a stress hormone. In addition, growth appears also to be stimulated by a root signal which is proportional to the flux of xylem sap (see also Beck, this volume, Chapter 5).
- The effects of increasing ABA transport in the xylem may be enhanced by blockage of phloem unloading under water stress, and it is compensated by light-dependent metabolization in the mesophyll (Fig. 22).
- There is no indication that loss in maximum rate of transpiration would increase the sensitivity of regulation (see Scheibe and Beck, this volume, Chapter 1). The maximum rate of water loss is determined by other physiological functions, e.g., the coupled flux of CO_2 and water vapor.

It appears that the interactions of feedback as a sensing mechanism, of feedforward between root and shoot, and of the plant internal circulation are important in stabilizing the water relations of higher plants. In homoiohydric higher plants (in contrast to poikilohydric organisms)

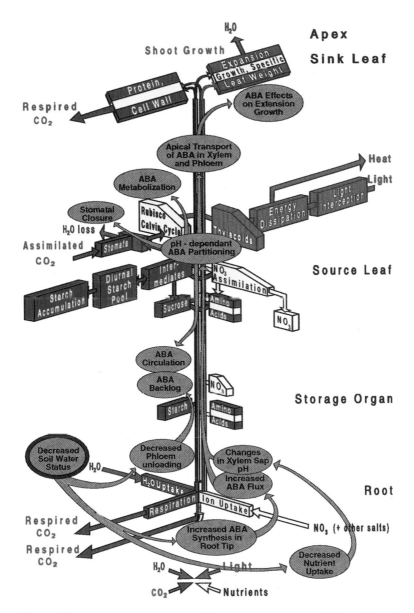

Figure 22 Whole-plant response (carbon, water, nutrient, and light relations) following a change in soil water status. Decreased soil water status (oval with black margin) will affect phloem unloading, which may cause a buildup of ABA. At the same time root tips will increase ABA sythesis and nutrient uptake will be reduced. This will cause a change in xylem ABA concentration of ABA in the leaf. Depending on the pH in leaf cell walls, ABA will be reloaded into the phloem and either recirculate or reach the leaf-growing region, it will be metabolized, or it will close stomata. The ABA, which reaches the growing tip via the xylem and the phloem, will reduce extension growth.

survival depends on the maintainence of hydration of living cells. A single regulating process would probably be too coarse and too risky for coping with the large variety of factors which may cause water stress. Besides drought in the atmosphere and in the soil, an imbalance between the biomass of root and shoot may cause such disturbance.

Acknowledgments

A large number of guest scientists and postdoctoral, doctoral, and master's degree students have joined together to complete this 12-year project. We acknowledge their cooperation and help. We especially thank our scientific guests J. Boyer, I. Cowan, J. Ehleringer, G. Farquhar, T. Hall, D. Hollinger, T. Hsiao, F. Kelliher, J. Milburn, S. Nagaraja, H. Nonami, J. Passioura, K. Shakel, N. C. Turner, and J. Zhang, as well as our doctoral students T. Gollan, U. Heckenberger, M. Küppers, M. Lauerer, U. Schurr, and A. Wartinger. We thank W. Gigl, R. Krug, R. Meserth, and M. Wartinger for technical assistance.

References

Atwell, B. J., and Newsome, J. C. (1990). Turgor pressure in mechanically impeded lupin roots. *Aust. J. Plant Physiol.* **17,** 49–56.

Bates, L. M., and Hall, A. E. (1981). Stomatal closure with soil water depletion not associated with changes in bulk leaf water status. *Oecologia* **50,** 62–65.

Blackman, P. G., and Davies, W. J. (1984). Age-related changes in stomatal response to cytokinins and abscisic acid. *Ann. Bot. (London)* [N.S.] **64,** 121–125.

Boyer, J. S., Cavalieri, A., and Schulze, E.-D. (1985). Control of the rate of cell enlargement: Excision, wall relaxation, and growth-induced water potentials. *Planta* **163,** 327–543.

Brinckmann, E., Hartung, W., and Wartinger, M. (1990). Abscisic acid of individual leaf cells. *Physiol. Plant.* **80,** 51–54.

Brouwer, R. (1983). Functional equilibrium: Sense or nonsense? *Neth. J. Agric. Sci.* **31,** 287–289.

Bunce, J. A. (1990). Abscisic acid mimics effects of dehydration on area expansion and photosynthetic partitioning in young soybean leaves. *Plant, Cell Environ.* **13,** 295–298.

Chapin, F. S., III, Schulze, E.-D., and Mooney, H. A. (1990). The ecology and economics of storage in plants. *Annu. Rev. Ecol. Syst.* **21,** 423–427.

Cleland, R. E. (1986). The role of hormones in wall losening and plant growth. *Aust. J. Plant Physiol.* **13,** 93–103.

Cosgrove, D. J. (1986). Biophysical control of plant cell growth. *Ann. Rev. Plant Physiol.* **37,** 377–405.

Coutts, M. P. (1981). Leaf water potential and control of water loss in droughted sitka spruce seedlings. *J. Exp. Bot.* **131,** 1193–1201.

Cowan, I. R. (1977). Stomatal behaviour and environment. *Adv. Bot. Res.* **4,** 117–228.

Cowan, I. R. (1993). As to the mode of action of the guard cells in dry air. *Ecol. Stud.* **100** (in press).

Cowan, I. R., Raven, J. A., Hartung, W., and Farquahr, G. D. (1982). A possible role for abscisic acid in coupling stomatal conductance and photosynthetic carbon metabolism in leaves. *Aust. J. Plant Physiol.* **9,** 489–498.

Davies, W. J., and Zhang, J. (1991). Root signals and the regulation of growth and development of plants in drying soil. *Annu. Rev. Plant Physiol. Plant Mol. Biol.* **42,** 55–76.

Ehleringer, J. R., Schulze, E.-D., Ziegler, H., Lange, O. L., Farquhar, G. D., and Cowan, I. R. (1985). Xylem tapping mistletoes: Primarily water or nutrient parasites? *Science* **227,** 1479–1481.

Farquhar, G. D. (1978). Feedforward responses of stomata to humidity. *Aust. J. Plant Physiol.* **5,** 787–800.

Farquhar, G. D., and Sharkey, T. D. (1982). Stomatal conductance and photosynthesis. *Annu. Rev. Plant Physiol.* **33,** 317–345.

Frensch, J., and Schulze, E.-D. (1988). The effect of humidity and light on cellular water relations and diffusion conductance of leaves of *Tradescantia virginiana* L. *Planta* **173,** 554–562.

Fry, S. C. (1989). Cellulases, hemicelluloses and auxin-stimulated growth: A possible relationship. *Physiol. Plant.* **75,** 532–536.

Glatzel, G. (1983). Mineral nutrition and water relations of hemiparasitic mitletoes: A question of partitioning. Experiments with *Loranthus euripaeus* on *Ouercus petraea* and *Ouercus robur. Oecologia* **56,** 193–201.

Gollan, T., Turner, N. C., and Schulze, E.-D. (1985). The responses of stomata and leaf gas exchange to vapour pressure deficits and soil water content. III. In the sclerophyllous woody species *Nerium oleander. Oecologia* **65,** 356–362.

Gollan, T., Passioura, J. B., and Munns, R. (1986). Soil water status affects the stomatal conductance of fully turgid wheat and sunflower leaves. *Aust. J. Plant Physiol.* **13,** 459–464.

Gollan, T., Schurr, U., and Schulze, E.-D. (1992). Stomatal response to drying soil in relation to changes in the xylem sap composition of *Helianthus annuus.* I. The concentration of cations, anions, amino acids in, and pH of, the xylem sap. *Plant, Cell Environ.* **15,** 551–559.

Hall, A. E., and Schulze, E.-D. (1980). Drought effects on transpiration and leaf water status of cowpea in controlled environments. *Aust. J. Plant Physiol.* **7,** 141–147.

Hartung, W. (1983). The site of action of abscisic acid at the guard cell plasmalemma of *Valerianella locusta. Plant, Cell Environ.* **6,** 427–428.

Hartung, W., and Davies, W. J. (1991). Drought-induced changes in physiology and ABA. *In* "Abscisic Acid, Physiology and Biochemistry" (W. J. Davies and H. G. Jones, eds.), pp. 63–79. Bios Scientific Publishers, Oxford.

Hartung, W., Heilmeier, H., Wartinger, A., Kettemann, I., and Schulze, E.-D. (1990). Ionic content and abscisic acid relations of *Anastatica hierochuntica* L. under arid conditions. *Isr. J. Bot.* **39,** 373–382.

Heckenberger, U. (1992). Stomatareaktion auf kontrollierte Veränderungen des Xylemsaftes mit Abscisinsäure. Diplomarbeit, University of Bayreuth.

Hollinger, D. Y., Kelliher, F. M., Schulze, E.-D., and Köstner, B. M. (1993). Coupling of tree transpiration to atmospheric turbulence. *Nature (London)* (in press).

Hornberg, C., and Weiler, E. W. (1984). High-affinity binding sites for abscisic acid on the plasmalemma of *Vicia faba* guard cells. *Nature (London)* **310,** 321–324.

Hsiao, T. C., and Jing, J. (1987). Leaf and root expansive growth in response to water deficits. *In* "Physiology of Cell Expansion During Plant Growth" (D. J. Cosgrove and D. P. Knievel, eds.), The American Society of Plant Physiologists, Rudeville pp 180–192.

Jarvis, P. G. (1992). Self-assembly of plant cell walls: Opinion. *Plant, Cell Environ.* **15,** 1–6.

Kappen, L., Oertli, J. J., Lange, O. L., Schulze, E.-D., Evenari, M., and Buschbom, U. (1975). Seasonal and diurnal courses of water relations of the arido-active plant *Hammada scoparia* in the Negev desert. *Oecologia* **21,** 175–192.

Kelliher, F. M., Köstner, B. M. M., Hollinger, D. Y., Beyers, J. N., Hunt, J. E., McSeveny, T. M., Meserth, R., Weir, P. L., and Schulze, E.-D. (1992). Evaporation, xylem sap flow, and tree transpiration in a New Zealand broad-leaved forest. *Agric. For. Met.* **62**, 53–73.

Körner, C. (1993). Maximum stomatal conductance of major vegetation types of the globe. *Ecol. Stud.* **100** (in press).

Körner, C., Scheel, J. A., and Bauer, H. (1979). Maximum leaf diffusive conductance in vascular plants. *Photosynthetica* **13**, 45–82.

Köstner, B. M. M., Schulze, E.-D., Kelliher, F. M., Hollinger, D. Y., Beyers, J. N., Hunt, J. E., McSeveny, T. M., Meserth, R., and Weir, P. L. (1992). Transpiration and canopy conductance in a prestine broad-leaf forest of *Nothofagus:* An analysis of xylem sap flow and eddy correlation measurements. *Oecologia* **92**, 236–241.

Lange, O. L., Lösch, R., Schulze, E.-D., and Kappen, L. (1971). Responses of stomata to changes in humidity. *Planta* **100**, 76–86.

Lauerer, M. (1991). Der Einfluss von Lichtintensität und Ribulose-1,5-bisphosphat-Carboxilase/Oxigenase auf Photosynthese, Wachstum und Blattanatomie bei Tabak (*Nicotiana tabacum* L.) transformiert mit antisense rbcS. Diplomarbeit, University of Bayreuth.

Lockhart, J. A. (1965). An analysis of irreversible plant cell elongation. *J. Theor. Biol.* **8**, 264–275.

Lösch, R., and Schenk, B. (1978). Humidity response of stomata and the potassium content of guard cells. *J. Exp. Bot.* **29**, 781–787.

Lösch, R., and Schulze, E.-D. (1993). Internal coordination of plant responses to drought and excess evaporational demand. *Ecol. Stud.* **100** (in press).

Maerker, U. (1965). Zur kenntnis der Transpiration der Schliesszellen. *Protoplasma* **60**, 61–78.

Maier-Maerker, U. (1983). the role of peristomatal transpiration in the mechanism of stomatal movement. *Plant, Cell Environ.* **6**, 369–380.

Mattyssek, R., Maruyama, S., and Boyer, J. S. (1991). Growth-induced water potentials may mobilize internal water for growth. *Plant, Cell Environ.* **14**, 917–923.

Meidner, H., and Mansfield, T. A. (1968). "Physiology of Stomata." McGraw-Hill, London.

Meinzer, F. C., Grantz, D. A., and Smit, B. (1991). Root signals mediate coordination of stomatal and hydraulic conductance in growing sugarcane. *Aust. J. Plant Physiol.* **18**, 329–338.

Mott, K. A., and Parkhust, D. F. (1991). Stomatal responses to humidity in air and helox. *Plant, Cell Environ.* **14**, 509–515.

Munns, R. (1992). A leaf elongation bioassay detects an unknown growth inhibitor in xylem sap from wheat and barley. *Aust. J. Plant Physiol.* **19**, 127–136.

Nagarajah, S., and Schulze, E.-D. (1983). Responses of *Vigna unguiculata* (L.) Walp. to atmospheric and soil drought. *Aust. J. Plant Physiol.* **10**, 385–394.

Neals, T. F., and McLeod, A. L. (1992). Do leaves contribute to the abscisic acid present in the xylem of "droughted" sunflower plants? *Plant, Cell Environ.* **14**, 979–986.

Noble, P. S. (1974). "Biophysical Plant Physiology and Ecology." Freeman, San Francisco.

Nonami, H., and Boyer, J. S. (1990). Primary events regulating stem growth at low water potentials. *Plant Physiol.* **93**, 1601–1609.

Nonami, H., and Schulze, E.-D. (1989). Cell water potential, osmotic potential, and turgor in the epidermis and mesophyll of transpiring leaves. Combined measurements with the cell pressure probe and nanoliter osmometer. *Planta* **177**, 35–46.

Nonami, H., Schulze, E.-D., and Ziegler, H. (1990). Mechanisms of stomatal movement in response to air humidity, irradiance and xylem water potential. *Planta* **183**, 57–64.

Oren, R., Werk, K. S., and Schulze, E.-D. (1986). Relationships between foliage and conducting xylem in *Picea abies* (L.) Karst. *Trees* **1**, 61–69.

Passioura, J. B. (1987). The use of the pressure chamber for continuously monitoring and controlling the pressure in the xylem sap of the shoot of intact transpiring plants. *Proc. Int. Conf. Meas. Soil Plant Water Status*, Logan, Utah, pp. 31–34.

Passioura, J. B., and Gardner, P. A. (1990). Control of leaf expansion in wheat seedlings growing in dry soil. *Aust. J. Plant Physiol.* **17**, 149–158.

Raschke, K. (1979). Movements of stomata. *Encycl. Plant Physiol., New Ser.* **7**, 383–441.

Raschke, K., and Kühl, U. (1970). Stomatal response to changes in atmospheric humidity and water supply: Experiments with leaf sections of *Zea mays* in CO_2-free air. *Planta* **87**, 36–48.

Schulze, E.-D., (1986a). Carbon dioxide and water vapor exchange in response to drought in the atmosphere and in the soil. *Annu. Rev. Plant Physiol.* **37**, 247–274.

Schulze, E.-D. (1986b). Whole-plant responses to drought. *Aust. J. Plant Physiol.* **13**, 127–141.

Schulze, E.-D. (1991). Water and nutrient interactions with plant water stress. *In* "Response of Plants to Multiple Stresses" (H. A. Mooney, W. E. Winner, and E. J. Pell. eds.), pp. 89–101. Academic Press, San Diego.

Schulze, E.-D., and Ehleringer, J. R. (1984). The effect of nitrogen supply on growth and water-use efficiency of xylem tapping mistletoes. *Planta* **162**, 268–275.

Schulze, E.-D., and Hall, A. E. (1982). Stomatal responses, water loss, and CO_2 assimilation rates of plants in contrasting environments. *In* Encyclopedia Plant Physiology N. S., Vol. 12B (Lange, O. L., *et al.,* eds.), pp. 181–230. Springer Verlag, Berlin.

Schulze, E.-D., and Küppers, M. (1979). Short-term and long-term effects of plant water deficits on stomatal response to humidity in *Corylus avellana* L. *Planta* **108**, 319–326.

Schulze, E.-D., Lange, O. L., Buschbom, U., Kappen, L., and Evenari, M. (1972). Stomatal responses to changes in humidity in plants growing in the desert. *Planta* **108**, 259–270.

Schulze, E.-D., Lange, O. L., Evenari, M., Kappen, L., and Buschbom, U. (1974). The role of air humidity and leaf temperature in controlling stomatal resistance of *Prunus armeniaca* L. under desert conditions. I. A simulation of the daily course of stomatal resistance. *Oecologia* **17**, 159–170.

Schulze, E.-D., Schilling, K., and Nagarajah, S. (1983). Carbohydrate partitioning in relation to whole plant production and water use of *Vigna unguiculata* (L.) Walp. *Oecologia* **58**, 169–177.

Schulze, E.-D., Turner, N. C., and Glatzel, G. (1984). Carbon, water and nutrient relations of two mistletoes and their hosts: A hypothesis. *Plant, Cell Environ.* **7**, 293–299.

Schulze, E.-D., Steudle, E., Gollan, T., and Schurr, U. (1988). Response to Dr. P. J. Kramer's article, "Changing concepts regarding plant water relations," Volume 11, Number 7, pp. 565–568. *Plant, Cell Environ.* **11**, 573–576.

Schulze, E.-D., Lange, O. L., Ziegler, H. and Gebauer, G. (1991). Carbon and nitrogen isotope ratios of mistletoes growing on nitrogen and non-nitrogen fixing hosts and on CAM plants in the Namib desert confirm partial heterotrophy. *Oecologia* **88**, 457–462.

Schurr, U. (1992). Die Wirkung von Bodenaustrocknung auf den Xylem- und Phloemtransport von *Ricinus communis* und deren Bedeutung für die Interaktion zwischen Wurzel and Sproß. Doktor-Thesis, Bayreuth, Germany.

Schurr, U., Gollan, T., and Schulze, E.-D. (1992). Stomatal response to drying soil in relation to changes in the xylem sap composition of *Helianthus annuus*. II. Stomatal sensitivity to abscisic acid imported from the xylem sap. *Plant, Cell Environ.* **15**, 561–567.

Semmel, H., Horn, R., Hell, U., Dexter, A. R., and Schulze, E.-D. (1990). The dynamics of soil aggregate formation and the effect on soil physical properties. *Soil Technol.* **3**, 113–129.

Seybold, A. (1961–1962). Ergebnisse und Probleme pflanzlicher Transpirationsanalysen. *Heidelb. Akad. Wiss., Jh.* **62**, 5–8.

Sheriff, D. W. (1979). Stomatal aperture and the sensing of the environment by guard cells. *Plant, Cell Environ.* **2**, 15–22.

Sheriff, D. W., and Meidner, H. (1974). Water pathways in leaves of *Hedera helix* L. and *Tradescantia virginiana* L. *J. Exp. Bot.* **25**, 1147–1156.

Slovik, S., and Hartung, W. (1992). Compartmental distribution and redistribution of abscisic acid in intact leaves. III. Analysis of the stress signal chain. *Planta* **187**, 37–47.

Smith, R. C., and Fry, S. C. (1991). Endotransglycosylation of xyloglucans in plant cell suspension cultures. *Biochem. J.* **279**, 529–535.

Spollen, W. G., and Sharp, R. E. (1991). Spatial distribution of turgor and root growth at low water potentials. *Plant Physiol.* **96**, 438–443.

Steudle, E. (1992). The biophysics of plant water: Compartmentation, coupling with metabolic processes, and flow of water in plant roots. *In* "Water and Life: Comparative Analysis of Water Relationships at the Organismic, Cellular, and Molecular Levels" (G. N. Somero, C. B. Osmond, and C. L. Bolis, eds.), pp. 173–204. Springer-Verlag, Berlin.

Tardieu, F., and Davies, W. J. (1992). Stomatal response to abscisic acid is a function of current plant water status. *Plant Physiol.* **98**, 540–625.

Tardieu, F., Zhang, J., Katerji, N., Bethenod, O., Palmer, S., and Davies, W. J. (1992). Xylem ABA controls the stomatal conductance of field-grown maize subjected to compaction or drying soil. *Plant, Cell Environ.* **15**, 193–199.

Tenhunen, J. D., Catarino, F. M., Lange, O. L., and Oechel, W. C. (1985). "Plant Responses to Stress: Functional Analysis in Mediterranean Ecosystems." Springer-Verlag, Heidelberg.

Trewavas, A. (1986). Understanding the control of plant development and the role of growth substances. *Aust. J. Plant Physiol.* **13**, 447–458.

Turner, N. C., and Jones, M. M. (1980). Turgor maintenance by osmotic adjustment. *In* "Adaptation of Plants to Water and High Temperature Stress" (N. C. Turner and P. J. Kramer, eds.), pp. 67–103. Wiley, New York.

Turner, N. C., Schulze, E.-D., and Gollan, T. (1984). Responses of stomata and leaf gas exchange to vapour pressure deficits and soil water content. I. Species comparisons at high soil water contents. *Oecologia* **63**, 338–342.

Turner, N. C., Schulze, E. D., and Gollan, T. (1985). The responses of stomata and leaf gas exchange to vapour pressure deficits and soil water content. II. In the mesophytic herbaceous species *Helianthus annuus*. *Oecologia* **65**, 348–355.

Waring, R. H., Schroeder, P. E., and Oren, R. (1982). Application of the pipe model theory to predict canopy leaf area. *Can. J. For. Res.* **12**, 556–560.

Wartinger, A., Heilmeier, H., Hartung, W., and Schulze, E.-D. (1990). Daily and seasonal course of leaf conductance and abscisic acid in the xylem sap of almond trees (*Prunus dulcis* (Miller) D. A. Webb) under desert conditions. *New Phytol.* **116**, 581–587.

West, D. W., and Gaff, D. F. (1976). The effect of leaf water potential, leaf temperature and light intensity on leaf diffusive resistance and the transpiration of leaves of *Malus sylvestris*. *Physiol. Plant.* **38**, 98–104.

Zhang, J., and Davies, W. J. (1990). Changes in the concentration of ABA in xylem sap as a function of changing soil water status can account for changes in leaf conductance and growth. *Plant, Cell Environ.* **13**, 277–287.

Zhang, J., Schurr, U., and Davies, W. J. (1987). Control of stomatal behaviour by abscisic acid which apparently originates in the root. *J. Exp. Bot.* **38**, 1174–1181.

8

The Regulation of Plant Water at the Cell, Tissue, and Organ Level: Role of Active Processes and of Compartmentation

E. Steudle

I. Introduction

Water is by far the most mobile substance in plants. The water permeability of a typical plant cell membrane is larger by 3 to 8 orders of magnitude than that of nutrient ions such as K^+ or NO_3^- or slowly permeating nonelectrolytes (sucrose, glucose). Because of its high mobility, water tends to distribute according to thermodynamic gradients (activity, pressure) within tissues and organs of higher plants, and water potential gradients will tend to vanish. Regulation of water potential in terms of maintaining a desired water potential in a certain tissue within a plant organ while others are at a different water potential is usually not possible. The situation is different at the boundaries of plants. Here, high hydraulic resistances of cuticles prevent water losses of shoots into the dry atmosphere and considerable gradients are usually maintained. In contrast, the relatively low hydraulic resistance of roots provides for sufficient uptake of water. The feedback regulation of transpirational water losses by stomata allows a regulation of the water status of the plant (as expressed by the water potential), if there is sufficient supply from the root or from internal water sources (see Schulze, this volume, Chapter 7).

Despite the fact that water tends to quickly reach thermodynamic equilibrium in plants, there are internal water movements which are caused by the water potential difference between soil and atmosphere, growth processes, active and passive solute transport, and the compartmentation of solutes. In the latter cases, water flow is coupled to solute flow. Components of water potential (turgor pressure, osmotic pressure) rather than water potential itself are the properties regulated such as during turgor or osmoregulation and growth. This is a direct consequence of the tendency of water potential to equilibrate rapidly within plants.

Turgor pressure is related to cell volume and water content by the elastic properties of cells and tissues and, hence, these parameters are directly related to processes governed by water potential. In the regulation of plant water, very often a homeostasis of turgor, volume, water content, or osmotic pressure is achieved. On the other hand, changes of the components of water potential will cause changes in water potential which are the driving force for a redistribution of water. During redistributions of water, very often metabolic processes interact with water. This allows some metabolic control or regulation of plant water status as is observed during extension growth or during the transfer of solutes between different compartments (e.g., during phloem transport; see section IV of this chapter and Komor, this volume, Chapter 6).

In this chapter, the regulation of the water status of plants at the levels of cells, tissues, and organs is examined. Emphasis is given to aspects of the coupling between water and solute relations and to compartmentation phenomena as they occur during expansion growth and in roots. For general aspects of plant water relations, the reader is referred to the basic reviews of Dainty (1963, 1976), Zimmermann and Steudle (1978), Boyer (1985), Tomos (1988), and Steudle (1989a, 1990, 1992, 1993). In addition, the reviews of Steudle (1985) and Cosgrove (1986, 1992) focus on the biophysics of growth and those of Pitman (1982), Weatherley (1982), Dainty (1985), Passioura (1988), Steudle (1989b,c), and Moreshet and Huck (1991) focus on transport across roots. The regulation of transpirational water losses via stomata is not considered because this is subject of a separate chapter in this book (Schulze, Chapter 7).

In this chapter, a comprehensive theory of the water relations of plant cells and tissues is given which extends the basic theories of Philip (1958) and Molz and Ikenberry (1974) for both stationary and dynamic water relations of tissues. The theory focuses on interactions between water and solute transport and on compartmentation phenomena. It is applied to growth and to the water uptake into roots in an attempt to work out general principles of the regulation of plant water.

II. Water Relations at the Cell Level

As a good approximation plant cells can be considered to behave like ideal (semipermeable) osmometers. In this case, the linear force/flow relations of irreversible thermodynamics yield for the water (volume) flow across the cell membrane (J_V; Steudle, 1989a, 1992, 1993):

$$J_V = -\frac{1}{A}\frac{dV}{dt} = Lp\{P - RT(C^i - C^o)\} = Lp \cdot \Delta\psi. \tag{1}$$

J_V is water-flow density in $m^3 \cdot m^{-2} \cdot s^{-1}$. A = cell surface area, V = cell volume, Lp = hydraulic conductivity, P = cell turgor, and C^i, C^o = concentration of internal and external solutes, respectively. The term in brackets on the right side of Eq. (1) represents the difference in water potential ($\Delta\psi$) which is the driving force under these conditions. Equation (1) is an Ohm's law analogue. In principle, it should only hold for small driving forces. However, practical experience shows that a linear relation between flow and force is also found under conditions of a considerable deviation from equilibrium.

To a large extent, the hydraulic conductivity (Lp) will control the water exchange between a cell and its immediate surroundings. During conditions of steady flow, Lp will be the only controlling factor besides the cell surface area. During dynamic processes, i.e., under conditions of changing water potentials, the time constant for the uptake or loss of water of cells (τ) becomes important, which incorporates additional parameters. For an isolated cell, the time constant is given by

$$\tau = \frac{1}{k_w} = \frac{T^w_{1/2}}{\ln(2)} = R_c \cdot C_c = \frac{1}{Lp \cdot A} \cdot \frac{V}{(\varepsilon + \pi^i)}. \tag{2}$$

Here, k_w and $T^w_{1/2}$ represent the rate constant and half-time of water exchange of the cell with its surroundings. R_c and C_c are the hydraulic resistance ($R_c = 1/(Lp \cdot A)$) and storage capacity of a cell for water ($C_c = V/(\varepsilon + \pi^i)$), respectively. ε is the elastic coefficient of the cell wall ($= V \cdot dP/dV$) and π^i the osmotic pressure of the cell sap ($= RT \cdot C^i$). It can be seen from Eq. (2) that, analogous to electrical circuits, τ is given as the product of a resistance and a capacity ($R_c \cdot C_c$). When changes of water potential occur, cells are "charged" or "depleted" of water via a hydraulic resistance (Fig. 1). The meaning of R_c is straightforward (inverse of a hydraulic conductance, $Lp \cdot A$). However, the storage capacity of a cell for water (C_c) is dependent not only on its volume, but also on its elastic extensibility ($1/\varepsilon$) and osmotic pressure (π^i). It will usually hold that $\varepsilon \gg \pi^i$.

The parameter k_w ($T^w_{1/2}$) in Eq. (2) can be measured with high precision. The vailidity of the relations given in Eqs. (1) and (2) has been verified

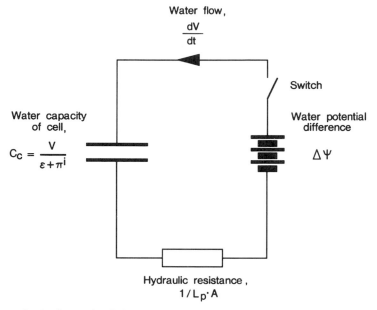

Figure 1 Analogue circuit for the water transport in an individual plant cell. The cell (protoplast) is represented by a hydraulic capacity (C_c) which is charged or depleted via a hydraulic resistance ($1/Lp \cdot A$). The force driving the water flow across the cell membrane is the water potential difference between cell and surroundings (see Eqs. (1) and (2)).

repeatedly using cell pressure probes (see reviews by Dainty, 1976; Zimmermann and Steudle, 1978; Tomos, 1988; Steudle, 1989a, 1990, 1992, 1993). The example shows that, for describing the regulation or control of plant water, electrical analogues may be useful. This will also hold for higher levels of organization (tissue, organ, entire plant; e.g., Molz and Ferrier, 1982; Steudle, 1989a, 1990, 1992; Nobel and Goldstein, 1992). However, when the modeling of plant water in terms of electric analogues is started at the level of individual cells, this would refine and extend the simplifying models in which steady flows across a plant are described as a flow across fairly large and heterogenous compartments (e.g., root, shoot, leaf) represented by resistors arranged in series or in parallel (van den Honert, 1948). However, rather complicated networks would have to be considered even if homogenous tissues are taken into account (see Section III.B).

A. Solute Flow Coupled with Water Flow

The situation of a "regulation" of cell water gets more complicated when water and solute flows are coupled to each other. This may occur in a "dynamic" way by a direct (frictional) interaction between water and solutes as they cross membranes or other osmotic barriers. However, the contribution of these direct interactions to overall flows should usually be small (Steudle and Tyerman, 1983; Rüdinger *et al.*, 1992). More important is the active and passive solute transport which changes osmotic gradients across membranes and creates additional forces. In plant tissues, the effects of these forces on water flow and water relations could be quite complex due to the presence of a network of hydraulic resistors and of osmotic barriers of different selectivity. In the presence of solute flows across a single membrane, we would have to take into account a coupling between the flow of water (J_V in $m^3 \cdot m^{-2} \cdot s^{-1}$) and that of solutes ($J_s$ in mol $m^{-2} s^{-1}$). For the sake of simplicity, we may consider only one permeating solute (subscript 's') crossing a membrane. Then the following basic relations hold (Steudle, 1989a, 1992, 1993):

$$J_V = -\frac{1}{A}\frac{dV}{dt} = Lp\{P - \sigma_s \cdot RT \cdot (C_s^i - C_s^o)\}, \tag{3}$$

and

$$J_s = -\frac{1}{A}\frac{dn_s^i}{dt} = P_s(C_s^i - C_s^o) + (1 - \sigma_s) \cdot \overline{C}_s \cdot J_V + J_s^*. \tag{4}$$

Compared with Eq. (1), Eq. (3) is modified by the reflection coefficient (σ_s) which is a measure of the frictional interaction between flows or a measure of the "passive selectivity" of the membrane. $\sigma_s = 1$ represents the ideally semipermeable membrane (for which the solute permeability is zero). On the other hand, a reflection coefficient of $\sigma_s = 0$ denotes a membrane exhibiting no selectivity for the given solute, "s." Equation (3) demonstrates that the reflection coefficient will modify the driving force which would then become different from the water potential difference which is usually used in a first approximation. In plants, this is very important in all cases where selective membranes are absent as in the apoplast, in the xylem, or during the longitudinal solution flow across the sieve plates of the phloem.

In Eq. (4), n_s^i is the number of moles of substance "s" in the cell and P_s the permeability coefficient of this solute. $\overline{C}_s$ = mean of solute concentration of cell and medium. The first term on the right side of the equation describes the diffusional (passive) permeability of the solute and the second term ($(1 - \sigma_s) \cdot \overline{C}_s \cdot J_V$) the solvent drag (direct interaction between flows; see above). J_s^* denotes the active component of solute flow which relates the passage of solute "s" across the membrane to a metabolic reaction (reaction flow caused, for example, by the action of an ATPase). It can be seen from Eqs. (3) and (4) that, by convention, the uptake of water and solutes into a cell ($dV/dt, dn_s^i/dt > 0$) will be denoted by a negative J_s and J_V and the export of water and solutes by positive values.

In the simple case, where a cell is operating at a constant rate of solute pumping ($J_s^* < 0$) and a certain leak rate, a steady turgor (P_o) will be established ("pump-leak model"). Under these conditions, J_s and J_V will be zero. It is easily verified from Eqs. (3) and (4) that the steady turgor, P_o, will then be

$$P_o = -\sigma_s \cdot RT \frac{J_s^*}{P_s}.$$ (5)

An increase of the active uptake of solutes will increase P_o as well as cell volume and water content. On the other hand, an increase of the leak (P_s) will decrease P_o. Equation (5) demonstrates that turgidity and homeostasis are dynamic rather than static processes resulting from the action of a pump (or pumps) and leak(s). The osmotic responses of the system (turgor, volume) are of interest, when a certain concentration of solute "s" is applied to the cell, and it is subjected to an "osmotic stress." At constant J_s^*, the system will generally react in this case in a biphasic response of turgor or volume due to a superposition of water and solute flows as shown in Fig. 2 for an isolated cell (*Chara* internode) and for an isolated leaf epidermis. Owing to the reduced water potential of the medium, turgor will decrease in a first rapid "water phase." However, it will be restored again during a phase of (passive) solute uptake which will equilibrate the solute concentration across the cell membrane. Unlike the examples shown in Fig. 2, the half-time of the second phase may be rather long for the solutes present in a cell vacuole. The behavior is predicted by Eqs. (3) and (4) or by integrated forms of these equations (Steudle and Tyerman, 1983; Steudle, 1989a; Rüdinger *et al.*, 1992). This has been exactly verified in experiments with isolated cells or tissue using rapidly permeating solutes which also proved the validity of the theory.

Figure 2 Measurement of water and solute transport in plant cells using the cell pressure probe. In (A), measurements are shown on an isolated giant internode of *Chara corallina* and in (B) on cells of the isolated leaf epidermis of *Tradescantia virginiana*. Hydrostatic experiments and experiments with a nonpermeating solute (mannitol, sucrose) gave "monophasic" responses in pressure from which the hydraulic conductivity could be evaluated (Eq. (2)). Permeating solutes (dimethyformamide, ethanol, n-propanol) produced "biphasic" responses from which permeability (P_s) and reflection (σ_s) coefficients of the solutes could be evaluated as well. The rate of the second phase was limited by the permeation of solute into the cell. Note that for *Tradescantia* the cells responded with an increase of turgor when subjected to higher concentrations of the low-molecular-weight alcohols ethanol and n-propanol, i.e., for these solutes the cells showed an "anomalous osmosis" (negative reflection coefficients). Effects were completely reversible, i.e., when changing back to the original solution (APW = artificial pond water), symmetrical responses occurred in the opposite direction (after Tyerman and Steudle, 1982; Steudle and Tyerman, 1983).

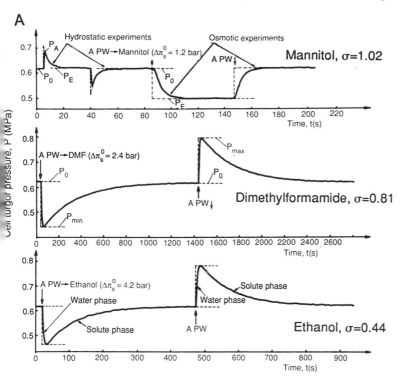

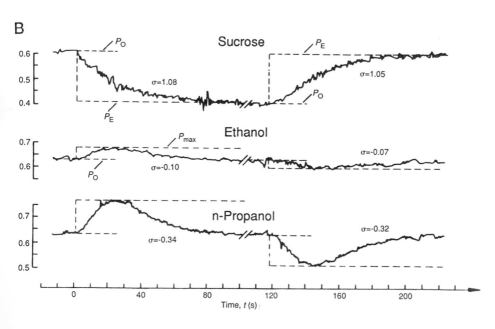

At first sight, the response may be interpreted in terms of a regulation of turgor or cell volume, but it is obvious that it is a simple adjustment due to an equilibration of water and solute flows. Since water is moving much faster than solutes, there will be a transient equilibration of water flow ($J_V = 0$) during the process followed by a phase dominated by solute flow. During the "solute phase," water will follow the solutes, and the water potential gradient (more precisely, $P - \sigma_s \cdot RT\ (C_s^i - C_s^o)$) across the membrane will be maintained close to zero. Figure 2 shows that similar osmotic responses are obtained when the concentration of solute "s" is lowered outside the cell. In this case, a transient maximum turgor is obtained. Figure 2B also shows that for certain solutes a different type of response may be obtained with plant cells. Turgor increased when the concentration of the solute was increased ("anomalous osmosis"). In this case, which would rarely occur with the solutes usually present in plants, the reflection coefficient is negative, which means that the solute entering the cell would drag water into the cell at a rate which is larger than that caused by osmotic water loss.

B. Active Processes

Changes of active solute flow (J_s^* in Eq. (4)) would have effects comparable to those of passive solute flow examined in the previous section. For example, if we assume that a cell (or tissue) which is able to undergo an osmoregulation or turgor regulation is subjected to a change in water potential, there will be a rapid phase of equilibration of water potential (water phase). This phase will be followed by a much slower phase of active solute flow caused by a change of J_s^* (Eq. (5)). If the regulation of turgor (or volume) is perfect, the original turgor will be reestablished. Unlike that of the passive equilibration of solutes, the process will require some feedback control of J_s^* or P_s or of both by turgor or volume or by some other parameter directly related to both which triggers J_s^* (or P_s) according to Eq. (5). Regulation of turgor may be more or less perfect. Such systems have been described repeatedly (for reviews, see Kauss, 1978; Zimmermann, 1978; Zimmermann and Steudle, 1978; Wyn Jones and Gorham, 1982; Kirst, 1991). Osmoregulatory processes will be important during the adaptation of plants to drought or high salinity and during extension growth. They also will be important during the regulation of stomatal movement or for pressure flow in the phloem. For some siphonale algae such as *Valonia* and *Acetabularia*, turgor pressure regulation has been intensively studied by measuring turgor directly (e.g., Zimmermann and Steudle, 1974, 1978; Zimmermann, 1978; Steudle *et al.*, 1977). These measurements have been combined with the measurement of ion fluxes (namely that of potassium) and with those of electrical properties of the membranes (membrane potential and resistance). For

Valonia, it has been shown that the uptake of potassium decreased with turgor which suggested a direct effect on the potassium pump. On the other hand, the efflux of potassium increased with increasing turgor pressure which showed that the passive efflux (P_s in Eq. (5)) was also affected. At a certain steady turgor, influx and efflux were balancing. Interestingly, the balancing turgor was a function of the cell size and increased with decreasing cell size. Since extension growth is driven by turgor (see section IV of this chapter), this would mean that small cells would be able to maintain a higher steady-state turgor and growth rate provided that the effect of turgor on wall tension was not completely compensated for by the smaller cell diameter. Thus, by the pressure dependence of potassium fluxes both the homeostasis of mature cells as well as the growth rate of growing cells could be regulated by a feedback control mechanism. Combined measurements of turgor and of electrical properties of the cell membrane supported the flux measurements. By measuring the dielectric breakdown of cell membranes, estimates were obtained of the compressibility of the membranes which suggested that compressible proteins in the membrane (carriers of channels) acted as the sensing element (the pressure-dependent switch) by which homeostasis was achieved (Coster and Zimmerman, 1976; Coster *et al.,* 1976).

For higher plants, the situation may be different. For example, water stress caused by high salinity or by freezing would induce the production of low-molecular-weight solutes in cells (compatible solutes; Wyn Jones and Gorham, 1982). However, knowledge about the precise mechanisms which explain homeostatic and other types of responses of higher plants to water stress is still poor. More detailed measurements are necessary, which would also require more advanced techniques. In some cases, the fact that mechanisms for the transduction of changes of water potential or its components are not known has led to the conclusion that water potential and its components are not important for sensing. These conclusions have been supported by experiments in which responses of shoots (closing of stomata) to soil drought (water content) were measured while the water potential in the soil was kept constant by applying a balancing gas pressure. Thus, the shoot was hydraulically uncoupled from the root (Passioura, 1980; Turner *et al.,* 1985; Gollan *et al.,* 1985). The results of these experiments suggested that biochemical (abscisic acid) rather than hydraulic signals were important for the root-to-shoot communication (Schulze, this volume, chapter 7). Water content or mechanical properties of the soil rather than the water potential appeared to be important for the sensing at the root (Passioura, 1988). However, these ideas seem to be somewhat premature because the experimental evidence for sensing soil water content rather than soil water potential and possible effects of soil structure are still weak. It is hard to see how variables other than the water potential could be sensed by root tissue. It appears to be necessary to integrate active (metabolic) processes into the concept of water potential which forms the solid basis for plant water relations (Schulze et al., 1988).

C. Active Water Flow and Electroosmosis

Stomatal regulation of transpiration provides an "active" mechanism for plants to maintain the water potential of shoots within certain limits (Schulze, 1986; Schulze, this volume, chapter 7). This type of a regulation of water relations at the level of entire plants is based on a strategy to avoid water losses. There appear to be no active mechanisms of a "direct pumping" of water in plants. These would be mechanisms in which water would be moved actively within the plant body to improve water status. The movement would have to be directly coupled to a metabolic reaction (e.g., to the splitting of ATP). Such mechanisms are highly unlikely, because any active pumping of water would be short-circuited by the high water permeability of plant cells tending to rapidly equilibrate gradients of water potential in tissues as soon as they occur (Steudle, 1989a, 1992).

For the same reason, it is also unlikely that electrokinetic phenomena such as electroosmosis provide suitable mechanisms to move water or a solution and to establish gradients, although the electroosmotic efficiency between ion and water flow may be high in membranes or suitable porous materials (Fensom and Dainty, 1963). This statement also refers to electroosmotic mechanisms which have been proposed for the solution flow in the phloem as an alternative or supplement to the Münch pressure flow (Spanner, 1975). Thus, because of its high mobility, movements of water linked to metabolic energy should be restricted to "secondary active processes" which affect osmotic gradients. These movements require selective membranes. The following sections emphasize this point and also take into account compartmentation which is an additional important factor in the water relations of plant tissues.

III. Regulation of Tissue Water

In principle, Eqs. (1) to (5) should hold for isolated cells as well as for cells positioned in tissues. However, in tissues the overall water and solute relations differ considerably from those of isolated cells because of the network formed by cells and because of the fact that tissues comprise the apoplast as an additional pathway for both water and solutes. For individual cells, the water relations of the wall compartment (apoplast) can be neglected usually. In tissues, as in isolated cells, we have to consider interactions between solutes and water. Effects of active or passive solute flows may be distinguished. In the apoplast only passive (diffusional or convective) solute flows are possible. These points have been incorporated into the integrated theory of tissue water relations outlined below. The theory considerably extends the established theories of Philip (1958) and of Molz and Ikenberry (1974). Steady-state and dynamic properties of tissues are considered. Interactions between water and solutes and between water and active processes are emphasized as well as effects of compartmentation (Steudle, 1989a, 1992).

A. Pathways and Steady-State Water Flow

With respect to stationary hydraulic properties of tissues, the situation appears to be fairly simple. If a gradient of pressure is applied across a tissue there will be three different pathways for water to move (Fig. 3): (a) the apoplasmic path around protoplasts, (b) the symplasmic path via plasmodesmata, and (c) the transcellular (vacuolar) path across cell membranes layer by layer.

The combined symplasmic and transcellular components may be sum-

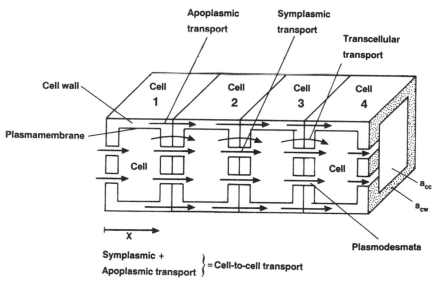

Figure 3 Schematic representation of the transport pathways across a tissue in one dimension (*x*). Only four cells are shown. There is an apoplasmic (cell wall), a symplasmic (via plasmodesmata), and a transcellular (vacuolar) pathway. The transcellular and the symplasmic path may be summarized as the cell-to-cell path. a_{cc} and a_{cw} represent the cross-sectional areas available for the two main pathways (cell-to-cell and apoplasmic). For further explanation, see text (after Molz and Ferrier, 1982).

marized as a "cell-to-cell" component. For solutes, it is usually thought that symplasmic transport is much faster than the transcellular transport because of the low permeability of membranes for solutes. However, for water the situation is different. The experimental distinction between the symplasmic and transcellular component is very difficult because of the high water permeability of cell membranes. To date, there are no techniques available to quantitatively open or close plasmodesmata and to measure the corresponding effects on water transport (hydraulic conductivity). Therefore, the two components (symplasmic and transcellular) should be combined in a cell-to-cell component. Thus, in a tissue only two parallel pathways would have to be considered which would contribute to the overall transport of water according to their hydraulic conductivities and cross-sectional areas. The overall hydraulic conductivity of a tissue (Lp_r) in one direction in $m \cdot s^{-1} \cdot MPa^{-1}$ would be given by:

$$Lp_r = \left[\gamma_{cc} \cdot \frac{Lp}{2} + \gamma_{cw} \cdot \frac{Lp_{cw}}{\Delta x} \right] \cdot \frac{\Delta x}{d}, \tag{6}$$

where Δx = cell thickness in x direction (Fig. 3); d = tissue thickness; Lp_{cw} = specific hydraulic conductivity of cell wall material in $m^2 \cdot s^{-1} \cdot MPa^{-1}$; and γ_{cw} and γ_{cc} are the fractional cross-sectional areas of the apoplasmic and cell-to-cell path, respectively. A factor of 2 is employed in the first term in the brackets on the right side of Eq. (6), because two membranes would have to be crossed per cell layer. The absolute value of Lp would incorporate plasmodesmata and cell membranes. Also, measured values of cell Lp would usually incorporate the hydraulic resistances of two membranes arranged in series, i.e., the plasmalemma and the tonoplast.

Equation (6) holds for hydrostatic gradients as they occur across plant tissues (e.g., when tensions are created in the xylem by transpiration). Difficulties will arise, however, when the equation is applied to osmotic water flow, since osmotic gradients will only cause small effective driving forces across the apoplasmic pathway. This is so, because the reflection coefficient of cell walls would be very low. Nearly no selectivity is expected for the apoplast. The effective force driving a water flow along the apoplast will be $\sigma_s \cdot \Delta \pi_s$ (see Eq. (3)). In contrast, during the exchange of water between apoplast and symplast (protoplasts), osmotic forces will be fully exerted. Thus, we expect complications in tissue water relations when osmotic forces are operating. This is briefly examined in the next section.

B. Dynamic Water and Solute Relations of Tissues: Integrated Theory

Philip (1958) was the first to quantitatively formulate dynamic changes of the water status of plant tissues. He neglected the apoplasmic component of water transport and calculated the rate of propagation of changes of the water potential of a tissue following a change of water potential in the surroundings by assuming that for each cell layer the water flow across the layer would be proportional to the water potential difference across the layer ($\Delta\psi$):

$$J_V = \frac{\text{Lp}}{2} \cdot \Delta\psi . \qquad (7)$$

This equation is analogous to Eq. (1) but considers that for every cell layer two membranes have to be crossed (cf. Eq. (6)). For the changes of water potential in space and time, Philip's approach resulted in a kinetics which was formally identical with those used for diffusion processes. A "diffusivity" (D_c) was worked out with the units of a diffusion coefficient ($m^2 \cdot s^{-1}$) in order to quantify the rate of a propagation per unit driving force and the dynamics of changes of water potential. Diffusivities are also used in soil physics to quantify the dynamics of soil water and, being a soil physicist, Philip simply extended this approach to plants. D_c was related to basic cell parameters such as Lp, ε, and cell dimensions (Steudle, 1989a, 1992):

$$D_c = \frac{\text{Lp} \cdot a_{cc} \cdot \Delta^2}{2 \cdot C_c} = \frac{a_{cc} \cdot \Delta x^2 \cdot \ln(2)}{2 \cdot A} \cdot \frac{1}{T^w_{1/2}} . \qquad (8)$$

It can be seen from Eq. (8) that D_c is inversely proportional to the half-time of water exchange of tissue cells (eq. (2)) which can be directly measured using the cell pressure probe. The Philip model as well as extended models (see below) can be experimentally tested. It predicts that, as in diffusion kinetics, the half-time of a propagation of changes of water potential in a tissue would be proportional to the square of the dimensions of the tissue in the direction of the propagation. This was verified with plant tissue to a good approximation (Westgate and Steudle, 1985). It should be noted that, although the diffusivity has the units of a diffusion coefficient, the processes characterized by the parameter are by no means related to the diffusion of water. What is transported are changes of water potential (free energy), turgor, or volume (water content) which are all linearly related to each other. Thus, we are dealing with bulk flows of water and the relation to diffusion is only a formal one.

Molz and Ikenberry (1974) extended the Philip model by introducing the parallel cell wall pathway. They assumed a rapid equilibration between a protoplast and the adjacent apoplast in a tissue which strictly coupled the two differential equations for water flow in the two pathways to each other. Because of the high mobility of water in plant tissue we know today (namely, from cell pressure probe data) that this assumption of a "local equilibrium" holds to a very good approximation. The diffusivity (D_t) calculated by Molz and Ikenberry extended the expression of Philip for D_c according to properties of the parallel apoplasmic path:

$$D_t = \frac{\text{Lp}/2 \cdot a_{cc} \cdot \Delta x^2 + \text{Lp}_{cw} \cdot a_{cw} \cdot \Delta x}{C_c + C_{cw}}. \tag{9}$$

It can be seen from this equation that the hydraulic conductances of the parallel pathways for water are additive in the numerator and D_t increases with increasing conductance. The storage capacities in the denominator (C_c for the protoplast and C_{cw} for the apoplast) are also additive but D_t decreases with increasing cell water storage ($C_c + C_{cw}$). This is expected, since the propagation of changes of ψ is damped by the storage of water in the parallel pathways. In terms of an electric analogue, the Molz/Ikenberry model describing the dynamic water relations of tissues is given by a network consisting of two parallel pathways with cellular resistances (apoplasmic and cell-to-cell) in series in each path and storage capacities in parallel (Fig. 4). In the model, small hydraulic resistances connect the parallel pathways so that the condition of local equilibrium is fulfilled. The storage capacities of cells (protoplasts) and of the per cell apoplast (C_{cw}) are given by (see Eq. (2); Steudle, 1989a, 1990, 1992):

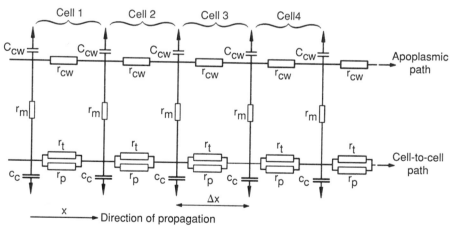

Figure 4 Analogous representation of the water transport across a plant tissue in terms of an equivalent electric circuit (one-dimensional). r_{cw}, r_t, and r_p denote the hydraulic resistances of the three possible pathways for water (apoplasmic, transcellular, and symplasmic; see Fig. 3). The parallel resistances r_t and r_p may be summarized as a resistance for the cell-to-cell path (r_{cc}). C_c and C_{cw} are the corresponding per cell capacities for water of the cells (protoplasts) and the apoplast which are arranged in parallel to the resistances. Note that the transcellular path (r_t) would imply crossing the cell walls between adjacent cells and two cell membranes to get from one vacuole (capacity) to the next.

$$C_c = \frac{dV}{d\psi} = \frac{V}{\varepsilon + \pi^i} \quad \text{and} \quad C_{cw} = \frac{dV_{cw}}{d\psi_{cw}} \approx \frac{\Delta V_{cw}}{\Delta \psi_{cw}}, \tag{10}$$

where V_{cw} = per cell water volume of apoplast. Experimentally, cell (protoplast) data (Lp, ε, $T^w_{2/2}$, C_c, etc.) are directly accessible using the cell pressure probe. In contrast, apoplast data (Lp_{cw} and C_{cw}) or the water content (per cell) of the apoplast (V_{cw}) are much harder to obtain (for references, see Steudle, 1989a; Peterson and Steudle, 1993). To date there have been only a few attempts to measure or estimate these values. For higher plants, this has been done usually by first measuring tissue properties (hydraulic conductance, diffusivity, etc.). Cell (protoplast) data were then determined separately using the cell pressure probe and apoplast properties were worked out by difference (Westgate and Steudle, 1985; Zhu and Steudle, 1991).

It has been already mentioned that, in principle, the kinetic model of Molz and Ikenberry is in agreement with experimental findings. However, there is also evidence that there are limitations of the model. The latter refer to interactions between water and solute flow in the tissue. Molz and Hornberger (1973) tried to incorporate one of the possible interactions by allowing for a diffusion of solutes along the cell-to-cell path. However, the contribution of this component of solute flow should be rather small because of the low permeability of cell membranes for solutes (see above). Apoplasmic flows and the active uptake of solutes into the symplast as well as exchanges of solutes between pathways should be much more important. At times, when data on rates of water exchange of individual tissue cells (i.e., half-times and diffusivities) were not available, the diffusion of solutes in the wall space has been considered to be rate limiting for osmotic swelling or shrinking of tissues. Today we know from data of the cell-to-cell diffusivity of water (D_c) obtained by cell pressure probe experiments that D_c usually dominates D_t because of the high mobility of water in tissues. However, some "colimitation" caused by the apoplasmic diffusion of solutes is possible.

The more important question is how apoplasmic solutes would contribute to the overall water relations in view of the low reflection coefficients of the apoplasmic path mentioned above. In order to quantify and to integrate these effects into an extended model of tissue water relations, a reflection coefficient of the wall material (σ_{cw}) has to be assumed. Furthermore, how changes of tissue water potential ($\Delta\psi$) cause changes in wall osmotic ($\Delta\pi_{cw}$) and nonosmotic components of ψ in the wall (pressure and matric components) needs to be known. This may be expressed by introducing a factor (F) for the apoplast which would be the fractional change in water potential in the apoplast (tissue) which would be osmotic in nature and not hydrostatic or matric. Thus, by using the factor F we would be able to express the special role of the osmotic component of water potential. F is defined by (Steudle, 1992):

$$F = \frac{d\pi_{cw}}{d\psi_{cw}} = \frac{d(P_{cw} - \tau_{cw})}{d\psi_{cw}} - 1 , \qquad (11)$$

since it holds that $\psi = P_{cw} - \pi_{cw} - \tau_{cw}$. Using Eq. (11) and also allowing for some cell-to-cell diffusion of solutes (permeability coefficient, P_s) and for a diffusion of solutes along the wall path (diffusion coefficient, D_s^{cw}), the tissue diffusivity, D_t, given in Eq. (9) can be rewritten as D_t' (Steudle, 1992):

$$D_t' = \frac{Lp/2 \cdot a_{cc} \cdot \Delta x^2 + Lp_{cw} \cdot a_{cw} \cdot \Delta x \{1 + F(1 - \sigma_{cw})\}}{C_c + C_{cw}}$$
$$+ \frac{a_{cw} \cdot \Delta x}{V_{cw}} D_s^{cw} + \frac{a_{cc} \cdot \Delta x}{V} \frac{P_s \cdot \Delta x}{2} . \qquad (12)$$

Besides the hydraulic effects, the diffusivity (D_t') given in Eq. (12) incorporates all possible effects of solutes except for active solute flow. It can be seen from Eq. (12) that osmotic gradients would contribute to the overall water flow in different ways depending on the absolute values of both F and σ_{cw}. F will range between -1 and 0 (Eq. (11)) and σ_{cw} between 1 and 0. For example, if $\sigma_{cw} = 0$ and the driving force in the apoplast is purely osmotic ($F = -1$), there will be no osmotic water flow in the apoplast since the second term in the numerator of Eq. (12) will vanish. On the other hand, if $F = 0$ (only hydrostatic and matric forces operating), water flow will be maximal regardless of the absolute value of σ_{cw}. As already mentioned, large differences between osmotic and hydrostatic water flow have been found for different types of plant tissue such as leaves, growing stem tissue, and roots (see Section V; Westgate and Steudle, 1985; Steudle and Boyer, 1985; Steudle and Frensch, 1989; Zhu and Steudle, 1991). These differences are readily explained by the present model. It demonstrates that the "osmotic hydraulic conductivity" of a tissue could be fairly small and, in fact, controlled by the cell Lp (cell-to-cell path; first term in the brackets on the right side of Eq. (6) or first term in the numerator of the first fraction on the right side of Eq. (12)). This could happen despite a high (potential) hydraulic conductivity of the cell wall (Lp$_{cw}$) if the water flow in the apoplast was purely osmotic in nature. On the other hand, the "hydrostatic hydraulic conductivity" could differ substantially from the osmotic and could be much larger depending on the absolute value of Lp$_{cw} \cdot a_{cw}$. This has been found repeatedly for quite different plant tissues (see below).

Equation (12) also incorporates the passive movement of solutes within the apoplast following a concentration gradient (diffusion coefficient of solutes in the apoplast, D_s^{cw}; second term on the right side of Eq. (12)).

Apoplasmic diffusion should be important in the root or during phloem loading and unloading (see below). Its contribution to D'_t will depend on the absolute value of D_s^{cw} which should be somewhat smaller than the diffusion coefficient of "s" in an aqueous bulk solution. On the contrary, the influence of cell-to-cell diffusion (permeability, P_s; last fraction on the right side of Eq. (12)) should be generally low compared with that of the other components of D'_t.

C. Active Solute Flow and Compartmentation

Interactions between water and solutes in tissues may be extended to active solute flow. However, in this case the kinetics are no longer of the diffusion type, since the active transport of solutes within the symplast or the uptake from the apoplast into the symplast usually follows a saturation type of kinetics (e.g., a Michaelis-Menten type of kinetics) and is not linearly related to concentration differences. Metabolic pumps operate actively against a gradient. Hence, the differential equations describing the flow at the level of individual cells would have to be integrated numerically which would require some knowledge of the pattern of active solute movement within a tissue. Two different transport models could be distinguished. In the first model, solutes are taken up at the tissue boundary into cells and are then transported symplastically across the tissue with the water following in both pathways (cell-to-cell and apoplasmic path) due to a reduced water potential. In the alternative model, solutes are taken up from the apoplast after traveling along this path for some time. The two different models play an important role during the processes of phloem loading and unloading. They should be also of considerable importance during the coupled uptake of water and solutes by roots and during growth. In case of a passive movement of solutes (Eq. (12)), it is evident that, depending on the absolute values of $P_s \cdot \Delta x/2$ and D_s^{cw}, quite different consequences for the water relations of a tissue may result if either an apoplasmic path or the symplasmic path would be preferred, since water would osmotically follow the solutes in either case in an exchange process between the parallel pathways. To date, these consequences have not been realized in phloem transport studies when focusing only on the movement of solutes (sugars) and neglecting the coupled water flow which may feed back to the solutes (sugars). For example, if solutes are taken up into the symplast (protoplasts), this would cause a per cell change of the solutes of Δn_s (in moles) which, in turn, would cause a change in water potential of (Steudle, 1992):

$$\Delta\psi = -\frac{C_c}{C_c + C_{cw}} \mathrm{RT} \frac{\Delta n_s}{V}. \tag{13}$$

Here, the factor of $C_c/(C_c + C_{cw})$ denotes that water will be shared in the tissue between the protoplast and apoplast according to their storage capacities. If solutes are taken up from the apoplast into the adjacent symplast, the water potential in the symplast becomes smaller whereas the water potential of the apoplast tends to increase. It can be shown that, under these conditions, the resulting change of tissue water potential would depend on the amount of solutes (Δn_s) transferred from the apoplast into the symplast:

$$\Delta\psi = \frac{1}{C_c + C_{cw}} RT \left[\frac{C_{cw}}{V_{cw}} - \frac{C_c}{V} \right] \Delta n_s . \tag{14}$$

It is interesting that, under these conditions, the direction of the change in water potential of the tissue could be positive, negative, or zero depending on the absolute values of the storage capacities and sizes of apoplast and symplast (term in brackets on the right side of Eq. (14)).

In an alternative approach, the effects of active solute movements on tissue water relations may be modeled assuming certain amounts of solutes generated either in the apoplast or in the symplast (per cell) and by again assuming local equilibrium for water. Changes of cell turgor and volume as well as of the tissue water potential would be of interest as a function of the amounts of solutes sequestered into the apoplast (Δn_s^{cw} in moles) or into the cell (protoplast, Δn_s). It has been shown that these changes (given as (small) differences compared with a reference state (subscript 'o')) will be (Steudle, 1989a, 1992):

$$P - P_o = \frac{\varepsilon}{\varepsilon + \pi^i} \frac{1}{C_c + C_{cw}} \left[\Delta V_{tissue} + RT \cdot C_{cw} \left(\frac{\Delta n_s}{V} - \frac{\Delta n_s^{cw}}{V_{cw}} \right) \right], \tag{15}$$

$$V - V_o = \frac{C_c}{C_c + C_{cw}} \left[\Delta V_{tissue} + RT \cdot C_{cw} \left(\frac{\Delta n_s}{V} - \frac{\Delta n_s^{cw}}{V_{cw}} \right) \right], \tag{16}$$

and

$$\psi - \psi_o = \frac{1}{C_c + C_{cw}} \left[\Delta V_{tissue} - RT \cdot \left(\frac{\Delta n_s}{V} C_c + \frac{\Delta n_s^{cw}}{V_{cw}} C_{cw} \right) \right]. \tag{17}$$

The equations also incorporate effects due to a water uptake of the tissue improving its water status (ΔV_{tissue}). ΔV_{tissue} would be the per cell uptake of water into a tissue which is the difference between the amount of water gained from the root and that lost by transpiration. In Eq. (15), the factor $\varepsilon/(\varepsilon + \pi^i)$ refers to dilution effects due to an uptake of water into the cells (protoplasts). It can be seen from Eqs. (15) and (16) that solutes accumulated or generated in the cells improve turgor and volume (water content), whereas, in the apoplast, they have the opposite effect. This would be expected, but in quantitative terms the effect in the cell (protoplast) would be much less than that in the wall space since the protoplast volume (V) is much larger than that of the (per cell) water volume of the apoplast (V_{cw}). The latter would be subject also to much bigger changes and high concentrations of solutes could be more easily maintained or changed quickly in the small compartment. Furthermore, it is important to recognize that changes in the osmotic pressure of cells $(RT \cdot \Delta n_s/V)$ would not be equivalent to changes in turgor $(P-P_o,$ Eq. (15)). Assuming that $\varepsilon/(\varepsilon + \pi^i) \approx 1$, then the factor $C_{cw}/(C_c + C_{cw})$ would be important. It could be as small as 1/3 to 1/10 (Steudle, 1989a, 1992). The reason osmotic changes are not directly transferred into equivalent changes in turgor (which should be important for any turgor-dependent process in plants) is that the water available in the tissue is shared between the two compartments according to their storage capacities. Accordingly, it can also be shown that if ΔV_{tissue} is sufficiently large (saturating supply with water), there will be a 1:1 change in turgor following a change in the osmotic pressure of the cell. This is verified from Eqs. (15) and (17) for water saturation, i.e., for $\psi = \psi_o$.

Equations (15) to (17) are important, since they define in quantitative terms the conditions under which a regulation or homeostasis of turgor, cell volume, or water content could be expected in plant tissues. For example, if solutes are taken up from the apoplast into the symplast $(\Delta n_s = - \Delta n_s^{cw} > 0)$ at $\Delta V_{tissue} = 0$ (water withheld), turgor and cell volume will increase. Conversely, they will decrease if solutes are sequestered from the protoplast into the apoplast. However, a supply of water to the tissue may compensate for this effect. At $\Delta V_{tissue} = 0$, cell volume and turgor would remain constant, if $\Delta n_s/\Delta n_s^{cw} = V/V_{cw}$. Thus, although water is quite mobile in plant tissues, the compartmentation of solutes has some influence on tissue water relations. Differences in the storage capacities of the apoplast and protoplast play a key role. As demonstrated, this holds at the cellular level. However, it should also hold for larger compartments such as during the distribution of water between different tissues or organs as long as the condition of water flow equilibrium between compartments is maintained.

The model calculations quantify the effects of stored solutes on the storage of plant water as it would occur during osmoregulation, phloem loading and unloading, growth, and other processes. During osmoregulation, changes of turgor or volume may feed back on Δn_s or Δn_s^{cw} or may result in an exchange between compartments. The model would also be useful for describing osmotic processes occurring during CAM (Steudle, 1989a). During the dark phase of CAM, changes in turgidity would depend on both the accumulation of malic acid in the vacuoles and on the supply of water. During phloem unloading rapid changes of Δn_s^{cw} may occur which should cause big changes in water potential and turgor which, in turn, could play an important role in the regulation of the process, if there are turgor-dependent processes involved. It is important to remember that the processes are dynamic rather than static and that movements of solutes and water are strictly coupled. In the following section, the model is applied to experiments with growing hypocotyls of castor bean (*Ricinus communis* L.). These experiments offer some evidence for the model's validity.

IV. Application of Integrated Theory: Extension Growth

A. The Lockhart Model of Extension Growth

During extension growth, the uptake of water into growing cells plays an important role. Nearly all of the volume changes of growing cells which can be 10%/h or even more are due to water uptake. However, this does not necessarily mean that the rate of water uptake would limit growth and that growth could be regulated by triggering certain water relations parameters (e.g., the hydraulic conductivity of cell membranes) or by endogenous (e.g., auxins) and exogenous factors (e.g., water potential). At the level of isolated cells exhibiting a very low hydraulic resistance for water uptake this is highly unlikely, but in tissues or organs of higher plants where quite a number of cell layers would have to be crossed by water to reach growing regions the situation could be different. The idea of a limitation of growth by water implies that growth-induced water potential gradients are created during growth as suggested by Boyer and co-workers (e.g., Molz and Boyer, 1978; Boyer, 1985, 1988; Nonami and Boyer, 1990; Matyssek *et al.*, 1991).

On the other hand, mechanical properties of the primary walls of growing cells may act as "mechanical resistances" that limit growth. The plastic or viscous properties of the walls of young cells could terminate the irreversible extension of the wall material. The idea of a limitation of growth by mechanical resistances has been accentuated by Cosgrove and co-workers (Cosgrove, 1986, 1987, 1993). For individual cells and tissues, it has been found that the rate of cell extension (the growth rate, r, in % of volume growth per unit time) is linearly related to the stress exerted by cell turgor, i.e.:

$$r = \frac{1}{V}\frac{dV}{dt} = m\{P(t) - Y\}. \tag{18}$$

The factor m describes the extensibility of the cell wall in units of a fluidity (= inverse of a viscosity) and Y a certain yield threshold of turgor or tension in the wall which is necessary to overcome chemical and other bonds between microfibrils in the wall. It has been shown that this is valid for isolated internodes of *Nitella,* where the growth rate was directly measured as a function of turgor, as well as for higher plant tissue (Green *et al.,* 1971). However, it has been shown also that both growth parameters *m* and *Y* can vary and can be under some metabolic control. Under steady-state conditions (constant growth rate), the combination of Eq. (18) with that for the water uptake (Eq. (1)) yields:

$$\left(\frac{1}{\text{Lp} \cdot A} + \frac{1}{V \cdot \text{m}}\right) \frac{dV}{dt} = (\Delta\pi - Y). \tag{19}$$

This equation is written in a form analogous to Ohm's law showing that the overall "growth resistance" is the sum of a hydraulic ($1/\text{Lp} \cdot A$) and a mechanical resistance ($1/V \cdot m$). It is also clear that the overall driving force in the system is the osmotic gradient between cell and surroundings ($\Delta\pi$). If the growing tissue were close to water saturation and the hydraulic resistance sufficiently low, this gradient would be close to cell turgor. The combined hydraulic and mechanical model anticipated by Lockhart (1965) can be drawn in a simple equivalent electric circuit with two resistors in series similar to that given in Fig. 1 (Steudle, 1985). A metabolic control of growth would be possible by an extrusion of protons into the wall space triggered by auxin which, in turn, would directly loosen the wall material, e.g., by loosening covalent and noncovalent bonds between microfibrils, or by activating a pH-dependent "wall-loosening factor" (acid growth hypothesis; Cleland, 1981). This hypothesis (in which metabolism acts on the wall extensibility, *m*) has been questioned on the grounds that (1) auxin (IAA) may have other functions during growth regulation besides triggering a K^+/H^+ exchange at the plasmalemma (Kutschera and Schopfer, 1985) and that (2) growth limitation could be limited only by the epidermis of stems rather than by the entire tissue of young plants (Kutschera, 1989). Nevertheless, the acid growth theory is, to date, the most accepted basis of the understanding of extension growth. Recently, Fry (1989) elaborated the acid growth hypothesis. Since cellulases have low pH optima, he suggested that the principal effect of auxin is an increase of the activity of these enzymes and that wall loosening is brought about by the cellulase-catalyzed cleavage of load-bearing hemicellulose molecules which interconnect microfibrils in the wall. An adjustment of the extensibility (*m*) rather than of the driving force or the yield threshold (*Y*) also appears to take place during root growth where the growth rate can be maintained despite lower turgor because of an increase of the extensibility (Pritchard *et al.,* 1991; Spollen and Sharp, 1991).

It is usually assumed that, during extension growth, $\Delta\pi$ remains constant over considerable periods of time and that the force driving the process does not limit growth during these time intervals. Effects of solute transport (e.g., the uptake of assimilates supplied from the phloem via the apoplast to the cells) usually have been excluded or explained away (Cosgrove, 1987). However, it is clear that, at least in the long-term, assimilates offered to a growing hypocotyl have a favorable effect on growth (Stevenson and Cleland, 1981; Schmalstig and Cosgrove, 1990; Meshcheryakov *et al.,* 1992).

Current growth models based on either tissue hydraulics or mechanical properties are controversial (see above). However, most of the discussion seems to miss the point by omitting or neglecting effects of solutes interacting with hydraulic and mechanical properties. This involves the possibility that there is a fairly direct metabolic regulation of growth besides that via wall extensibility (see above). The fact that, in the absence of an external solute supply, growth continues for some time without a change in growth rate does not mean that solute flow could not have an influence on growth if internal solute movements, water, and mechanical properties of cell walls were tightly coupled to each other. This idea extends existing concepts. The concept of a water/solute coupling put forward in the previous section for mature cells has been applied to growing cells. It provides a model of the regulation of extension growth which integrates metabolism into the Lockhart concept via active solute flow and compartmentation. It provides a more realistic model of growth. In the following, the concept is discussed in relation to recent measurements on young seedlings of castor bean (*R. communis* L.).

B. Extension of the Growth Theory: Osmotic Compartmentation Model

According to Eq. (19) the ultimate driving force for growth is a difference of osmotic pressure maintained between the cell (protoplast) and the immediate surrounding tissue (apoplast). In order to evaluate this force, we have to calculate changes of $\Delta\pi$ caused by active or other movements of solutes in the tissue. Furthermore, we have to incorporate dilution effects into the approach. As for mature tissue (see Section III. C) changes in turgor, cell volume, and water potential are important. The procedure to derive these relations is analogous to the approach used for mature tissue (see Eqs. (15) to (17)), since the condition of local equilibrium of water potential also holds for immature tissue. Effects of the osmotic compartmentation of solutes on turgor, cell volume, and water potential may be described as follows (Meshcheryakov *et al.,* 1992):

$$P - Y = \frac{1/\mathrm{Ex}}{1/\mathrm{Ex} + \pi_c^i\, C_c^g + C_{\mathrm{cw}}} \left[\Delta V_{\mathrm{tissue}} + \mathrm{RT} \cdot C_{\mathrm{cw}} \left(\frac{\Delta n_s}{V} - \frac{\Delta n_s^{\mathrm{cw}}}{V_{\mathrm{cw}}} \right) \right], \quad (20)$$

$$V - V_c = \frac{C_c^g}{C_c^g + C_{cw}} \left[\Delta V_{tissue} + RT \cdot C_{cw} \left(\frac{\Delta n_s}{V} - \frac{\Delta n_s^{cw}}{V_{cw}} \right) \right], \quad (21)$$

and

$$\psi - \psi_c = \frac{1}{C_c^g + C_{cw}} \left[\Delta V_{tissue} - RT \cdot \left(\frac{\Delta n_s}{V} C_c^g - \frac{\Delta n_s^{cw}}{V_{cw}} C_{cw} \right) \right]. \quad (22)$$

As in Eqs. (15) to (17), small changes of P, V, and ψ are considered using the yield threshold (Y) and the corresponding values of the volume and water potential as the reference at which the growth rate is zero (Eq. (18)). Except for using the total extensibility (Ex) instead of the elastic modulus and for assuming that the storage capacity of the protoplast (C_c^g) would increase during growth, the equations are identical. The total (viscous plus elastic) extensibility of the cell (protoplast) is given by:

$$Ex = m \cdot \Delta t + 1/\varepsilon, \quad (23)$$

and the capacities of the protoplast (C_c^g) and apoplast (C_{cw}) by

$$C_c^g = \frac{V}{1/Ex + \pi_c^i} \quad \text{and} \quad C_{cw} = \frac{dV_{cw}}{d\psi}. \quad (24)$$

It is evident that during growth, capacities and extensibilities would change with time. Therefore, Eqs. (20) to (22) would only be valid for short (differential) time intervals. In order to arrive at the time dependence of pressure, volume, or water potential in growing tissue, the equations must be integrated which would require some knowledge of the parameters (Ex, capacities etc.) and their changes with time. As in the equations given for mature tissue, ΔV_{tissue} represents the net uptake of water (per cell) into the growing tissue (supply by the root minus transpiration). ΔV_{tissue} may be controlled by water, mechanical parameters, or both.

It is clear from Eqs. (20) to (22) that besides water and mechanical properties changes of solute contents of the apoplasmic and protoplasmic compartments could become rate limiting. For example, at a given water supply to the growing tissue, changes of the solute content of the apoplast (Δn_s^{cw}) and of the symplast (Δn_s) would affect turgidity and growth in a different way. An increase of solutes in the protoplast would improve turgor and growth. A decrease would reduce it. The effects would be quite different for the same amount of solutes sequestered into the different compartments, since $V_{cw} \ll V$. Furthermore, effects due to an increase of the osmotic pressure of the symplast would only in part be reflected by an increase in cell turgor and subsequent growth (factor of $C_{cw}/(C_c{}^g + C_{cw})$ in Eq. (20)) analogous to mature cells. As for mature cells, the supply with solutes would reduce the water potential regardless of whether solutes are supplied to the symplast or the apoplast. It can be seen from Eq. (22) that the water potential of a growing tissue is maintained if an equivalent water supply is given to the tissue. Thus, the ratio or balance between water and solute supply is important for the water status and would be crucial for the regulation of growth. The processes of mechanical extension, water flow, and solute supply are tightly linked to each other. For example, there are experiments in which water was withheld from growing tissue and a relaxation of turgor and growth rate was measured in order to evaluate both the extensibility and the yield threshold. It was concluded that the time constant of the process was controlled by mechanical properties of the cell wall (m and ε; Cosgrove, 1985). However, it could also have been controlled by solute movements which tended to reduce ψ and turgor so that growth ceased. This could have resulted from a supply of solutes to the apoplast at a reduced supply of water to the same compartment which could easily cause the effects. Similarly, if negative water potentials are created during growth even under conditions of high water supply (Boyer, 1985), it cannot be easily concluded that water transport is the limiting factor. Also in this case, it would be necessary to prove that effects of solutes or of interactions between solutes and water really can be excluded. This has never been done because solute compartmentation does not appear in the simplified growth equations traditionally used to describe growth.

C. Measurements on *Ricinus*

Evidence for a regulation of extension growth by an interaction between water and solutes has been demonstrated recently with *Ricinus* seedlings grown in hydroponics (Meshcheryakov *et al.*, 1992). When plants were grown with roots and cotyledons in 0.5 mM CaSO$_4$, the hypocotyl grew quite well when fixed horizontally in a chamber for the measurement of turgor in the hypocotyl (Fig. 5). Those parts of the plant which were not in 0.5 mM CaSO$_4$ solution were covered with wet Kleenex to keep the water potential at $\psi = 0$. Under these conditions, the cells in the growing region of the hypocotyl had a high turgor (5 to 10 bar). However, steep gradients occurred in the cortex, from the inside (i.e., from an area around the vascular bundles) to the periphery, which were as large as 0.5 to 0.7 MPa (5 to 7 bar) over a distance of only 470 μm (Fig. 6A). Measurements of the osmotic pressure in individual cells across the cortex (nanoliter osmometer) showed that $P = \pi$. Hence, there was no gradient of water potential under these conditions across the tissue. However, when the water supply was reduced by cutting off either the root or the

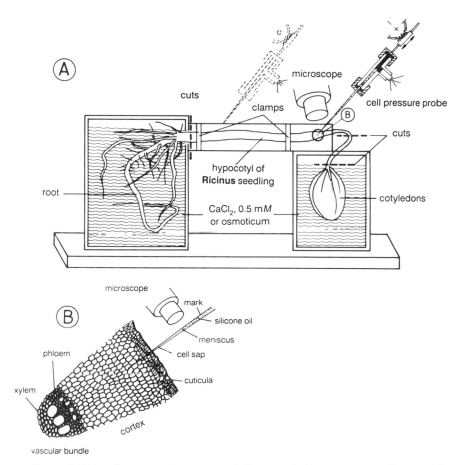

Figure 5 Setup for measuring turgor and of water relations parameters of cortex cells in the growing hypocotyl of young seedlings of *Ricinus communis* L. using the cell pressure probe (A). Seedlings were fixed horizontally under a stereo microscope. The tip of the microcapillary of the probe was introduced in defined layers of the cortex (B) to measure turgor and other water relation's parameters (Lp, ε, $T^{w}_{1/2}$, etc.) either in growing or in mature parts of the hypocotyl. To avoid water losses from the tissue, the hypocotyl was covered with wet Kleenex. As indicated, parts of the seedling could be cut off (roots, cotyledons, etc.), but the cut surfaces were kept moist also with wet Kleenex. Osmotic solutes could be applied to the medium bathing the root or the cotyledons (0.5 mM $CaCl_2$) to change the water supply. Changes of turgor gradients across the cortex were recorded following a change of water supply or a supply of assimilates from the phloem, when cotyledons or the hook were removed (after Meshcheryakov *et al.*, 1992).

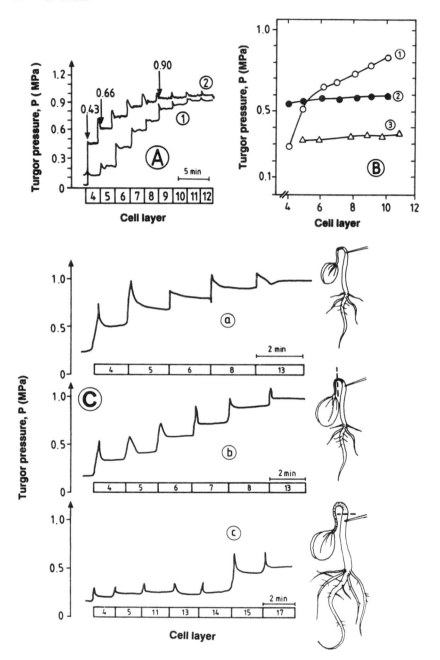

Figure 6 (A) Typical traces of cell turgor vs time in different cell layers (layes 4 to 12) of the hypocotyl of *Ricinus communis* L. as measured with the setup shown in Fig. 5. Trace 1 refers to the mature and trace 2 to the growing region of hypocotyls. Cell layers are

cotyledons in a wet atmosphere, the gradients of turgor pressure quickly disappeared (Fig. 6B) and a negative water potential built up in the tissue. Gradients also disappeared when all the tissue in the hook which contained starch was removed, thus, removing the supply of solutes (sugars) via the phloem (Fig. 6C).

The experiments suggested that the gradients in pressure or solute content were due to a supply of assimilates from the phloem which were used in the cells for extension growth and other processes (storage and syntheses) along the cortex. In the presence of sufficient water supply, growth was maintained. However, at a reduced supply the balance between water and solutes was no longer maintained as shown theoretically in the previous section. This reduced gradients of turgor as well as absolute values of water potential and turgor and stopped growth according to Eqs. (18) and (20). The compartmentation of solutes appeared to be very important. If the roots were first cut off to stop growth and to level off gradients, and the xylem of the cut stump was then connected to a water pipe and pressurized with water, gradients and growth could be reestablished (Fig. 7A). Transient changes in turgor as large as 1.2 MPa were measured in cortical cells under these conditions which were due to a rapid dilution or sweeping away effect in the apoplast where assimilates piled up during reduced water supply (Fig. 7B). Thus, it appeared that, to a considerable extent, rapid changes of the apoplasmic solute concentration took place which could be estimated in this type of experiment. Obviously, both the relation between water and solute supply to the growing tissue as well as the effect of compartmentation were important as predicted from the osmotic compartmentation theory.

The work on *Ricinus* shows that apoplasmic concentrations are a dynamic rather than a static parameter as has been suggested previously (Cosgrove and Cleland, 1983). They depend on both the water and the solute status, but should also be influenced by the fact that water potentials

---

numbered toward deeper layers. Steady-state values of turgor measured 0.5 min after puncturing a cell were equal to the osmotic pressures of cell sap determined with the nanoliter osmometer (data added by arrows to curve 2). Thus, when the growing tissue was well supplied with water, there were no gradients of water potential across the tissue. (B) Changes of profiles of turgor pressure after cutting off the root. Trace 1 refers to standard conditions (root and cotyledons bathed in 0.5 m*M* CaCl$_2$), and trace 2 to the situation 0.5 h after the roots were cut off. When the supply with water was reduced by cutting, negative water potentials developed in the cortex (data not shown). (C) Recorder traces of profiles of turgor as in A after the cotyledons right at the hook (b; time: 1 h) or at the beginning of the growing zone (c) were cut off. When the hook was completely removed, sources of assimilates for supplying the growth of the hypocotyl were also removed. It can be seen that this reduced both the absolute value of turgor and gradients in the cortex (after Meshcheryakov *et al.*, 1992).

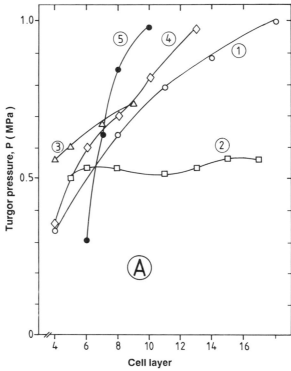

Figure 7 (A) Changes of gradients of turgor across the hypocotyl of growing seedlings of castor bean caused by an application of hydrostatic pressure to the xylem of seedlings after the root was cut off. Trace 1 represents the control and trace 2 the situation 1 h after the root was cut off. Trace 3, application of a hydrostatic pressure of 0.25 MPa for 10 min. Trace 4, application of 0.4 MPa for 10 min. Trace 5, 0.5 MPa for 10 min. It can be seen that steep gradients of turgor can be reestablished by pressurizing the xylem. (B) Same experiment as in A, but changes of turgor were followed during the application of pressure in a certain cell (layer No. 7) using the cell pressure probe. Roots were removed 2.5 h before pressurizing which caused turgor to drop to about 0.2 MPa. When the xylem was pressurized by 0.25 MPa, turgor rapidly increased to peak values of 1.2 and 1.35 MPa. When it was removed again, turgor declined slowly to reach a steady value after about 7 min. Cutting the seedling at the hook or removing part of the cotyledon caused a rapid decline of turgor to a value close to the original. It is remarkable that the absolute values of the rapid changes of turgor upon pressurizing the xylem were much larger than those of the pressure applied. This indicated that the supply of water via the xylem considerably diluted the apoplast concentration where solutes should have accumulated after excising the root (see Eq. (20)) (after Meshcheryakov *et al.*, 1992).

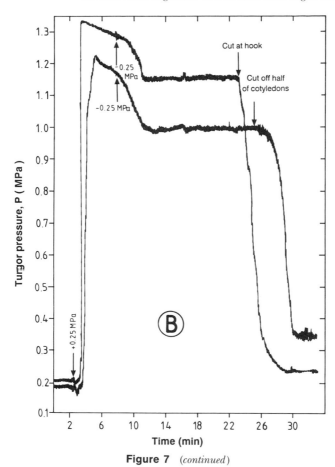

Figure 7 (*continued*)

rapidly equilibrate across the cell membrane (local equilibrium). These interactions (besides technical difficulties) explain the fact that, by centrifugation, infiltration, and other techniques, quite different results have been obtained in the past for the apoplasmic solute concentration (for literature, see Meshcheryakov *et al.*, 1992).

D. Advantages of the Osmotic Compartmentation Model of Growth

The new extended growth model given here incorporates solute effects (transport, production, compartmentation) into the classical Lockhart concept, which only considers hydraulic and mechanical effects. It may be termed an "osmotic compartmentation model." By incorporating solutes, turgor and growth rate are effectively linked to metabolism in addition to the water and the mechanics of extensible cell walls. The new model could be useful for solving some of the discrepancies and contradictions which have been discussed in growth physics in the past few years when the focus was only water or mechanical properties. In the osmotic compartmentation model, links between solute levels in the apoplasmic and protoplasmic compartments and metabolism could be numerous and could provide an efficient and low-cost mode of regulating growth. For example, active pumping (J_s^* in Eq. (5)) may be regulated to adjust turgor and growth rate (Δn_s in Eq. (20)). Alternatively and even more efficiently, there could be a regulation of the passive permeability of cell membranes (P_s in Eq. (5)) which would mainly affect the solute level of the apoplast (Δn_s^{cw} in Eq. (20)). In both cases, effects of turgor, concentration, or some other variables related to them on the activity of ATPases (pumps) or on the opening or closing of ion channels (leaks) would be required for regulation. In addition, the closing or opening of plasmodesmatal connections caused by pressure gradients in tissues could play a role (Oparka *et al.*, 1992). At present, it is not clear which of the different mechanisms of a coupling between metabolism and growth rate are important. In principle, they could be analyzed using pressure probe techniques by measuring turgor-dependent transport of solutes and solute compartmentation in addition to the water relations as demonstrated for *Ricinus* seedlings in this section.

V. Regulation of Root Water and Solute Relations

Interactions between water and solutes also play a role in roots during the uptake of nutrients and water. This type of interaction is well known from guttation and other root pressure phenomena. However, the couplings between water and solute flows during absorption processes in the root should be somewhat different from those occurring during growth, since solutes usually do not accumulate in root cells but pass radially across the root cylinder to the xylem where they are conveyed to the shoot in the transpiration stream. At low rates of transpiration the accumulation of solutes in the xylem contributes substantially to the overall driving force for water uptake. The crucial point for consideration in studies of root water relations is that the barrier for the radial uptake of water is rather complicated. Different parallel pathways (apoplast, symplast, transcellular path; see Figs. 3 and 8) may contribute differently to the overall flow. Furthermore, water and solutes are transported across different tissues arranged in series in the root (epidermis, cortex, stele). Therefore, transport models such as those developed in previous sections have to be applied. In a developing structure such as the root, the contribution of different root zones to the overall hydraulic properties will

rhizodermis   cortex   endodermis   stele

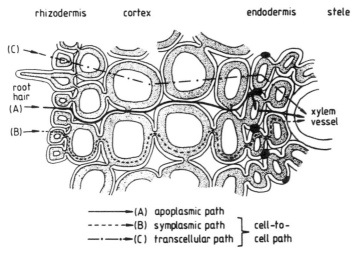

————→(A) apoplasmic path
------→(B) symplasmic path ⎤ cell-to-
—·—·—·—(C) transcellular path ⎦ cell path

Figure 8 Radial pathways for water solutes in roots. The apoplasmic path (A) is within the cell walls around the protoplasts of the rhizodermis, cortex, and stele. In the endodermis, the Casparian band interrrupts the apoplasmic flow causing a membrane-bound transport step. The symplasmic transport (B) is via plasmodesmata, and the transcellular path (C) across protoplasts, whereby two membranes have to be crossed per cell layer. The symplasmic and the transcellular components are summarized as the cell-to-cell path (see also Fig 3).

vary in time and space. This interesting aspect is also important for the overall efficiency of roots in taking up water and nutrients. In addition, the radial components of flow are in series with the longitudinal or axial components which also contribute to the overall hydraulic resistance of roots.

All these factors determine the ability of roots to supply the shoot with water under conditions of changing demands. It is thought that, given a certain root structure and some transpirational demand, water flow will be along a gradient of water potential. It is shown below that this simple model has to be extended. An extension is necessary for two main reasons: (i) interactions between water and solute transport in the two different pathways (cell-to-cell and apoplasmic path) and (ii) the special role of osmotic pressure as a driving force in the plant apoplast which already has been discussed in Section III of this chapter. Over longer periods, a regulation or limitation of water uptake may, of course, also occur through variation in rates of root growth or by the suberization of older parts of roots. These processes result either in an increase of the absorbing surface area or in a decrease of the hydraulic conductivity of the root (Lp_r). In order to quantify the capacity or efficiency of roots to take up water and solutes, transport models are required which are based on detailed measurements of the transport of water and nutrient salts. The root pressure probe technique (Fig. 9) has been employed in the past few years to measure these processes in order to model root transport. The technique which is based on the measurement of root pressure of excised roots has been applied to root segments as well as to root systems. Radial transport across the root cylinder as well as longitudinal transport have been studied. In this technique, it is important that roots or root systems are tightly connected to the probe. On one hand, the sealing should prevent any leakages across the sealing area. On the other hand, it has to be ensured that conducting xylem is not interrupted or blocked. There are physical criteria as well as staining experiments which are used to test these requirements (Figs. 10A and 11).

In order to work out models, the technique has been combined with the conventional cell pressure probe from which it originated. In the following section, some of the results obtained with the root pressure probe are summarized. The discussion focuses on transport models which are the prerequisite for working out the capacity of roots to take up water and solutes.

A. Transport Models of Roots

The traditional viewpoint of water relations of roots or more precisely of the water transport across roots is that the endodermis represents the main hydraulic and osmotic barrier. It is usually thought that, due to the existence of the Casparian band, this structure provides a nearly semipermeable barrier. The root is regarded as a fairly perfect osmometer (Weatherley, 1982) or as a structure exhibiting a single-membrane-equivalent barrier. Therefore, equations analogous to Eqs. (1) to (5) have been applied to roots to describe radial water and solute flow (Fiscus, 1975; Dalton *et al.*, 1975; Steudle and Jeschke, 1983; Miller, 1985; Dainty, 1985; Steudle *et al.*, 1987; Steudle, 1989a,b,c, 1990, 1992, 1993). There is evidence that this approach of describing uptake processes in terms of a two-compartment system (xylem and soil solution separated by a membrane-like structure) may be justified as a first approximation. Effects of unstirred layers appear to be smaller than one would expect at first glance (Steudle and Frensch, 1989). Usually, in root models the reflection coefficient of the osmotic barrier has been assumed to be close to unity for nutrient salts (as for a perfect osmometer), and their permeability is assumed to be very low to provide a small leakage which prevents a loss of nutrients by diffusional backflow into the soil when nutrients have been accumulated in the xylem.

Deviations from the simple two-compartment model may occur for structural reasons, namely, due to the existence of more than two com-

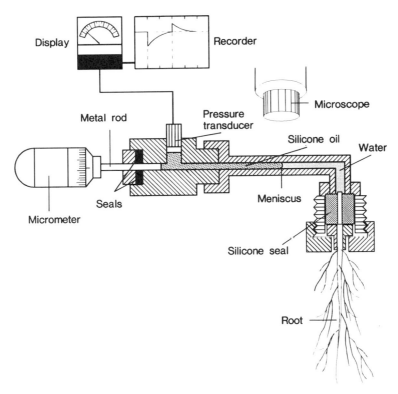

Figure 9 Root pressure probe for measuring water and solute relations of roots (schematical). The root segment or root system is tightly connected to the root pressure probe by silicone seals so that the root pressure develops in the measuring system which consists of the root, the seal, a measuring capillary, and a small pressure chamber. Half of the system is filled with silicone oil and the other one by 0.5 mM CaCl$_2$ solution so that a meniscus forms in the measuring capillary. This meniscus is used as a point of reference during the measurements. When a steady root pressure is established, water flows can be either induced by changing the pressure in the system using a metal rod (hydrostatic experiments) or by changing the osmotic pressure of the medium (osmotic experiments). The resulting changes of pressure are analyzed to work out water flows once the elasticity of the system is also measured. Furthermore, since the forces which drive the water are known, the hydraulic conductivity of the root (Lp$_r$) can be determined if the root surface area is known. If segments of roots are used which are open at the end, the hydraulic conductivity of the xylem can be evaluated (L_x). By measuring Lp$_r$ and L_x along a developing root, a complete picture of the hydraulic properties of roots is obtained. From the osmotic experiments (see text), reflection and permeability coefficients of solutes can be measured as well, analogous to the experiments with isolated cells (Fig. 2). Typical experiments with roots are shown in Figs. 11 and 12.

partments which may create deviations from linear force/flow relations (three-compartment model; Newman, 1976; Dainty, 1985). In other models, standing gradients of solutes in the apoplast have been proposed which would be created by a successive uptake of solutes from the cortical apoplast into the symplast or by an extrusion of these solutes into the stelar wall space (apoplast canal model; Taura *et al.*, 1988; Katou and Taura, 1989).

Structurally, there is also a basis for deviations from the simple model because of the existence of different parallel pathways for water and solutes such as the apoplasmic and cell-to-cell path or root zones of different permeability patterns. The existence of an exodermis with an additional Casparian band (or bands) would provide a serial arrangement of resistances (Peterson, 1988; Fig. 10B). This would enhance the composite character of roots already mentioned by the presence of different tissues arranged in series. Recent measurements of the hydraulic conductivity of the lateral walls of xylem vessels of young corn roots showed that their hydraulic resistance, although not rate limiting for radial water flow, would nevertheless contribute to the overall hydraulic resistance (Peterson and Steudle, 1993; see Section V.H). Despite these and other findings, it has been claimed that it would be appropriate to use the simple single-membrane-equivalent model, mostly because of its simplicity (Dainty, 1985). Deviations would be interpreted then in terms of some variability of transport coefficients. However, there is now considerable experimental and theoretical evidence which indicates that the model has to be extended. The evidence is based on

- the finding of low root reflection coefficients (σ_{sr}) and apoplasmic by-passes of solutes,
- apparent differences between osmotic and hydrostatic root Lp_r,
- the effect of water flow on the absolute value of root Lp_r,
- effects of longitudinal hydraulic resistance on the overall resistance,
- changes in transport coefficients (Lp_r, σ_{sr}, P_{sr}) along the developing root, and
- effects of external factors on root water relations.

These findings are directly related to basic functions of roots. They are important for the limitation and control of water and solute flows across roots. Changes in root Lp_r and σ_{sr} will directly affect the efficiency of roots for taking up water. The efficiency will be also affected by variation in the tissue properties along the developing root. In the following, some of the parameters that limit and control root transport will be discussed briefly in terms of a new root model which has been termed the "composite transport model of the root." The model integrates the different experimental findings listed above as well as well-known ana-

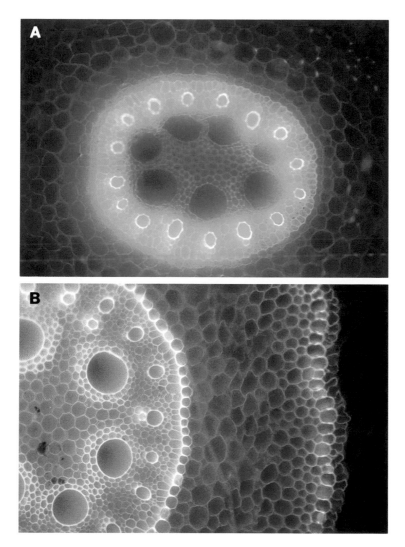

Figure 10 (A) Cross sections of young maize roots which had been sealed to the root pressure probe and through which Cellufluor, a fluorescent dye which binds to cellulose, has been forced. Regularly, all vessels of early metaxylem were stained which indicated good hydraulic contact between root pressure probe and root xylem (after Peterson and Steudle, 1993). (B) Casparian bands in the endodermis and exodermis of a maize root section stained with the fluorescent dye, berberine sulfate, according to the procedure of Brundrett *et al.* (1989) (courtesy of C.A. Peterson, University of Waterloo, Ontario, Canada).

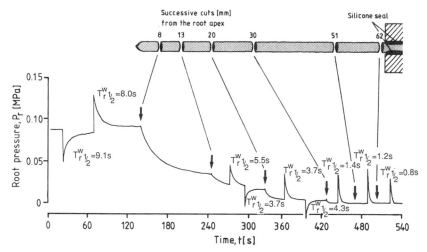

Figure 11 Cutting experiment on an excised maize root sealed to the root pressure probe. Prior to the experiment, root pressure relaxations were performed which revealed a half-time of 8 to 9 s. Successively excising root segments from the apex reduced root pressure as soon as mature vessels were hit. Due to the leaks produced by cutting, half-times successively decreased until they differed from the intact system by an order of magnitude when the cut was made right at the seal. This proved that the effect of the hydraulic resistance of the seal was negligible in the measurement of water flows across the root. Note that upon the first cut, root pressure did not completely decline to zero because the leak was still rather small (xylem not yet completely mature at a distance of 8 mm from the apex) and was compensated for by the solute pumping of the root (Eq. (5)). It should be noted also that from the type of cutting experiment shown in the figure the longitudinal hydraulic conductivity of the root could be evaluated (after Steudle and Frensch, 1989).

tomical features of roots. The root is looked at as a composite structure of different membrane-like barriers arranged either in series (such as different tissues: rhizodermis, cortex, etc.) or in parallel (such as the apoplasmic and cell-to-cell path; Figs. 3 and 8).

B. Low Reflection Coefficients and Apoplasmic Bypass of Solutes

Root pressure probes have been used in the past few years to test conventional root models and to work out basic transport coefficients of roots (Lp_r, σ_{sr}, P_{sr}). The experiments have been performed using root segments as well as root systems grown in hydroponics and in soil. The experiments strongly suggest that the osmotic barrier of roots exhibits a $\sigma_{sr} < 1$ that may range between 0.5 and 0.8 for nutrient salts (e.g., KNO_3 and KCl). Under these conditions cell membranes would exhibit a σ_s of unity (Table I). The findings have been carefully checked for unstirred layers which would reduce σ_{sr}, but this did not explain completely the low σ_{sr} (Steudle

Table I   Reflection (σ_{sr}) and Permeability (P_{sr}) Coefficients of Roots Determined by the Root Pressure Probe and Other Techniques

Root	Solute	Reflection coefficient, σ_{sr}	Permeability coefficient, $P_{sr} \cdot 10^{10}(m \cdot s^{-1})$	Reference
Lycopersicon esculentum	Nutrients	0.76	—	a
Glycine max	Nutrients	0.90	—	b
Zea mays	Nutrients	0.85	—	c
	Ethanol	0.3	60–190	d–f
	Mannitol	0.4–0.7	—	d–f
	Sucrose	0.54	30	d–f
	PEG 1000	0.64–0.82	—	d–f
	NaCl	0.5–0.6	60–140	d–f
	KCl	0.53	—	d–f
	KNO_3	0.5–0.7	<10–80	d–f
	NH_4NO_3	0.5	30–320	d–f
Phaseolus coccineus	Methanol	0.16–0.34	27–62	g
	Ethanol	0.15–0.47	44–73	g
	Urea	0.41–0.51	11	g
	Mannitol	0.68	1.5	g
	KCl	0.43–0.54	7–9	g
	NaCl	0.59	2	g
	$NaNO_3$	0.59	4	g
Phaseolus vulgaris	Nutrients	0.98	22	h

References: (a) Mees and Weatherley, 1957; (b) Fiscus, 1977; (c) Miller, 1985; (d) Steudle *et al.*, 1987; (e) Steudle and Frensch, 1989; (f) Zhu and Steudle, 1991; (g) Steudle and Brinckmann, 1989; (h) Fiscus, 1986.

and Frensch, 1989). There are few data for root σ_{sr} from other experiments. Using an experimental setup in which root exudation was measured as a function of hydrostatic pressure applied to the root, Fiscus (1977) worked out a $\sigma_{sr} = 0.90$ for nutrient salts in roots of soybean, but for other roots his estimates of σ_{sr} were close to unity (Fiscus, 1986). Miller (1985) has used a root manometer applied to the stumps of excised maize roots to carefully measure the reflection coefficient of the solutes in the root xylem. He found a $\sigma_{sr} = 0.85$ which is close to the range found in root pressure probe experiments (Table I).

The root pressure probe allowed the measurement of the passive permeability of solutes applied to the medium from biphasic response curves of root pressure analogous to those measured for cells (Fig. 12; see also Fig. 2). It was found that, for the solutes usually present in the root xylem, P_{sr} was rather small despite the low σ_{sr} which, at a first glance, would suggest a rather permeable root. Some values of σ_{sr} and P_{sr} obtained with the root pressure probe are listed in Table I and are compared with those obtained by other techniques. From the finding of low values for both σ_{sr} *and* P_{sr} we may conclude that the relation between reflection and permeability coefficients (as it is known for simple, homogeneous barriers such as cell membranes; Nobel, 1991) is quite different in a complex structure such as the root than it is in a cell.

The findings suggest that conventional root models in which the root is looked at as a nearly perfect osmometer with the endodermis as the main osmotic barrier (Weatherley, 1982) have to be reconsidered. The composite root model mentioned above explains the differences and is in accordance with osmotic properties of roots known from the literature. In the composite root model, a series arrangement results from the fact that structures of different permeation properties such as the exodermis, cortex, endodermis, and stele have to be crossed in a root. In fact, each cell layer may be considered an individual barrier. However, more important than the series elements are, probably, the different parallel components or pathways of transport. Parallel elements exist in roots for two different reasons: (i) there is an apoplasmic and a cell-to-cell path as already discussed in earlier sections, and (ii) there are different root zones exhibiting parallel arrays with different transport coefficients and also different rates of active pumping of nutrient salts. This should lead to characteristic changes of the overall transport properties. Irreversible thermodynamics shows that the overall transport properties of composite membranes would not be simply additive even if different cross-sectional areas available for transport are taken into account (Kedem and Katchalsky, 1963a,b). In a parallel arrangement of two membrane-like components *a* and *b*, the overall σ_{sr} would be given by that of the individual arrays (pathways; e.g., apoplasmic and cell-to-cell path). The overall σ_{sr} would result as a weighted mean of the individual parameters (σ_s^a and σ_s^b) according to the hydraulic conductivity on each path:

$$\sigma_{sr} = \gamma^a \cdot \frac{Lp^a}{Lp_r} \sigma_s^a + \gamma^b \cdot \frac{Lp^b}{Lp_r} \sigma_s^b . \tag{25}$$

$Lp^{a,b}$ = hydraulic conductivity of pathways a and b. $\gamma^{a,b}$ = fractional contributions of cross-sectional areas of pathways to the overall root area. Lp_r = overall hydraulic conductivity. It is evident from Eq. (25) that, provided the apoplasmic passage across the root exhibits a fairly low σ_{sr} (e.g., $\sigma_s^a \approx 0$) and a fairly high Lp^a, this component could still contribute to the overall σ_{sr} although the cross-sectional area is small. The physical reason for the effect is that, in the presence of two parallel pathways exhibiting different values of σ_s and Lp, opposing water flows would be created in the root which would result in circulation flow of water. For example, if the accumulation of nutrients were to create a root pressure at zero transpiration, there would be an osmotic water flow along the cell-to-cell path tending to equilibrate at $P_r \approx \Delta\pi_s$. However, an increase

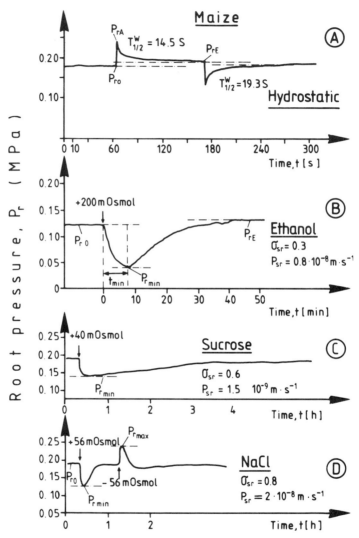

Figure 12 Typical root pressure relaxations of roots of maize (*Zea Mays* L.) and bean (*Phaseolus coccineus*) measured using the root pressure probe. Either hydrostatic (upper traces) or osmotic pressure gradients have been applied. In osmotic experiments, biphasic responses of root pressure were obtained from which reflection (ρ_{sr}) and permeability (P_{sr}) coefficients of the roots could also be determined in addition to half-times of water exchange ($T_{1/2}^w$) and hydraulic conductivities (see also Fig. 2).

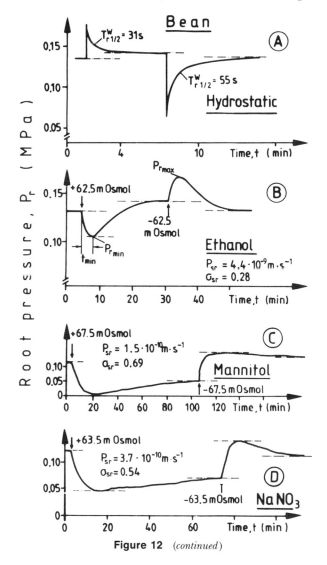

Figure 12 (*continued*)

of P_r would cause a back flow of solution along the nonselective apoplasmic path where $\sigma_s^a \approx 0$. Thus, a smaller stationary root pressure would be obtained at $J_{Vr} = 0$ than expected from the osmotic pressure difference. At $J_{Vr} = 0$, the opposing volume flows would exactly cancel. Since $P_r > \Delta\pi_s$ at $J_{Vr} = 0$, σ_{sr} would be smaller than unity. The function of the Casparian band in the root endodermis would be to decrease Lp^a and increase σ_s^a, thus tending to also increase the overall σ_{sr}. Permeability

(P_{sr}) would be lowered at the same time. In summary, the root exhibits basic functions of an osmometer, but there are considerable deviations such as apoplasmic bypasses and other properties which affect the absorption of water and nutrients.

Pressure probe data indicate apoplasmic bypasses of solutes and water in the root apoplast. This means that the Casparian band in the endodermis is not completely impermeable for solutes and water, at least during certain stages of development or at positions where primordia of secondary roots emerge (Peterson *et al.*, 1981). In order to test the hypothesis of apoplasmic bypasses of water and solutes, pressure probe experiments have been combined with X-ray microprobe measurements (EDX; Frensch *et al.*, 1992). Roots were treated with NaCl solutions (as the permeating solute) and "biphasic root pressure relaxations" were measured. During the relaxations, water tended to equilibrate across the root when it was subjected to an increased osmotic pressure ("water phase"). However, since osmotic solutes were permeating across the root cylinder, root pressures tended to reestablish due to an equilibration of the osmotic concentration. The situation was analogous to that of an isolated cell (Fig. 2). The same roots used in the osmotic experiments were analyzed with EDX, and results indicated a significant uptake of Na^+ into the xylem. When NaCl solutions were replaced by isoosmotic Kcl, Na^+ was released again (Fig. 13; Azaizeh and Steudle, 1991). Thus, the process of the uptake of sodium salts into the xylem was reversible. Radial profiles of ion concentrations (Na^+, K^+, Cl^-) showed that Na^+ appeared in both early and late metaxylem elements, whereas the latter were not mature in the young roots (Frensch *et al.*, 1992). Na^+ occurred in the vessels before it was detectable in the cell vacuoles of the root cortex. Thus, the EDX measurements confirmed the findings obtained by the root pressure probe. In experiments in which roots were connected to a root pressure probe and cell turgor was measured in individual cortex cells at the same time using the cell pressure probe, no biphasic responses of turgor were measured in the cells when solutes were used for which cell membranes were impermeable. However, the root pressure usually responded with two separate phases (Azaizeh *et al.*, 1992). This is expected since root cells did behave like nearly perfect osmometers. However, it was found in these experiments that the cell σ_s varied depending on the position in the cortex (Zhu and Steudle, 1991). It was close to unity right at the surface (epidermis) and decreased toward the root interior. Theoretically, this would be expected for a root being attached to the root pressure probe because of the opposing (circulation) flow of water mentioned above. However, when excised roots, not fixed to the root pressure probe, were measured the cell σ_s was found to be about unity throughout the tissue which is also expected theoretically (Azaizeh *et al.*, 1992).

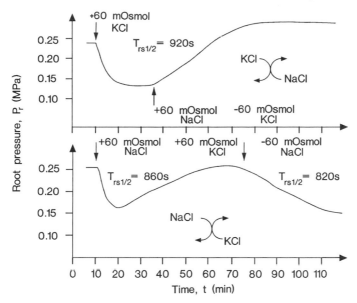

Figure 13 Osmotic experiments with maize roots demonstrating reversible exchange of solutes. In the upper part, roots fixed to the root pressure probe were first treated with 60 mOsmol kg^{-1} KCl solution which resulted in a monophasic response since KCl did not passively enter the root. The isoosmotic change to a solution of 60 mOsmol kg^{-1} NaCl resulted in an increase of root pressure caused by an uptake of NaCl into the xylem (a water phase was missing!). In the lower part, the sequence of the application of solutions was reversed. This resulted in a biphasic response of root pressure for NaCl. However, when the NaCl solution was exchanged by isotonic KCl solution, there was also no water phase and pressure decreased again. Thus, the uptake of the permeating solute (NaCl) was reversible as expected for a passive process (after Azaizeh and Steudle, 1991). $T_{rs1/2}$ describes the halftime of the solute transport.

It is astonishing that Na$^+$ salts entered the root xylem more easily than K$^+$ salts. The same rapid uptake usually occurred with ammonium salts. The finding is explained by the fact that potassium is actively accumulated in the xylem and the electrochemical gradient for K$^+$ (including the electrical trans-root potential) will prevent further uptake. Thus, the driving force for a passive uptake of K$^+$ is low and no solute phase is obtained even when it is offered in the medium at a high concentration. For Na$^+$ the situation is different since the Na$^+$ concentration is low in the xylem and the negative trans-root potential which is mainly a potassium diffusion potential (de Boer, 1989) favors the uptake even more, so that in some cases an overshoot in root pressure has been observed when Na$^+$ salts are offered to the root (Steudle and Frensch, 1989). This explanation for the differences between K$^+$ and Na$^+$ salts observed in maize and other roots was supported by the finding that under anoxia, when the root pressure and potassium concentrations in the xylem were lowered by switching off active pumps, potassium salts did enter the xylem passively and biphasic responses in root pressure were obtained (Birner and Steudle, 1993). These findings as well as the fact that the biphasic osmotic responses in root pressure were reversible (see above) strongly suggest a passive process.

Apoplasmic bypasses in roots are also suggested from other studies. For example, Hanson *et al.* (1985) used detopped root systems of red pine (*Pinus resinosa* Alt.) to which they applied pressures in hydroponics. They followed the uptake of the fluorescent apoplasmic marker PTS (= 3-hydroxy-5,8,10-pyrenetrisulfonate) into the exuding xylem sap at different pressures which would indicate an apoplasmic pathway in the root of the negatively charged dye anion. The measurements showed that the contribution of the apoplasmic path was small. Less than 1% of the total amount of solution was taken up along the apoplast. However, during anoxia up to 50% of the total flux was apoplasmic, and this change was reversible. Anaerobic conditions caused an increase of the root's hydraulic resistance resulting in the change in the relative contribution of the apoplasmic path. However, as discussed in the previous paragraph, there also could have been changes of the driving forces in response to anoxia. Yeo *et al.* (1987) report a bypass of PTS as well as of NaCl in roots of rice (*Oryza sativa*). It was confirmed that PTS uptake was confined to a compartment no greater than the root apoplast. For NaCl, it was concluded that the bypass flow would be a major contribution to the NaCl uptake in rice under saline conditions. For red pine, the results may be interpreted in terms of a rather tight root system, whereas for rice the bypass appears to be a substantial amount.

It should be noted that the existence of apoplasmic bypasses measured with the root pressure probe and quantified as permeability coefficients (P_{sr}) does not mean that roots are leaky and tend to easily release the solutes once taken up by active processes. In fact, the absolute values of permeability coefficients measured with the root pressure probe (Table I) were of an order similar to that of plant cell membranes. Hence, the rates of salt (nutrient) leakage could be easily balanced by the rates of pumping observed in roots which exhibit low values of P_{sr} despite relatively low σ_{sr} (Birner and Steudle, 1993).

C. Variable Hydraulic Conductivity of Roots

Further support for the composite transport model comes from the hydraulic properties of roots and from the fact that the apparent radial hydraulic conductivity depends on the nature of the driving force (hydrostatic or osmotic), at least in some species. For example, in maize, onion, and beech roots, the hydrostatic Lp_r was considerably larger than the osmotic Lp_r (in beech and in onion roots by 1 to 2 orders of magnitude; Steudle *et al.*, 1987; Steudle and Frensch, 1989; Frensch and Steudle, 1989; Zhu and Steudle, 1991; Melchior and Steudle, 1993; Heydt and Steudle, 1993). In others, the values were similar (barley and bean; Steudle and Jeschke, 1983; Steudle and Brinckmann, 1989). These differences are in agreement with the composite transport model provided that there are apoplasmic bypasses including the Casparian bands and provided that the apoplasmic hydraulic conductivity (Lp_{cw}) is high (Zhu and Steudle, 1991; Peterson and Steudle, 1993). In this case, an osmotic gradient would be very inefficient in the apoplast for driving a water flow because of the low reflection coefficient of this structure. On the other hand, hydrostatic gradients would be fully effective. The differences observed between different species would reflect structural differences. The development and tightness of the Casparian band would be crucial. The latter should depend on the species and the growing conditions. One would expect no differences in osmotic and hydrostatic Lp_r, if Casparian bands were completely tight or the conductance of the cell-to-cell path much larger in comparison to that of the apoplast (high cell Lp). This may be true for roots of barley and bean. Changes in Lp_r during root development have been found in onion roots grown hydroponically in hydrostatic experiments (Melchior and Steudle, 1993). For this species, Lp_r decreased significantly with increasing distance from the root apex which could have been due to either the development of the endodermis or that of the exodermis which was found in this species at a distance of 40 to 50 mm from the apex. In maize, where no exodermis developed in hydroponics, Lp_r remained fairly constant at a distance of 30 to 130 mm from the apex (Frensch and Steudle, 1989).

In order to work out the effects of variable Lp_r and of the different parallel pathways in the root cylinder, comparisons have also been made between the Lp of cortex cells and the hydrostatic and osmotic root Lp_r. Some of the data are listed in Table II. In barley and bean the results were consistent with the view of a substantial cell-to-cell transport for both hydrostatic and osmotic gradients. In maize, onion, and beech, however, the main pathway for water was around cells (apoplasmic) when a hydrostatic gradient was applied. In the presence of an osmotic gradient, it was substantially from cell-to-cell. The findings are in line with the composite transport model of the root.

It is interesting that Radin and Matthews (1989) estimated, for roots of young cotton seedlings, that the root Lp_r was larger by a factor of 2 than that of the cell Lp (measured with the cell pressure probe). Since Lp_r was measured by a pressure/flux method (hydrostatic gradient), the result is understandable. It has been interpreted in terms of a predominant apoplasmic water flow in the root. It points to a hydraulic response

Table II   Hydraulic Conductivity of Entire Roots (Lp_r) and of
Individual Root Cells (Lp)

Species	Root $Lp_r \times 10^8$ $(ms^{-1} MPa^{-1})$	Root cell $Lp \times 10^8$ $(ms^{-1} \cdot MPa^{-1})$	Techniques used	References
Hordeum distichon				
osm.flow	0.5–4.3	—		
hydr.flow	0.3–4.0	12		
Zea mays			Root and cell pressure	a–e
osm./hydr.	1.4/10	—/24	probe,	
Phaseolus coccineus				
osm./hydr.	(2–8)/(3–7)	—/190		
Triticum aestivum			Osmotic stop flow and	f
osm./hydr.	(1.6–5.5)/—	—/12	cell pressure probe	
Z. mays				
osm./hydr.	(0.9–5)/—	—/12		
Gossypium hirsutum			Pressure-flow and cell	
osm./hydr.	—/23	—/12	cell pressure probe	g

References: (a) Steudle and Jeschke, 1983; (b) Steudle *et al.*, 1987; (c) Zhu and Steudle, 1991; (d) Peterson and Steudle, 1993; (e) Steudle and Brinckmann, 1989; (f) Jones *et al.*, 1988; (g) Radin and Matthews, 1989.
Water flows were induced by either hydrostatic or osmotic gradients.

of cotton roots similar to that of maize, onion, or beech. Also using the cell pressure probe and an osmotic stop-flow technique (root level), Jones *et al.* (1983) found a predominant cell-to-cell transport in roots of wheat, although later experiments showed that the contribution of the apoplasmic path was larger than originally thought (Jones *et al.*, 1988). Unlike these measurements in which the root Lp_r and cell Lp were measured at the same time, there are other examples of hydraulic measurements in the literature which point to differences between osmotic and hydrostatic flow, although the distinction between the different types of flow is sometimes difficult to make because of the techniques used (for a discussion, see Steudle *et al.*, 1987 and Cruz *et al.*, 1992).

Root pressure probe experiments performed with bean root systems grown in soil support the findings reported above which mainly refer to roots of grasses (Lütgenau and Steudle, 1989; E. Steudle and R. Lütgenau, unpublished). In these experiments, the roots were kept in a pressure chamber (Fig. 14) such that, besides the usual procedure, hydrostatic pressure gradients could be changed by applying a pneumatic pressure to the roots sitting in soil or in a solution. It was found that half-times of water exchange were similar in both types of hydrostatic

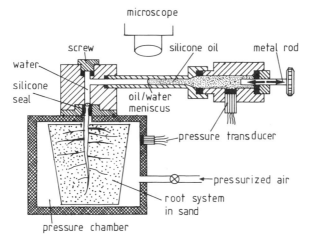

Figure 14 Combination of root pressure probe and pressure chamber. The root is fixed to a root pressure probe and sealed into a pressure chamber. Pneumatic pressures can be applied to the soil solution to create hydrostatic pressure gradients across the root cylinder and to induce water flows which are evaluated from the response in root pressure (Fig 15). Data of root hydraulic conductivity (Lp_r) obtained are compared with those obtained by the standard technique (see Fig. 12).

experiments. There was a nearly 1:1 response in the root pressure when pneumatic pressure was applied to the root system (Figs. 15C and 16). In the osmotic experiments, Lp_r was similar to that found for root segments of bean (Steudle and Brinckmann, 1989) and to the values from hydrostatic experiments. However, in the osmotic experiments it was difficult to change solutions in the soil rapidly enough to not delay the exchange process, and unstirred layer effects in the soil may have affected the values of reflection coefficients which were low (Fig. 15B).

D. Effects of Water Flow on Root Hydraulic Conductivity

There is evidence (for references, see Weatherley, 1982) that the hydraulic resistance of roots decreases with increasing water flow (transpiration). The effect has been known for more than 50 years and has been interpreted as a favorable adjustment of root hydraulics to the needs of the shoot. Different mechanisms and hypotheses have been discussed in the literature to account for the effect which is, in fact, rather variable (Weatherley, 1982). Fiscus (1975) has proposed that it may be apparent and caused by the dilution of xylem sap (i.e., by changes of the driving force) rather than by a change of the root Lp_r. In pressure/flow experiments with excised root systems he found a constant Lp_r when correcting for the dilution of xylem sap. Using intact plants of barley and lupin, Munns and Passioura (1984) reported an offset of the driving force

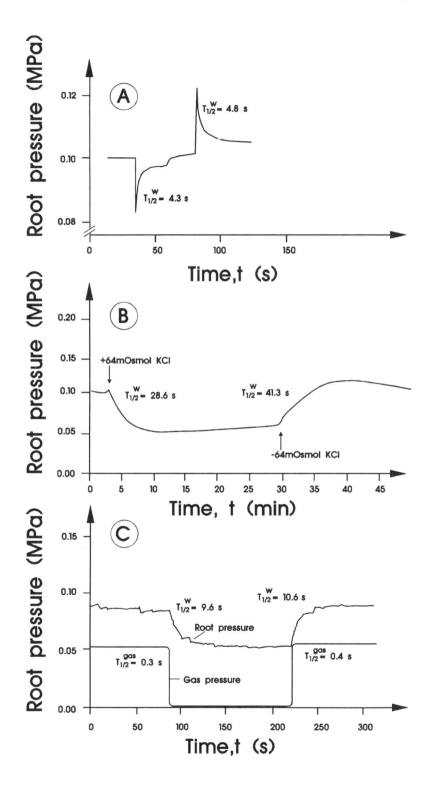

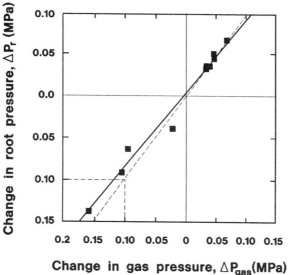

Change in gas pressure, ΔP_{gas}(MPa)

Figure 16 Pressure chamber experiment with a bean root (*Phaseolus coccineus*) sealed to the pressure chamber as shown in Fig. 14. Stationary changes in root pressure (ΔP_r) are plotted against changes of the pneumatic pressure (ΔP_{gas}). It can be seen that the responses were linear and nearly 1:1 (dotted line). Slope = 0.86 and r^2 = 98 (n = 10).

(pressure applied to root system) at zero water flow (transpiration) which was estimated by extrapolating the linear relationship between pressure and water flow (transpiration) back to zero flow. Thus, there was an apparently high hydraulic resistance in the root at low driving forces and flows. Passioura (1988) interpreted the offset in terms of a valve-like mechanism which he speculated to be located in the plasmodesmata of root cortex cells. Due to the structure of plasmodesmata and, namely, due to their possible blockage by membranes, he expected that, at a certain threshold of the pressure gradient driving the flow, the pores

---

Figure 15 Pressure chamber experiment with a bean root (*Phaseolus coccineus*) sealed to the pressure chamber as shown in Fig. 14. (A and B), the usual hydrostatic and osmotic experiments with the root sitting in sand and KCl as the osmotic solute. (C) the root was first pressurized with compressed air by about 0.5 bar followed by step changes of pneumatic pressure. It can be seen that changes in air pressure within the chamber were much faster than those of root pressure indicating that the processes observed were limited by the water exchange between soil solution and root xylem. Half-times of water exchange in the hydrostatic experiments (external (C) or internal (A) change of hydrostatic pressure) were similar, whereas in the osmotic experiments half-times were somewhat higher which, in part, could have been due to difficulties in exchanging the soil solution rapidly enough so that this process would not limit or colimit the water exchange across the root (after Lütgenau and Steudle, 1989).

would open allowing for a rapid passage of water at higher pressures or pressure gradients applied across the root. Recently, valve-like mechanisms related to plasmodesmata have been proposed also by Oparka *et al.* (1992).

The Passioura model still appears to be rather speculative, since there are no real data or ultrastructural evidence that a reversible valve-like mechanism could operate in plasmodesmata. Furthermore, because of the high water permeability of the root cell membranes, the contribution of the plasmodesmata to the overall water flow should be rather low (Weatherley, 1982; Cosgrove and Steudle, 1981). In terms of the composite transport model of the root, a dependency of Lp_r on the rate of water flow would be expected if there were a cell-to-cell and an apoplasmic path for water operating in parallel. At low rates of water flow, this would result in opposite (i.e., circulating) water flows which, in part, would cancel. Hence, an apparently high hydraulic resistance or offset would result. However, as soon as the flows in both pathways are in the same direction (i.e., at higher rates of transpiration and at low or negative xylem pressure), this would change, and the hydraulic resistance would decrease. Thus, the composite membrane model could readily explain the findings of Fiscus, Passioura, and others about variable root resistances. However, it would be difficult to check the hypothesis experimentally using the root pressure probe. The rate of water flows would have to be varied over large ranges. This would imply considerable reductions of root pressures so that tensions or negative pressures would have to be created in the xylem to drive large flows. In principle, the latter is possible with the probe, but could be difficult because of the danger of cavitations at higher tensions in the experimental system. For end segments of maize roots, it has been shown that the hydraulic conductivity as well as the reflection and permeability coefficients of the roots did not change significantly when the pressure in the root pressure probe to was as low as -0.3 MPa (Heydt and Steudle, 1991). Perhaps the application of external pressures rather than changes of root pressure would be more suitable. On the other hand, for testing the hypothesis, it would also be important to estimate opposing water flow which is proposed by the model. Experimentally, this should be difficult.

E. Longitudinal Hydraulic Resistance: Hydraulic Architecture of Roots

So far, the radial hydraulic conductivity or its inverse, the radial hydraulic resistance, has been discussed with the hydraulic resistance in the longitudinal (axial) direction, i.e., the resistance to water flow in the vessels of the xylem, being neglected. There are only a few detailed measurements in which both radial and axial resistances have been measured in roots (e.g., Frensch and Steudle, 1989; North and Nobel, 1991; Melchior and Steudle, 1993). In these cases, it has been shown that axial resistances are usually much smaller than radial. This is, in fact, a prerequisite for the simple two-compartment model used for roots. However, the hydraulic resistance of the root xylem (R_x) is dependent on the position along the root and the axial component could become dominant close to the root tip where xylem vessels are not yet mature. In this zone, which may be as long as 20 mm in hydroponically grown maize, R_x can be much larger than the radial resistance (R_r) (Frensch and Steudle, 1989). This would hydraulically isolate the root tip from the rest of the root. The length of this zone depends on the growing conditions (Azaizeh and Steudle, 1991). Simulations of the root efficiency for taking up water were made with computer models using measured hydraulic conductivities (Lp_r and L_x) and their dependency on the position in the root. These simulations showed how the efficiency of water uptake would be affected (Frensch and Steudle, 1989). The model calculations used the basic theory of Landsberg and Fowkes (1978). Provided that sufficient information is available the efficiency or hydraulic architecture could also be modeled for root systems. Quantification of the contribution of different parts of roots to the overall water uptake and the effects of factors such as the geometry and state of development of roots would be very valuable. The information is needed in the overall water budget of plants where a detailed hydraulic mapping of the root is still missing for technical reasons. For example, most scientists working in the field tend to neglect the contribution of older suberized parts of roots to the overall water uptake. However, this has been questioned (Kramer, 1983). In order to get the information, the root pressure probe would have to be applied to extensive root systems as well as to parts of roots in soil. This is possible.

F. Changes of Transport along Developing Roots

There is evidence that certain root zones would contribute to the overall uptake of water more than others (see above, and Rosene, 1937; Graham *et al.*, 1974). It is generally thought that water would be preferentially absorbed by roots when the xylem is already developed, but suberin lamellae are not yet formed in the endodermis. Evidence for this comes from measurements using potometers which have been applied to roots of herbaceous and woody plants (Sanderson, 1983; Häussling *et al.*, 1988). To date, the question of the contribution of different root zones cannot be completely answered because detailed measurements of the root Lp_r depending on the position as well as on the driving forces in root systems are still missing. The root pressure probe could be used to fill this gap and to provide data to work out the hydraulic architecture of roots. Such an experiment has already been performed with end segments of maize roots. For these roots, it was shown that σ_{sr} increased and P_{sr} decreased along the developing root, whereas Lp_r remained constant (Frensch, 1990).

G. Effects of External Factors on Root Water: High Salinity and Anoxia

The most interesting effects on root water and solute transport which could be directly evaluated using the root pressure probe technique would be those of low temperature, salinity, drought, mycorrhiza, and anoxia. To date, effects of salinity and of anoxia have been studied in more detail. Excessive soil salinity is an environmental stress that inhibits growth and development of plants in many regions of the world. The precise mechanism(s) of the inhibition at the root is still not fully understood, but may include osmotic effects as well as the toxicity caused by certain ions (Epstein, 1985; Cheeseman, 1988). Reductions of growth could be caused by an inhibition of water uptake due to a reduced water permeability of root membranes (Lp) which in turn may result in a reduced overall hydraulic conductivity of the roots (Lp_r). It is known that effects of salinity could in part be reversed by calcium (Cramer *et al.*, 1986; Lynch *et al.*, 1987). It is thought that at high salinity Na^+ competes for Ca^{2+} at the cell membranes and that this has a deleterious effect on the proper function and integrity of the membranes. In order to test this and to work out effects on the water relations of roots, root pressure probe experiments have been performed using young maize roots. The hydraulic conductivity of the cell membranes of cortex cells was measured as well (Azaizeh and Steudle, 1991; Azaizeh *et al.*, 1992). Maize seedlings grown in a nutrient solution plus 100 mM NaCl showed an Lp_r which was reduced by 30 to 60% compared with the control (1/5 Hoagland solution; 0.5 mM Ca). Increased levels of Ca (10 mM) had an ameliorative effect on the Lp_r of salinized roots so that the water permeability increased again, thus, increasing the availability of water for the plant. It was shown that at the cell level the effects were much larger than those at the root level. High salinity (100 mM NaCl) caused a reduction of the cell Lp by a factor of 3 to 6 which was reversed by a factor of 2 to 3 in the presence of calcium. The differences in the effects of NaCl on either the root Lp_r or the cell Lp were due to the fact that in the roots the water flow was mainly around cells under hydrostatic conditions (see above). The findings of a reduced root Lp_r are in line with those of Munns and Passioura (1984) and O'Leary (1969) for bean and lupin. Conversely, salinity had no effect on Lp_r in barley, tomato, and sunflower (Munns and Passioura, 1984; Shalhevet *et al.*, 1976). The findings are at variance with those of Tyerman *et al.* (1989) who found no change of the root cell Lp of *Nicotiana tabacum* at high salinity, i.e., for a plant which can grow reasonably well at high salinity (Flowers *et al.*, 1986).

As was found for increased salinity, anoxia also had a negative effect on root Lp_r in root pressure probe experiments with maize roots (Birner and Steudle, 1993). The decrease in Lp_r was probably due to a decrease of the hydraulic conductivity at the level of cell membranes (Lp). In addition to the changes in water transport reported in the literature (e.g., Everard and Drew, 1987; Zhang and Tyerman, 1991), the results obtained with the root pressure probe allowed the quantification of changes in root pressure, the permeability of roots to nutrients (P_{sr}), and the selectivity of roots (σ_{sr}). The steady-state root pressure of the maize roots used in the studies strongly decreased upon anoxia to reach a much lower value after several hours (Fig. 17). Since ion leakage of the roots (permeability coefficient, P_{sr}) was also reduced at a constant σ_{sr}, this indicated that the active uptake of solutes (nutrient salts) was inhibited (Eq. (5)). Remarkably, roots treated in anoxia for several hours did not recover as quickly as they changed their transport properties. Great care was taken to characterize the physiological status of the roots by performing osmotic experiments and other tests which proved that, despite anoxia, the roots were still properly functioning as osmometers. As already mentioned, potassium salts showed biphasic responses for roots during anoxia which was interpreted as a consequence of the reduced potassium concentration in the root xylem under these conditions (see above).

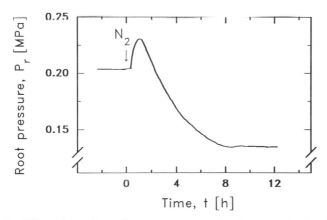

Figure 17 Effect of anoxia on the root pressure of maize roots. At the time marked by the arrow, the control medium of the root containing 8.9 mg O_2 kg^{-1} was exchanged for a medium containing only 0.2 mg O_2 kg^{-1}. After a transient increase of root pressure, this caused the stationary root pressure to decline to a much lower value. When the new stationary root pressure was established, the hydraulic conductivity of the root (Lp_r) as well as the permeability coefficient (P_{sr}) was decreased (data not shown). This proved that anoxia affected the active solute uptake ("root pumping") rather than making the root leaky for nutrients (see Eq. (5)) (after Birner and Steudle, 1993).

The two examples demonstrate that the root pressure probe may be used to determine changes of both water and solute uptake of roots in response to environmental stresses such as pollutants, low pH, high concentrations of aluminium, and heavy metals. Combining the approach with measurements at the cell level using the cell pressure probe should reveal deeper insight into mechanisms. Using these techniques one may envisage that roots could be used also as biosensors to detect changes in the soil which are harmful to roots either on the long-term or on the short-term (Rüdinger *et al.*, 1992). The absolute value of root pressure could be used as a direct sensor of the metabolic activity or roots in addition to monitoring changes of transport coefficients.

H. Work with Modified Roots: Root Steaming, Dissecting, and Puncturing

The work with modified roots provides a tool for getting a deeper insight into the mechanisms of water flow in roots and for locating the major barriers to water and solute movement in roots. Root transport properties may be modified by removing part of the cortex by scraping or by micro-dissection (Peterson *et al.*, 1993). Parts of roots may be also killed by steaming to remove the entire barrier for radial water flow except the lateral walls of mature vessels (Peterson and Steudle, 1993). Another possibility for modifying roots is the puncturing of the endodermis with needles (e.g., with the tip of a cell pressure probe) to create small holes of some 10-μm in diameter (Steudle *et al.*, 1993).

The puncturing of the endodermis offers the possibility of creating additional "apoplasmic bypasses" as small as only 10^{-2} to $10^{-3}\%$ of the entire surface area of the endodermis. It may be expected that this would affect the hydraulic conductivity, solute permeability, and reflection coefficients of roots and allow extrapolation to the possible size of bypasses in the intact system in terms of the composite membrane model. The experiments indicated that only small bypasses (leaks) were required to considerably reduce root pressure and to substantially lower σ_{sr} and increase P_{sr} at a fairly constant hydraulic conductivity. It should be noted that usually puncturing did not result in a zero root pressure. This means that, according to the pump/leak model of the root (Eq. (5)), the passive leak of ions could be compensated for by active pumping. Thus, the bypasses present in the intact system such as those across the root primordia (Peterson *et al.*, 1981), across the root tip, or some leakiness of Casparian bands could be sufficient to cause a $\sigma_{sr} < 1$ as found. On the other hand, the fact that the root hydraulic conductivity (half-time of water exchange) did not increase significantly upon puncturing could be simply explained by the high overall Lp_r of the intact system. It indicates that the radial hydraulic resistance was more evenly distributed across the root

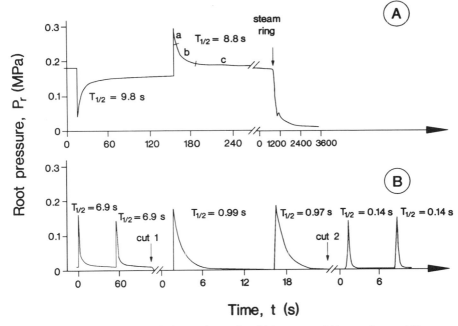

Figure 18 Graph of a typical experiment in which a zone of 11 mm along a 108-mm-long young maize root fixed to the pressure probe was steam ringed to kill the tissue and to make the endodermis leaky. Results are continuous from A to B. After steam ringing, root pressure dropped to a value close to zero (A) and half-times of water exchange decreased significantly although they did not become very short. Cutting off the root distal to the steamed area (cut 1) and cutting off the root at the seal of the root pressure probe (cut 2) resulted in drastic reductions of the half-time indicating that the radial rather than the longitudinal resistance to water flow limited the process and that the hydraulic resistance of the seal was sufficiently small during the measurement. From the changes caused in the rate of water exchange by steam ringing, the lateral hydraulic conductivity of the walls of root xylem vessels has been calculated (see Section V.H; after Peterson and Steudle, 1993).

tissue and not located at the endodermis. The puncturing experiments support the composite membrane model of the root. They do indicate that, at least for maize roots which lack a suberin lamella, the endodermis is a rather effective barrier for ions (low P_{sr}), but not for water. The former would be in line with the traditional view of the function of the endodermis, whereas the latter would be different. It is interesting that the holes produced in the endodermis did seal by a healing process after a few hours and root pressure recovered.

Experiments in which roots were scraped or dissected support this view. As long as the endodermis was not damaged by the manipulations, root pressure remained high or recovered after healing of the endodermis. The hydraulic conductivity of the modified roots only increased substantially when considerable amounts of the cortical tissue were removed.

In the steaming experiments the manipulation was the most severe. Killing of a few millimeters of a young maize root with a total length of about 100 mm resulted in an immediate decrease of root pressure to nearly zero and in an increase of the overall Lp_r from which the hydraulic conductivity of the steamed zone could be evaluated as well (Fig. 18). The Lp_r of the steamed zone was a factor of 3 to 13 larger than that of the intact root. In the steamed zone, the remaining radial hydraulic resistance could be considered as that of the walls of mature xylem vessels. This allowed the measurement of the lateral hydraulic conductivity of the walls of the vessels (Lp_x). It turned out that, although Lp_x was much higher than Lp_r, the contribution of the lateral hydraulic resistance to the overall radial resistance was not negligible.

Experiments with modified roots are being conducted to get more detailed information about the transport properties of the endodermis (water, solutes). They provide a good basis for elucidating the aggregated pathways used for transport in roots and will allow verification and further development of the composite membrane model of the root.

VI. Conclusions and Summary

At first glance, it may be assumed that the regulation of plant water should be much easier to describe than for other substances such as nutrient salts or assimilates. The reasons for this assumption are (i) water moves in plants along gradients of free energy or water potential, (ii) water is highly mobile within plants which would tend to rapidly equilibrate flows and gradients, and (iii) there is no direct coupling with metabolism, i.e., there is no active water flow. Unfortunately, this way of looking at plant water relations is not sufficient mainly because of couplings between water and solutes. Solutes act via the osmotic component of water potential (the osmotic pressure) which may be subject to metabolic influence or may even be regulated by active processes. Osmotic solutes are usually compartmented which leads to an "osmotic compartmentation of water" (elevated water content, high turgidity).

Another interaction between water and metabolism concerns the anatomy of plants during development. For example, the water storage capacity of plants involves physical properties of cell walls (plastic and elastic extensibility) which are under metabolic control such as the expanding walls of cells during growth. Other parameters would be the hydraulic conductivity of membranes or the size of compartments. Thus, there are different types of possible interactions between water and metabolism which have to be taken into account when attempting to understand water flows in plants. In this chapter, the most important parameters have been integrated into a general theory of tissue water which also includes growth.

From this theory, several general principles may be derived which govern plant water relations:

1. Water status is adequately described by water potential; gradients of water potential usually drive water transport.
2. There is a strict coupling between water flow and active and passive solute flow.
3. Compartmentation of solutes results in a compartmentation of water.
4. The apoplasmic compartment or transport path plays a special role in the interactions between water and solute relations of plants.

The first point states that the water potential concept provides the basis for defining both the water status of plants and the driving forces. During the regulation of water status, components of water potential such as turgidity, volume, or water content also play an important role because water potential gradients tend to vanish. Both the first and the second principle are important for a homeostasis of turgor and cell volume (water content) in many processes (osmoregulation, extension growth, stomatal regulation, etc.). It has been readily shown that during these processes parameters such as pressure (turgor), volume, and concentration act as set points for regulation. The specific mechanisms of regulation may require biochemical responses depending on signals of pressure, volume, or concentration. Physical quantities derived from the basic parameters (pressure, volume) such as the elastic extensibility of cell walls are involved in these mechanisms.

The third principle states that the compartmentation of solutes (e.g., in the apoplast or protoplasts) would cause concomitant movements of water which are well known (e.g., phloem loading and unloading and growth). In this chapter, a new physical model has been presented (osmotic compartmentation model) which shows how water would be shared between compartments and how this would influence the water status (water potential, volume, water content). In quantitative terms, the storage capacity of compartments for water plays a key role. Thus, a regulation of plant water, turgidity, water content, etc. would be possible by the compartmentation of certain solutes into either the apoplast or the symplast. This appears to be of great importance not only for growth processes, but also for phloem transport. The model represents an alternative and, in terms of energy consumption, a more effective concept than that of futile cycles (see this volume, chapters 1, 5–7). During expansion growth, the solute flow supplied to the growing tissue is strictly coupled to the water supply. In addition to compartmentation of solutes into apoplast and symplast, the ratio between flows is important for regulating tissue turgidity and the rate of growth. Thus, the present debate about a regulation of extension growth which adhears to the limitations of the Lockhart model and focuses only on either water transport or mechanical properties as limiting factors could be misleading. Couplings with solute relations have to be integrated into the classical Lockhart model not only theoretically, but also by experiments performed on the growing hypocotyl of the castor bean seedling.

Hydraulic and other transport properties of the apoplast (principle 4) have been shown to be of special importance in the water relations of plants. This aspect of plant water relations, which considers the existence of the apoplasmic compartment as the main difference between plant and animal tissue, has been largely overlooked in the past. This is so because of the lack of a theoretical background and techniques for measuring water relations parameters of the apoplast. The theory given in this chapter integrates water and solute relations of tissues. Although it may still be incomplete, it shows that the specific influence of the apoplast in the overall water and solute relations of plant tissues and organs largely originates from the fact that the osmotic component or water potential is quite ineffective in the apoplast as a driving force for water. However, these forces will be fully exerted across cell membranes, and local equilibria will be maintained between compartments. Depending on the contribution of the osmotic component to the overall water potential difference this should cause smaller driving forces across tissues than those expected from the water potential difference. The reflection coefficient of the apoplast and the absolute value of the osmotic component would be important. The effect has been readily demonstrated for leaf and growing tissue and for roots which exhibit an overall reflection coefficient of $\sigma_{sr} < 1$ for the same reason. Also, the dependence on the nature of driving forces for water (osmotic and hydrostatic) observed in the hydraulic conductivity of roots and other tissue would be due to the special properties of the apoplast just mentioned. The differences in the hydraulic properties refer to both dynamic and stationary water transport.

The root data reviewed in this chapter provide another example of a coupling between water and solute flow. In the root, couplings appear to be most important at low rates of transpiration where the osmotic component of the xylem water potential is important. The interaction between flows and the "composite structure" of the root tissue (apoplasmic vs cell-to-cell path) appears to be the reason for nonlinearities between forces and water flows in the root which have been observed by several authors. Compared with cells, the relatively low selectivity of roots expressed by reflection coefficients of less than unity is readily explained by the composite transport model of the root. This implies some apoplasmic bypass of water and solutes in roots including the endodermis. Again, the compartmentation between water and solutes plays an important role. Due to the existence of parallel pathways of different selectivity in the root, the model would propose some circulation flow of water across the root cylinder at low rates of transpiration. Thus, common views of the transport across roots have to be extended by a modified pattern of water and solute transport. The endodermis with the Casparian band especially is no longer considered to act as a nearly semipermeable structure. Nevertheless, the results show that the proper function of the root in the acquisation of nutrients and water will be ensured, because the passive permeability (leakage) of the endodermis is sufficiently low. This is in accordance with traditional views as long as the endodermis is not looked at as a semipermeable barrier. The validity of the composite transport model of the root is strongly supported by combined measurements of water transport at level of roots and cells (cell pressure probe). With the latter, the contribution of the cell-to-cell path to the overall water transport has been evaluated. This also allowed quantification of the apoplasmic component of radial water transport by comparing the cell Lp and the root Lp_r and considering the root anatomy. The validity of the composite transport model is, furthermore, supported by experiments with "modified roots" (steamed, dissected, and punctured roots), which allow characterization of transport properties of the endodermis.

Changes of transport coefficients (water, solutes) along developing roots indicate that the contribution of different root zones varies during root development. Besides the radial transport, the longitudinal transport has to be taken into account when the efficiency of roots and root system in taking up water is evaluated as shown in model calculations. In these calculations, measured values of the radial and longitudinal hydraulic conductivities were used to integrate the uptake of water by a given root. The approach can be applied also to root systems provided that there is a sufficient amount of detailed data which are available from root pressure probe experiments. With the information obtained for the cell, tissue, and organ level, models can be checked readily for their quantitative validity. Since the transport of water across roots is affected by external factors such as salinity, anoxia, or mycorrhiza, the approach also will be used in the future to evaluate the efficiency of roots with varying demands from the shoot and external factors, i.e., changes of the hydraulic architecture of roots under these conditions will be followed.

Acknowledgments

Between 1987 and 1992 the work reviewed in this chapter was, in part, supported by the Deutsche Forschungsgemeinschaft, Sonderforschungsbereich 137. The contributions of several guest scientist (Drs. Hassan Azaizeh, Technion, Israel Institute of Technology, Haifa, Israel; Benito Gunse, Universidad Autonoma de Barcelona, Spain; Anatoli B. Meshcheryakov, Russian Academy of Sciences, Moscow; Carol A. Peterson, University of Waterloo, Canada; Guo-Li Zhu, Agricultural University of Peking, China) are gratefully acknowledged as well as those of different co-workers (T. Birner, Dr. J. Frensch, H. Heydt, R. Lütgenau, W. Melchior, M. Murrmann, M. Rüdinger,). The author is also grateful to Dr. S. W. Hallgren, Oklahoma State University, for reading the manuscript and to Burkhard Stumpf for his expert technical assistance.

References

Azaizeh, H., and Steudle, E. (1991). Effects of salinity on water transport of excised maize (*Zea mays* L.) roots. *Plant Physiol.* **97**, 1136–1145.

Azaizeh, H., Gunse, B., and Steudle, E. (1992). Effects of NaCl and $CaCl_2$ on water transport across root cells of maize (*Zea mays* L.) seedlings. *Plant Physiol.* **99**, 886–894.

Birner, T. P., and Steudle, E. (1993). Effects of anaerobic conditions on water and solute relations and active transport in roots of maize (*Zea mays* L). *Planta* **190**, 474–483.

Boyer, J. S. (1985). Water transport. *Annu. Rev. Plant Physiol.* **36**, 473–516.

Boyer, J. S. (1988). Cell enlargement and growth-induced water potentials. *Physiol. Plant.* **73**, 311–316.

Brundrett, M. C., Enstone, D. E., and Peterson, C. A. (1989). A. berberine-aniline blue fluorescent staining procedure for suberin, lignin, and callose in plant tissue. *Protoplasma* **146**, 133–142.

Cheeseman, J. M. (1988). Mechanisms of salinity tolerance in plants. *Plant Physiol.* **87**, 547–550.

Cleland, R. E. (1981). Wall extensibility: Hormones and wall extension. *Encycl. Plant Physiol., New Ser.* **13B**, 225–276.

Cosgrove, D. J. (1985). Cell wall yield properties of growing tissues. Evaluation by *in vivo* stress relaxation. *Plant Physiol.* **78**, 347–356.

Cosgrove, D. J. (1986). Biophysical control of plant cell growth. *Annu. Rev. Plant Physiol.* **37**, 377–405.

Cosgrove, D. J. (1987). Linkage of wall extension with water and solute uptake. *In* "Physiology of Cell Expansion During Plant Growth" (D. J. Cosgrove and D. P. Knievel, eds), Proc. 2nd Symp. Plant Physiol., pp. 88–100. Am. Soc. Plant Physiol., Rockville, MA.

Cosgrove, D. J. (1993). Wall extensibility: Its nature, measurement, and relationship to plant growth. *Tansley Rev., New Phytol.* (in press).

Cosgrove, D. J., and Cleland, R. E. (1983). Solutes in the free space of growing stem tissues. *Plant Physiol.* **72**, 326–331.

Cosgrove, D. J., and Steudle, E. (1981). Water relations of growing pea epicotyl segments. *Planta* **153**, 343–350.

Coster, H. G. L., and Zimmermann, U. (1976). Transduction of turgor pressure by cell membrane compression. *Z. Naturforsch. C* **31C**, 461–463.

Coster, H. G. L., Steudle, E., and Zimmermann, U. (1976). Turgor pressure sensing in plant cell membranes. *Plant Physiol.* **58**, 636–643.

Cramer, G. A., Epstein, E., and Läuchli, A. (1986). Effect of NaCl and $CaCl_2$ on ion activities in complex nutrient solutions and root growth of cotton. *Plant Physiol.* **81,** 792–797.

Cruz, R. T., Jordan, W. R., and Drew, M. C. (1992). Structural changes and associated reduction of hydraulic conductance in roots of *Sorghum bicolor* L. following exposure to water deficit. *Plant Physiol.* **99,** 203–212.

Dainty, J. (1963). Water relations of plant cells. *Adv. Bot. Res.* **1,** 279–326.

Dainty, J. (1976). Water relations of plant cells. *Encycl. Plant Physiol., New Ser.* **2,** Part A, 12–35.

Dainty, J. (1985). Water transport through the root. *Acta Hortic.* **171,** 21–31.

Dalton, F. N., Raats, P. A. C., and Gardner, W. R. (1975). Simultaneous uptake of water and solutes by plant roots. *Argron. J.* **67,** 334–339.

de Boer, A. H. (1989). Xylem transport. *In* "Methods of Enzymology" (S. Fleischer and R. Fleischer, eds.), Vol. 174, pp. 277–287. Academic Press, San Diego.

Epstein, E. (1985). Salt tolerant crops: Origins, development, and prospects of the concept. *Plant Soil* **89,** 183–198.

Everard, J. D., and Drew, M. C. (1987). Mechanisms of inhibition of water movement in anaerobically treated roots of *Zea mays* L. *J. Exp Bot.* **38,** 1154–1165.

Fensom, D. S., and Dainty, J. (1963). Electro-osmosis in *Nitella. Can. J. Bot.* **41,** 685–691.

Fiscus, E. L. (1975). The interaction between osmotic- and pressure-induced water flow in plant roots. *Plant Physiol.* **55,** 917–922.

Fiscus, E. L. (1977). Determination of hydraulic and osmotic properties of soybean root systems. *Plant Physiol.* **59,** 1013–1020.

Fiscus, E. L. (1986). Diurnal changes in volume and solute transport coefficients of Phaseolus roots. *Plant Physiol.* **80,** 752–759.

Flowers, T. J., Flowers, S. A., and Greenway, H. (1986). Effects of sodium chloride on tobacco plants. *Plant, Cell Environ.* **99,** 615–645.

Frensch, J. (1990). Wasser- und Teilchentransport in Maiswurzeln. Dissertation, University of Bayreuth, Germany.

Frensch, J., and Steudle, E. (1989). Axial and radial hydraulic resistance to roots of maize (*Zea mays* L.). *Plant Physiol.* **91,** 719 726.

Frensch, J., Stelzer, R., and Steudle, E. (1992). NaCl Uptake in Roots of *Zea mays* Seedlings: Comparison of Root Pressure Probe and EDX Data *Ann. Bot. (London)* **70,** 543–550.

Fry, S. C. (1989). Cellulases, hemicelluloses and auxin-stimulated growth: A possible relationship. *Physiol. Plant.* **75,** 532–536.

Gollan, T., Passioura, J. B., and Munns, R. (1985). Soil water status affects the stomatal conductance of fully turgid wheat and sunflower leaves. *Aust. J. Plant Physiol.* **13,** 459–464.

Graham, J., Clarkson, D. T., and Sanderson, J. (1974). Water uptake by roots of marrow and barley plants. *Annu. Rep.—Agric. Res. Counc. Letcombe Lab.,* pp. 9–12.

Green, P. B., Erickson, R. O., and Buggy, J. (1971). Metabolic and physical control of cell elongation rate: In vivo studies in *Nitella. Plant Physiol.* **47,** 423–430.

Hanson, P. J., Sucoff, E. I., and Markhart, A. H. (1985). Quantifying apoplastic flux through red pine root systems using trisodium, 3-hydroxy-5,8,10-pyrenetrisulfonate. *Plant Physiol.* **77,** 21–24.

Häussling, M., Jorns, C. A., Lehmbecker, G., Hecht-Buchholz, C. H., and Marschner, H. (1988). Ion and water uptake in relation to root development in Norway spruce (*Picea abies* (L.) Karst). *J. Plant Physiol.* **133,** 486–491.

Jones, H., Tomos, A. D., Leigh, R. A., and Wyn Jones, R. G. (1983). Water-relation parameters of epidermal and cortical cells in the primary root of *Triticum aestivum* L. *Planta* **158,** 230–236.

Jones, H., Leigh, R. A., Wyn Jones, R. G., and Tomos, A. D. (1988). The integration of whole-root and cellular hydraulic conductivities in cereal roots. *Planta* **174,** 1–7.

Katou, K., and Taura, T. (1989). Mechanism of pressure-induced water flow across plant roots. *Protoplasma* **150**, 124–130.

Kauss, H. (1978). Osmotic regulation in algae. *In* "Progress in Phytochemistry" (L. Reinhold, J. B. Harborn, and T. Swain, eds.), pp. 1–27, Pergamon, Elmsford, NY.

Kedem, O., and Katchalsky, A. (1963a). Permeability of composite membranes. Part 2. Parallel elements. *Trans. Faraday Soc.* **59**, 1931–1940.

Kedem, O., and Katchalsky, A. (1963b). Permeability of composite membranes. Part 3. Series array of elements *Trans. Faraday Soc.* **59**, 1941–1953.

Kirst, G. O. (1991). Salinity tolerance of eukaryotic marine algae. *Annu. Rev. Plant Physiol. Plant Mol. Biol.* **41**, 21–53.

Kramer, P. J. (1983). "Water Relations of Plants." Academic Press, Orlando, FL.

Kutschera, U. (1989). Tissue stresses in growing plant organs. *Physiol. Plant.* **77**, 157–163.

Kutschera, U., and Schopfer, P. (1985). Evidence against the acid-growth theory of auxin action. *Planta* **163**, 483–493.

Landsberg, J. J., and Fowkes, N. D. (1978). Water movement through plant roots. *Ann. Bot. (London)* [N.S.] **42**, 493–508.

Lockhart, J. A. (1965). An analysis of irreversible plant cell elongation. *J. Theor. Biol.* **8**, 264–276.

Lütgenau, R., and Steudle, E. (1989). Root pressure measurements on intact root systems: Combination of root pressure probe and pressure chamber. *In* "Plant Membrane Transport: The Current Position" (J. Dainty, M. I. de Michelis, E. Marre, and F. Rasi-Caldogno, eds.), pp. 579–580. Elsevier, Amsterdam.

Lynch, J., Cramer, G. R., and Läuchli, A. (1987). Salinity reduces membrane-associated calcium in corn root protoplasts. *Plant Physiol.* **83**, 390–394.

Matyssek, R., Maruyama, S., and Boyer, J. S. (1991). Growth-induced water potentials may mobilize internal water for growth. *Plant, Cell Environ.* **14**, 917–923.

Mees, G. C., and Weatherley, P. E. (1957). The mechanism of water absorption by roots. I. Preliminary studies on the effects of hydrostatic pressure gradients. *Proc. R. Soc. London, Ser. B.* **147**, 367–380.

Melchior, W., and Steudle, E. (1993). Water transport in onion (*Allium cepa* L.) roots. Changes of axial and radial hydraulic conductivity during root development. *Plant Physiol.* **101**, 1305–1315.

Meshcheryakov, A., Steudle, E., and Komor, E. (1992). Gradients of turgor, osmotic pressure, and water potential in the cortex of the hypocotyl of growing *Ricinus* seedlings. Effects of the supply of water from the xylem and of solutes from the phloem. *Plant Physiol.* **98**, 840–852.

Miller, D. M. (1985). Studies of root function in *Zea mays.* III. Xylem sap composition at maximum root pressure provides evidence of active transport in the xylem and a measurement of the reflection coefficient of the root. *Plant Physiol.* **77**, 162–167.

Molz, F. J., and Boyer, J. S. (1978). Growth-induced water potentials in plant cells and tissues. *Plant Physiol.* **62**, 423–429.

Molz, F. J., and Ferrier, J. M. (1982). Mathematical treatment of water movement in plant cells and tissues: A review. *Plant, Cell Environ.* **5**, 191–206.

Molz, F. J., and Hornberger, G. M. (1973). Water transport through plant tissues in the presence of a diffusable solute. *Soil Sci. Soc. Am. Proc.* **37**, 833–837.

Molz, F. J., and Ikenberry, E. (1974). Water transport through plant cells and walls: Theoretical development. *Soil Sci. Soc. Am. Proc.* **38**, 699–704.

Moreshet, S., and Huck, M. C. (1991). Dynamics of water permeability. *In* "Plant Roots: The Hidden Half" (Y. Waisel, A. Eshel, and U. Kafkai, eds.), pp. 605–626. Dekker, New York.

Munns, R., and Passioura, J. B. (1984). Hydraulic resistance of plants. III. Effects of NaCl in barley and lupin. *Aust. J. Plant Physiol.* **11**, 351–359.

Newman, E. I. (1976). Interaction between osmotic- and pressure-induced water flow in plant roots. *Plant Physiol.* **57**, 738–739.

Nobel, P. S. (1991). "Physicochemical and Environmental Plant Physiology." Academic Press, San Diego.

Nobel, P. S., and Goldstein, G. (1992). Desiccation and freezing phenomena for plants with large water capacitance—cacti and espeletias. *In* "Water and Life: Comparative Analysis of Water Relationships at the Organismic, Cellular, and Molecular Levels" (G. N. Somero, C. B. Osmond, and C. L. Bolis, eds.), pp. 240–257. Springer-Verlag, Berlin.

Nonami, H., and Boyer, J. S. (1990). Primary events regulating stem growth at low water potentials. *Plant Physiol.* **93**, 1601–1609.

North, G. B., and Nobel, P. S. (1991). Changes in hydraulic conductivity and anatomy caused by drying and rewetting roots of *Agave desertii* (Agavaceae). *Am. J. Bot.* **78**, 906–915.

O'Leary, J. W. (1969). The effect of salinity on permeability of roots to water. *Isr. J. Bot.* **18**, 1–9.

Oparka, K. J., Wright, K. M., and Prior, D. A. M. (1992). Regulation of symplasmic transport in a storage sink. *J. Exp. Bot.* **43S**, 11.

Passioura, J. B. (1980). Transport of water from soil to shoot in wheat seedlings. *J. Exp. Bot.* **31**, 333–345.

Passioura, J. B. (1988). Water transport in and to the root. *Annu. Rev. Plant Physiol. Plant Mol. Biol.* **39**, 245–265.

Peterson, C. A. (1988). Exodermal Casparian bands: Their significance for ion uptake in roots. *Physiol. Plant.* **72**, 204–208.

Peterson, C. A., and Steudle, E. (1993). Lateral hydraulic conductivity of early metaxylem vessels in *Zea mays* L. roots. *Planta* **189**, 288–297.

Peterson, C. A., Emanuel, G. B., and Humphreys, G. B. (1981). Pathway of movement of apoplastic fluorescent dye tracers through the endodermis at the site of secondary root formation in corn (*Zea mays*) and broad bean (*Vicia faba*). *Can. J. Bot.* **59**, 618–625.

Peterson, C. A., Murrmann, M., and Steudle, E. (1993). Location of the major barrier(s) to the movement of water and ions in young roots of *Zea mays* L. roots. *Planta* **190**, 127–136.

Philip, J. R. (1958). Propagation of turgor and other properties through cell aggregations. *Plant Physiol.* **33**, 271–274.

Pitman, M. G. (1982). Transport across plant roots. *Q. Rev. Biophys.* **15**, 481–554.

Pritchard, J., Wyn Jones, R. G., and Tomos, A. D. (1991). Turgor, growth and rheological gradients of cereal roots and the effect of osmotic stress. *J. Exp. Bot.* **42**, 1043–1049.

Radin, J. W., and Matthews, M. E. (1989). Water transport properties of cells in the root cortex of nitrogen- and phosphorus-deficient cotton seedlings. *Plant Physiol.* **88**, 264–268.

Rosene, H. F. (1937). Distribution of the velocities of absorption of water in the onion root. *Plant Physiol.* **12**, 1–19.

Rüdinger, M., Hierling, P., and Steudle, E. (1992). Osmotic biosensors. How to use a characean internode for measuring the alcohol content of beer. *Bot. Acta* **105**, 3–12.

Sanderson, J. (1983). Water uptake by different regions of the barley root. Pathways of the radial flow in relation to development of the endodermis. *J. Exp. Bot.* **34**, 240–253.

Schmalstig, J. G., and Cosgrove, D. J. (1990). Coupling of solute transport and cell expansion in pea stems. *Plant Physiol.* **94**, 1625–1633.

Schulze, E.-D. (1986). Carbon dioxide and water vapor exchange in response to drought in the atmosphere and in the soil. *Annu. Rev. Plant Physiol.* **37**, 247–274.

Schulze, E.-D., Steudle, E., Gollan, T., and Schurr, U. (1988). Response to Dr. P. J. Kramer's article, "Changing concepts regarding plant water relations," Volume 11, Number 7, pp. 565–568. *Plant, Cell Environ.* **11,** 573–576.

Shalhevet, J., Maass, E. V., Hoffmann, G. J., and Ogata, G. (1976). Salinity and the hydraulic conductance of roots. *Physiol. Plant.* **38,** 224–232.

Spanner, D. C. (1975). Electroosmotic flow. *Encycl. Plant Physiol., New Ser.* **1,** 301–327.

Spollen, W. G., and Sharp, R. E. (1991). Spatial distribution of turgor and root growth at low water potentials. *Plant Physiol.* **96,** 438–443.

Steudle, E. (1985). Water transport as a limiting factor in extension growth. *In* "Control of Leaf Growth" (N. R. Baker, W. J. Davies, and C. K. Ong, eds)., pp 35–55. Cambridge Univ. Press, Cambridge.

Steudle, E. (1989a). Water flow in plants and its coupling to other processes: An overview. *In* "Methods of Enzymology" (S. Fleischer and R. Fleischer, eds.), Vol. 174, pp. 183–225. Academic Press, San Diego.

Steudle, E. (1989b). Water transport in roots. *In* "Structural and Functional Aspects of Transport in Roots" (B. C. Loughman, O. Gasparikova, and J. Kolek, eds.), pp. 139–145. Kluwer Academic Publishers, Amsterdam.

Steudle, E. (1989c). Water transport in roots. *In* "Plant Water Relations and Growth Under Stress" (M. Tazawa, M. Katsumi, M. Masuda, and Y. Okamoto, eds.), pp. 253–260. Yamada Science Foundation, Osaka, and Myu, K. K. Tokyo.

Steudle, E. (1990). Methods for studying water relations of plant cells and tissues. *In* "Measurement Techniques in Plant Sciences (Y. Hashimoto, P. J. Kramer, H. Nonami, and B. R. Strain, eds.), pp. 113–150. Academic Press, San Diego.

Steudle, E. (1992). The biophysics of plant water: Compartmentation, coupling with metabolic processes, and water flow in plant roots. *In* "Water and Life: Comparative Analysis of Water Relationships at the Organismic, Cellular, and Molecular Levels" (G. N. Somero, C. B. Osmond, and C. L. Bolis, eds.), pp. 173–204. Springer-Verlag, Berlin.

Steudle, E. (1993). Pressure probe techniques: Basic principles and application to studies of water and solute relations at the cell, tissue, and organ level. *In* "Water Deficits: Plant Responses from Cell to Community" (J. A. C. Smith and H. Griffith, eds.). Bios Scientific Publishers, Oxford (in press).

Steudle, E., and Boyer, J. S. (1985). Hydraulic resistance to radial water flow in growing hypocotyl of soybeans measured by a new pressure perfusion technique. *Planta* **164,** 189–200.

Steudle, E., and Brinckmann, E. (1989). The osmometer model of the root: Water and solute relations of *Phaseolus coccineus. Bot. Acta* **102,** 85–95.

Steudle, E., and Frensch, J. (1989). Osmotic responses of maize roots. Water and solute relations. *Planta* **177,** 281–295.

Steudle, E., and Jeschke, W. D. (1983). Water transport in barley roots. *Planta* **158,** 237–248.

Steudle, E., and Tyerman, S. D. (1983). Determination of permeability coefficients, reflection coefficients and hydraulic conductivity of *Chara corallina* using the pressure probe: Effects of solute concentratins. *J. Membr. Biol.* **75,** 85–96.

Steudle, E., Zimmermann, U., and Lelkes, P. I. (1977). Volume and pressure effects on the potassium fluxes of *Valonia utricularis. In* "Transmembrane Ionic Exchanges in Plants" (M. Thellier, A. Monnier, M. Demarty, and J. Dainty, eds.), Vol. 258, pp. 123–132. CNRS, Paris.

Steudle, E., Oren, R., and Schulze, E.-D. (1987). Water transport in maize roots. *Plant Physiol.* **84,** 1220–1232.

Steudle, E., Murrmann, M., and Peterson, C. A. (1993). Transport of water and solutes across maize roots modified by puncturing the endodermis: Further evidence for the composite transport model of the root. *Plant Physiol.* (in press).

Stevenson, T. T., and Cleland, R. E. (1981). Osmoregulation in the *Avena* coleoptile in relation to auxin and growth. *Plant Physiol.* **67,** 749–753.

Taura, T., Iwaikawa, Y., Furumoto, M., and Katou, K. (1988). A model for radial water transport across plant roots. *Protoplasma* **144,** 170–179.

Tomos, A. D. (1988). Cellular water relations of plants. *Water Sci. Rev.* **3,** 186–277.

Turner, N. C., Schulze, E.-D., and Gollan, T. (1985). The responses of stomata and leaf gas exchange to vapour pressure deficits and soil water content. II. In the mesophyllic herbaceous species *Helianthus annuus. Oecologia* **65,** 348–355.

Tyerman, S. D., and Steudle, E. (1982). Comparison between osmotic and hydrostatic water flow in a higher plant cell: Determination of hydraulic conductivities and reflection coefficients in isolated epidermis of *Tradescantia virginiana. Aust. J. Plant Physiol.* **9,** 461–479.

Tyerman, S. D., Oats, P., Gibbs, J., Dracup, M., and Greenway, H. (1989). Turgor-volume regulation and cellular water relations of *Nicotiana tabacum* roots grown in high salinities. *Aust. J. Plant Physiol.* **16,** 517–531.

van den Honert, T. H. (1948). Water transport as a catenary process. *Discuss. Faraday Soc.* **3,** 146–153.

Weatherley, P. E. (1982). Water uptake and flow in roots. *Encycl. Plant Physiol. New Ser.* **12B,** 79–109.

Westgate, M. E., and Steudle, E. (1985). Water transport in the midrib tissue of maize leaves: Direct measurement of the propagation of changes in cell turgor across a plant tissue. *Plant Physiol.* **78,** 183–191.

Wyn Jones, R. G., and Gorham, J. (1982). Osmoregulation. *Encycl. Plant Physiol., New Ser.* **12C,** 35–58.

Yeo, A. R., Yeo, M. E., and Flowers, T. J. (1987). The contribution of an apoplastic pathway to sodium uptake by rice roots in saline conditions. *J. Exp. Bot.* **38,** 1141–1153.

Zhang, W. H., and Tyerman, S. D. (1991). Effect of low O_2 concentration and azide on hydraulic conductivity and osmotic volume of the cortical cells of wheat roots. *Aust. J. Plant Physiol.* **18,** 603–613.

Zhu, G. L., and Steudle, E. (1991). Water transport across maize roots: Simultaneous measurement of flows at the cell and root level by double pressure probe technique. *Plant Physiol.* **95,** 305–315.

Zimmermann, U. (1978). Physics of turgor- and osmoregulation. *Annu. Rev. Plant Physiol.* **29,** 121–148.

Zimmermann, U., and Steudle, E. (1974). The pressure-dependence of the hydraulic conductivity, the membrane resistance, and membrane potential during turgor pressure regulation in *Valonia utricularis. J. Membr. Biol.* **16,** 331–352.

Zimmermann, U., and Steudle, E. (1978). Physical aspects of water relations of plant cells. *Adv. Bot. Res.* **6,** 45–117.

III

Flux Control at the Soil–Organism Interface

9

Patterns and Regulation of Organic Matter Transformation in Soils: Litter Decomposition and Humification

W. Zech and I. Kögel-Knabner

I. Introduction

Soil organic matter represents a major component of the world surface carbon reserves. The total carbon in dead organic matter in the forest floor and in the underlying mineral soil has been estimated to be 1450×10^9 t C, exceeding the amount stored in living vegetation by a factor of 2 or 3 (Schlesinger, 1977; Meentemeyer et al., 1982; Jenkinson, 1988). Therefore, mineralization, litter decomposition, and humification represent important processes in the terrestrial carbon cycle. Schlesinger (1990) estimated that about 0.7% of the annual terrestrial net primary production is sequestered into carbon of refractory humic substances. Organic matter of forest soils is composed of a mixture of above- and belowground plant residues (primary resources), microbial residues (secondary resources), and humic compounds (Swift et al., 1979). Humic compounds are formed concomitantly during the microbial decomposition of primary and secondary resources. The conceptual view of soil organic matter in the present study is that of a continuum ranging from fresh plant litter to humic substances, the final products of humification. A complete separation of plant remains and humic compounds is not possible, due to the fact that organic matter at all stages of degradation and humification is present simultaneously in natural soils, albeit in different amounts. Undisturbed soils have a characteristic distribution of undecomposed and decomposed litter and humus (Jenkinson, 1988). As the horizon sequence in forest humus is directly related to the degree of humification the formation pathway of humic substances can be delineated by following the chemical evolution observed concurrently in plant-derived macromolecules and in humic substances with depth. It is generally accepted now that humification in soils involves a number of processes and that more than one humification process is active in a certain soil. The predominance of specific formation pathways of humic substances in a certain environment presumably depends on the type of precursor material and on the environmental conditions (Ertel *et al.*, 1988; Oades, 1989). The objective of the present study is to delineate the major decomposition and humification processes operating in forest soils.

II. The Soil Carbon Cycle

The parent material for the decomposition and humification processes is derived from plant residues (primary resources) and microbial and animal products (secondary resources). The composition of forest humus layers depends on the input and composition of primary and secondary resources and on the decomposition rate of the different compound classes. The entire process of litter decomposition involves leaching, comminution, and catabolism (Swift *et al.*, 1979). Figure 1 summarizes the different processes involved in litter decomposition. Fragmentation of plant litter by soil animals strongly affects the release of dissolved organic carbon (Gunnarsson *et al.*, 1988). The individual processes mentioned above are reiterated at different stages of decomposition (Eijsackers and Zehnder, 1990), leading to the accumulation of organic matter in distinctive forest floor horizons. They represent various stages of decomposition and humification, from fresh litter material (L horizon) and transitional horizons like the fermentation layer (Of) to completely humified horizons (Oh, Ah). Depending on the decomposition rate one of the forest humus types mull, moder, mor, or tangelmor develops (for definitions of profiles see Figs. 2A and B). Carbon turnover rates are controlled by three main groups of factors: (1) the site-specific environmental factors like temperature and water regime, (2) the interactions with the soil matrix which result in a definite resource quality (chemical composition of litter), and (3) the nature of the decomposer community (Swift *et al.*, 1979; Anderson, 1988).

With depth in the forest floor the organic matter is gradually mineralized and humified resulting in a mixture of macromorphologically identifiable plant residues and morphologically unstructured humic compounds. These morphological changes are directly related to the degree

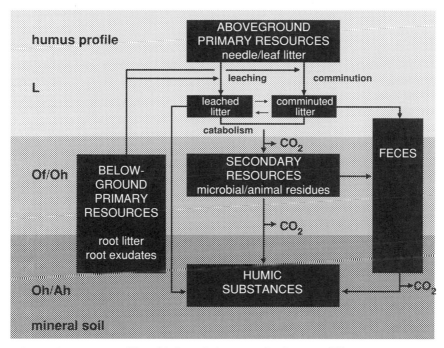

Figure 1 Processes of litter biodegradation operating between different compartments of carbon in forest soils.

of humification (Angehrn-Bettinazzi *et al.*, 1988). However, chemical transformation and transport of the organic material are possible through the action of the soil fauna, especially earthworms, which are of major importance in mull humus profiles (Anderson, 1988; Schaefer and Schauermann, 1990). Additionally, one must account for the input of root litter and the transport of organic material into the uppermost mineral soil horizon, the A horizon (Blume, 1965). In deeper humus horizons and in the mineral soil the contribution of root litter is significant. Organic matter is transported in the mineral soil horizon by the action of animals and also by leaching and precipitation of water-soluble organic matter. Babel (1975) and Ohta *et al.* (1986) point out that the transport of particulate organic matter into the A horizon is of importance in some forest soils.

Numerous definitions exist for humic substances (Waksman, 1938; Laatsch, 1957; Stevenson, 1982; Oades, 1988; Hayes *et al.*, 1989a; Theng *et al.*, 1989). Most of them are operational definitions, such as the definition based on solubility characteristics of the fulvic acid, humic acid, and humin isolates. Other definitions require humic substances to be of a

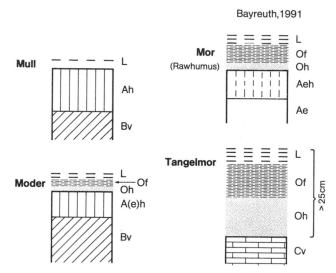

Bayreuth,1991

Figure 2A (A) Morphological description of the main forest humus types in Germany (L, litter; Of, fermentation layer; Oh, completely humified surface layer; Ah, humic mineral surface layer; Ae, albic horizon; A(e)h and Aeh, transitions between Ae and Ah; Bv, cambic horizon; Cv, weathered parent material).

definite structure (e.g., aromatic, phenolic-rich), because they are formed according to a presumed humification process, or of a nondefined structure. The basis for these definitions may be the result of limitations in analytical techniques suitable for the characterization of humic substances. The number of definitions reflects the problems associated with the separation of plant, animal, or microbial residues from humic substances. In the present work humic substances are viewed as an integrated component of soil organic matter. According to this concept, only two major compartments of carbon exist in soil. One is the "living" compartment, consisting of plant roots, animals, and microorganisms, and the other is the "dead" compartment or soil organic matter compartment. The soil organic matter compartment itself is composed of plant, animal, and microbial residues at different stages of decomposition and of humic substances, but excluding live roots, animals, and microorganisms. With increasing depth and decomposition in forest soils, the relative amount of morphologically and chemically identifiable detritus decreases and the amount of amorphous material formed during humification increases.

(Kögel, 1986; Ziegler *et al.*, 1986; Kögel *et al.*, 1988b). Hydroxy fatty acids derived from cutin and suberin can be determined after saponification or transesterification (Kögel-Knabner *et al.*, 1989; Ziegler and Zech, 1990; Riederer *et al.*, 1993). Other chemical degradative methods that have been used for the structural investigation of humic substances, e.g., permanganate oxidation or sodium amalgam reduction, do not permit an unambiguous structural characterization. The major problems associated with these methods are low yields of degradation products and generation of artifacts (Norwood, 1988). Therefore, the significance of aromatic and phenolic structures as building blocks of humic substances has been overemphasized.

Analytical pyrolysis has found wide application in structural studies of plant litter and soil organic matter. A thermal energy pulse applied to a macromolecule causes fracture of weaker bonds and yields pyrolysis products characteristic of the original structure. For studies of humic substances a widely used technique is flash pyrolysis with gas chromatographic separation of the pyrolysis products before identification with electron impact (EI) mass spectrometry (Py-GC-MS) (Saiz-Jimenez and De Leeuw, 1986). In pyrolysis-MS the products pass directly into a mass spectrometer ion source operating with low EI energies (10–15 eV) or with field ionization (Py-FIMS), so that mainly molecular ions are recorded (Bracewell and Robertson, 1987; Hempfling *et al.*, 1988). Secondary reactions are avoided by pyrolysis in a rapid stream of inert gas or in vacuum. Pyrolysis is especially useful for compound classes that are difficult to analyze by chemical degradative methods (e.g., cutan, suberan, lignin) and may be applied to whole soils as well as to isolates (Saiz-Jimenez and De Leeuw, 1986; Schnitzer, 1990; Hempfling *et al.*, 1991). However, the results obtained from analytical pyrolysis are essentially qualitative and should be combined with other analytical techniques in order to avoid misleading conclusions (Nip *et al.*, 1987).

Recent developments in ^{13}C nuclear magnetic resonance (NMR) have provided the possibility of obtaining well-resolved, chemically informative spectra of solid samples. The technique of cross-polarization magic-angle-spinning (CPMAS) has overcome the problems associated with line broadening and low signal intensity in solid-state NMR. The technique which is carried out nondestructively on dry ground samples provides an estimate of bulk soil organic matter chemical composition. The problems associated with lack of resolution and a low signal-to-noise ratio for CPMAS ^{13}C NMR of mineral soil samples can be overcome by de-ashing with HCl/HF without major structural changes in the organic chemical composition of the carbon-enriched samples (Preston *et al.*, 1989). If the conditions for the CPMAS experiment in relation to the relaxation behavior of the ^{13}C nuclei are chosen carefully, a semiquantitative interpretation of the NMR data seems possible (Wilson, 1987; Voelkel, 1988).

The relaxation behavior can also be exploited for structural investigations using the dipolar dephasing technique. With this pulse sequence the contribution of protonated and nonprotonated carbons or of mobile and rigid alkyl-carbon structures to the signal intensity of ^{13}C NMR spectra can be calculated (Wilson, 1987). The dipolar dephasing technique has been successfully used for the structural characterization of complex plant biopolymers, like lignin and tannins (Wilson and Hatcher, 1988), and forest soil humic acids (Kögel-Knabner et al., 1991).

IV. Composition and Distribution of the Input to Humification

A. Primary Resources

The basis of structural investigations of forest soil organic matter is a detailed knowledge of the composition of the parent litter material in morphological and molecular terms. Forest litter layers are composed almost exclusively of aboveground plant litter including leaves or needles, twigs, fruits, and buds. In deeper humus horizons and in the mineral soil the contribution of root litter is significant. In some ecosystems belowground inputs from fine root turnover may contribute more to the organic matter decomposition cycle than aboveground litterfall (Raich and Nadelhoffer, 1989). According to McClaugherty et al. (1984) fine root litter input, especially significant in the Oh and Ah horizons, is similar in magnitude to foliar litter production. Vogt et al. (1986) estimated the input of root litter to be 20 to 50% of the total C input to temperate forest soils. High amounts of partially decomposed root residues were found in different forest soils after density fractionation (Beudert et al., 1989; Preston, 1992). Woody debris is also an important component of forest ecosystems, comprising 24 to 39% of the total organic matter input (Harmon et al., 1986).

Results from ^{13}C NMR spectroscopy show that total carbon in forest litter layers consists of about 20% alkyl carbon (chemical-shift range 0–50 ppm), 55% O-alkyl carbon (50–110 ppm), 20% aromatic carbon (110–160 ppm), and 5% carboxyl carbon (160–200 ppm; Fig. 4). The ^{13}C NMR spectra of litter from different tree species and different sites and the overall chemical composition derived from the spectra are remarkably similar (Wilson et al., 1983; Hempfling et al., 1987; Kögel et al., 1988a). The signal intensity in the O-alkyl region is mainly due to polysaccharides (cellulose and hemicelluloses) and lignin side chains. The aromatic signal intensity can be attributed mainly to lignin and also to

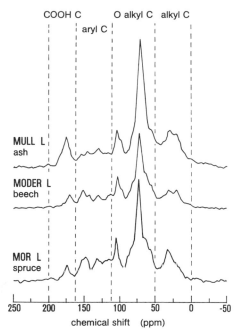

Figure 4   CPMAS ^{13}C NMR spectra of the bulk samples of three different litter layers under European ash (*Fraxinus excelsior* L.; mull), European beech (*Fagus sylvatica* L.; moder), and Norway spruce (*Picea abies* (L.) Karst.; mor).

tannins. Lignin is a complex three-dimensional polymer biosynthesized by dehydrogenative polymerization of three phenylpropane monomers (coniferyl, sinapyl, and *p*-coumaryl alcohol). The monomers are linked together by several different carbon–carbon and ether linkages, most of which are not readily hydrolyzable. The relative proportions of each monomeric unit, commonly referred to as guaiacyl/vanillyl, syringyl, and *p*-hydroxyphenyl units, in the lignin of a particular plant species depend on its phylogenetic origin (Sarkanen and Ludwig, 1971). The lignin of hardwoods such as beech consists of about equal proportions of guaiacyl and syringyl monomers; softwood lignin is composed mainly of guaiacyl units. Extractable lipids account for about 50% of the alkyl carbon (Ziegler, 1989; Ziegler and Zech, 1989). About 60% of the total carbon signal intensity of the NMR spectra of the litter layers can be identified. The carbon unaccounted for by wet-chemical methods is found mainly in the alkyl-carbon region and is most likely due to aliphatic biomacro-molecules of the protective layers in higher plants (cuticle and periderm), i.e., the polyesters cutin and suberin, and cutan and suberan, the highly aliphatic biomacromolecules discovered recently in plant cuticles (Tegelaar *et al.*, 1989). Cutin consists of characteristic long-chain (mainly C_{16} and C_{18}) hydroxy and epoxy fatty acids, which are cross-linked to form a three-dimensional amorphous polyester-type biomacromolecule. Suberin is composed of a mixture of hydroxy fatty acids and phenolic constituents (Kolattukudy, 1981). Woody debris consists mainly of O-alkyl carbon from polysaccharides and aromatic carbon from lignin (Preston *et al.*, 1990). The bark of woody debris is rich in suberin and tannins. Although considerable progress has been made in recent years, data on the chemical composition of plant litter are still lacking for many tree species and quantitative data on composition of roots are lacking almost completely.

B. Secondary Resources

Secondary resources are composed mainly of fungal and bacterial cell wall remains. Animal residues are quantitatively of minor importance (Swift *et al.*, 1979). Microbial standing biomass contributes 1–4% of the organic carbon in soils (Kassim *et al.*, 1981). Microbial residues are even more complex in composition than are plant remains. Bacterial cell walls are composed of peptidoglycan (murein), lipids, and lipopolysaccharides, containing a variety of unusual monomers (Rogers *et al.*, 1980). Many bacteria produce extracellular polysaccharides consisting of neutral or acidic sugar monomers (Hepper, 1975). The cell walls of fungi contain chitin, chitosan, cellulose, and a variety of noncrystalline polysaccharides, mainly mannans and glucans. Despite the heterogeneity of fungal cell walls with respect to the variety of macromolecules present, they can be subdivided into (1) an inner layer of chitin, glucans, or cellulose, forming the skeletal wall components, which are embedded in various matrix polymers (mainly glucans) and (2) an outer layer, composed of noncrystalline polysaccharides (Peberdy, 1990). It should be pointed out that the outer layer is soluble in dilute alkali leaving the inner layer as a residue (Wessels and Sietsma, 1981). Recent studies indicate that microbial residues also contain macromolecular alkyl-carbon structures (Zelibor *et al.*, 1988; Tegelaar *et al.*, 1989) in addition to extractable lipids. Melanins are minor cell wall components in many fungi but play a significant role for protection against lysis by enzymes and irradiation (Bell and Wheeler, 1986; Peberdy, 1990). Information on the chemical composition of secondary resources is mainly qualitative. Baldock *et al.* (1990) isolated mixed bacterial and fungal cultures from an agricultural soil and found that the bacterial materials contained more alkyl and carboxyl carbon but less O-alkyl and acetal carbon than the fungal materials. The microbial materials isolated from a forested Typic Dystrochrept were similar in composition to the bacterial materials described by Baldock *et al.* (1990), suggesting that mainly bacterial material was isolated (Bell and Wheeler,

1986). Certainly, more information is needed on the chemical composition of secondary resources.

V. Decomposition and Humification Processes

The current concepts on formation of humic substances can be divided into abiotic condensation models and biopolymer degradation models. A review of both types of models is given by Hedges (1988) and Hatcher and Spiker (1988). In the degradative schemes, recalcitrant plant and microbial polymers are viewed as the precursors of humic substances, which are formed progressively via the humin, humic acid, and finally fulvic acid step. The biopolymer degradation model assumes that recalcitrant plant and microbial biomacromolecules are selectively preserved during biodegradation, while the labile components of plants and microorganisms are completely mineralized. The recalcitrant biomacromolecules, which form the humin fraction, are further oxidized to form humic and fulvic macromolecules. The condensation models propose that humic substances evolve from the polymerization of low-molecular-weight organic precursors, generated during biodegradation of plant and microbial residues. In this type of model the formation pathway progresses from fulvic acids via humic acids to humin. The abiotic condensation models proposed for soils include the polyphenol model and the melanoidin or browning reaction. In the polyphenol model a variety of phenols of plant or microbial origin are assumed to be oxidized to quinones, which condense with each other or with amino acids and ammonia to form humic macromolecules. Simple sugars and amino acids are the initial monomers for the melanoidin reaction. They condense to form dark nitrogen-rich humic-like polymers. The melanoidin hypothesis has been criticized for several reasons. The precursor molecules are present in soils only in very low quantities and the reaction proceeds very slowly at acid or neutral conditions and natural temperatures. Therefore the precursor molecules are not likely to persist long enough for a reaction to take place. Also, the structural characteristics of synthetic melanoidins as determined by ^{13}C NMR spectroscopy are different from natural humic substances (Hedges, 1988). Both types of models assume an increase of the molecular weight from fulvic acids, through humic acids to humin. It is obvious that these concepts, which are based exclusively on the conventional classification according to solubility in acids and bases, are not well in line with the previous considerations for the "soil organic matter continuum."

The litter input to forest soils is mineralized in a two-stage process. A rapid initial phase of plant litter (primary resources) decomposition and transformation is followed by a second phase. The preferential decomposition of easily mineralizable materials leads to the selective preservation of refractory plant or microbial components (Hatcher and Spiker, 1988). In the second phase the microbial biomass and its metabolic products (secondary resources) which were built up during the initial phase and recalcitrant selectively preserved compounds are decomposed (Swift *et al.*, 1979; Jenkinson, 1988). The biodegradation of plant or microbial constituents can be restricted because of several mechanisms. A certain compound may possess an intrinsic recalcitrance due to its chemical structure. Another possibility is that the compound is protected from microbial attack by recalcitrant compounds, such as lignin or melanins. These selectively preserved materials can be incorporated directly in humic compounds or undergo further transformation reactions. Once reaching the mineral soil, the partly degraded organic matter can be stabilized due to precipitation or sorption to the mineral soil matrix. Organic matter in aggregated soils can be physically protected from microbial attack (Oades, 1989). Transformation of organic matter to humic compounds may also be promoted due to a catalytic effect of soil minerals. Carbon from a variety of different plant and microbial sources contributes to the formation of humic matter. The formation of forest humus is viewed here as the result of a combination of different processes which can be summarized as resynthesis by microorganisms, selective preservation, and direct transformation (Fig. 5). However, one should bear in mind that the intensity of the individual processes can be different depending on the type of soil environment. The sum of these individual humification processes is expressed in the morphology of different forest humus types.

Several forest soil profiles with the humus types mull, moder, mor, and tangelmor have been investigated in great detail by a combination of wet-chemical degradative methods and ^{13}C NMR spectroscopy (Hempfling *et al.*, 1987; Kögel-Knabner *et al.*, 1988, 1990; Zech *et al.*, 1985, 1990a,b, 1992; Ziegler *et al.*, 1992). Selected results are shown in Fig. 6 referring to forest litter derived from European ash (*Fraxinus excelsior* L.), European beech (*Fagus sylvatica* L.), and Norway spruce (*Picea abies* (L.) Karst.) ($n = 7$). The L horizons which show a mean C-to-N ratio of 43 consist of 56% O-alkyl C, 18% aryl C, 19% alkyl C, and 7% carboxyl C as identified by ^{13}C NMR spectroscopy. With progressive decomposition from the L to the A horizons (C-to-N of 21) O-alkyl C decreases to 42% of total organic carbon, and alkyl C (to 30%) as well as carboxyl C (to 11%) increase, whereas aromatic C remains more or less constant over the whole soil profiles (Fig. 7). Hydrolyzable polysaccharides exhibit a decomposition pattern similar to O-alkyl C, but at a lower level, decreasing from 40 to 14% of organic matter. CuO lignin

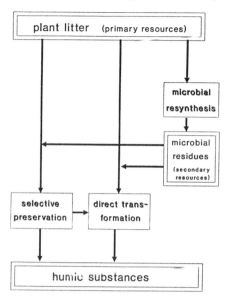

Figure 5 Schematic representation of the different humification processes operating on the transformation of plant litter to humic compounds.

amounts to 17% in the L and only 6% in the A horizons, thus behaving quite differently from the aromatic C as determined by CPMAS ^{13}C NMR spectroscopy. This indicates that with progressive decomposition an increasing part of total aryl C present is accounted for by nonlignin aromatic structures. The mineralization of lignin is accompanied by a chemical alteration of the remnant lignin molecule. The acid-to-aldehyde ratios of the vanillyl unit $(ac/al)_V$ are indicative of the degree of side-chain oxidation in the lignin molecule, because cleavage of C_α-C_β bonds and/or oxidation of C_α by white-rot fungi lead to an increased production of phenolic acids with respect to the aldehydes obtained from CuO oxidation of decayed lignin. As shown in Fig. 8, $(ac/al)_V$ increases considerably with depth in the humus profiles; the increase from Oh to the mineral Ah being always more spectacular than that from L to Oh. This general pattern of lignin degradation is observed regardless of the type of parent litter material and site conditions, although a high biological activity in soils in combination with the presence of clay leads to the incorporation of less modified plant materials into soil organic matter and thus to significantly lower $(ac/al)_V$ ratios in mull as compared to moder and mor humus type A horizons. This finding holds true for the plant residues and organomineral complexes as well (Fig. 9).

CPMAS ^{13}C NMR spectroscopy shows that total carbon in forest soil

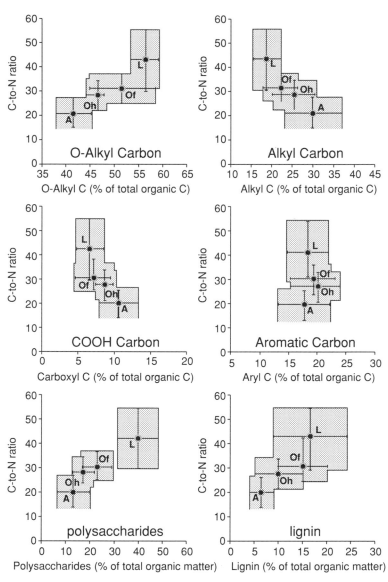

Figure 6 Chemical changes of organic matter with progressive decomposition from the L to the A horizons, expressed as the mean of seven Bavarian forest soils under ash, beech, and Norway spruce with mull, moder, and mor humus types.

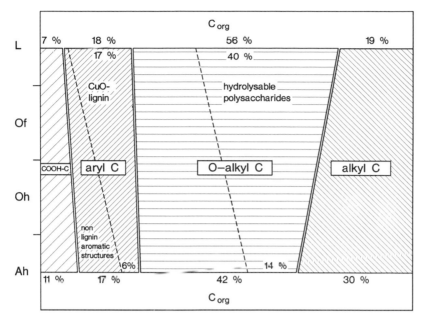

Figure 7 Average changes in the carbon species distribution with progressive decomposition from the L to the Ah horizons in Bavarian forest soils ($n = 7$) under ash, beech, and Norway spruce.

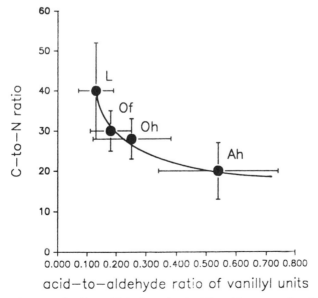

Figure 8 Changes of acid-to-aldehyde ratios (ac/al)$_V$ with progressive decomposition from the L to the A horizons, expressed as the mean of seven Bavarian forest soils under ash, beech, and Norway spruce with mull, moder, and mor humus types.

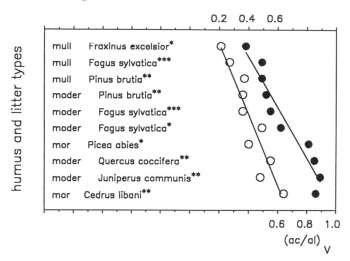

Figure 9 Acid-to-aldehyde ratios (ac/al)$_V$ of the residual lignin in plant residues and organomineral complexes as determined for the A horizons of different mull, moder, and mor humus profiles that have developed in the temperate and mediterranean climate, and under controlled laboratory conditions. *Temperate climate. **Mediterranean Climate. ***Laboratory experiments. ○, Plant residues; ●, organomineral complexes.

organic matter contains about 10–30% aromatic carbon (Hempfling *et al.*, 1987; Kögel-Knabner *et al.*, 1988; Zech *et al.*, 1992). The major aromatic components of plant litter are lignin and tannins. Evidence of the fate of lignin during decomposition in soils is obtained from chemical degradation and determination of the monomers released, e.g., CuO oxidation (Ertel and Hedges, 1984, 1985; Hedges *et al.*, 1988), from analytical pyrolysis (Hempfling *et al.*, 1987; Hempfling and Schulten, 1990; Saiz-Jimenez and De Leeuw, 1984), and from ^{13}C NMR spectroscopy (Zech *et al.*, 1987b). Lignin is attacked preferentially at the ether linkages (Fig. 10). Carbon–carbon-linked lignin structural units, such as pinoresinol, phenylcoumaran, and biphenyl units, are more resistant to biodegradation than the ether-linked structures (Haider *et al.*, 1985). In most cases woody angiosperm lignin is decomposed at a higher rate compared to coniferous lignin. Biodegradation of lignin in forest soils leads to a modification of the remnant lignin polymer, which has a lower content of intact lignin moieties due to ring cleavage and a higher degree of side-chain oxidation (Kögel-Knabner *et al.*, 1988; Ziegler *et al.*, 1986; Kögel *et al.*, 1988b). This is reflected by changes in the yields of CuO-oxidation products as a measure of intact lignin moieties and their acid/aldehyde ratios, which can be used as an indication for the degree of side-chain oxidation of the remnant lignin with depth in forest soils (Ertel and Hedges, 1984, 1985; Hedges *et al.*, 1988; Kögel-Knabner *et al.*, 1988;

Bayreuth,1992

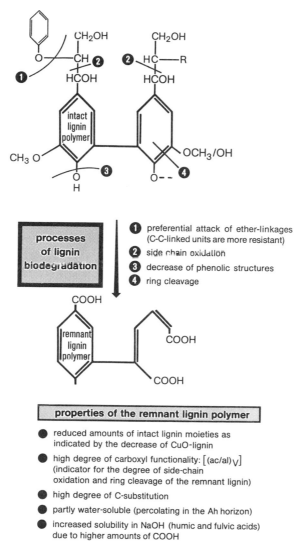

Figure 10 Processes of lignin biodegradation and properties of the remnant lignin polymer.

Ziegler *et al.*, 1986). Pyrolysis–field ionization–mass spectrometry shows that the more recalcitrant C-C-linked moieties are selectively preserved compared to the ether-linked moieties in forest soils (Hempfling and Schulten, 1989, 1990). Chemically altered, polymeric, water-soluble lignin fragments produced during microbial degradation of lignin (Ellwardt *et al.*, 1981; Seelenfreund *et al.*, 1990) contribute significantly to the aromatic-carbon structures in mineral soil horizons. In the mineral horizons of forest soils lignin structural units characterized by a strong chemical alteration accumulate, due to the precipitation or sorption of highly decomposed water-soluble lignin fragments (Kögel *et al.*, 1988b).

The humic acid fraction isolated from fresh litter of European beech and Norway spruce contains mainly aromatic C derived from lignin and tannin structures which are both partly extractable by alkaline solvents. The most prominent feature of the NMR spectra is the decrease of phenolic C and methoxyl C with increasing degree of humification. Simultaneously the signal intensity at 130 ppm in the ^{13}C NMR spectra increases. Detailed structural assignments for this signal can be obtained by measuring the percentage of signal intensity from protonated and nonprotonated C by dipolar dephasing ^{13}C NMR spectroscopy. Then the percentage of phenolic C (aryl-O) and C-substituted aromatic C (aryl-C) can be calculated from the NMR data. The percentage of nonprotonated aromatic C remains constant and the C-substituted aromatic-C fraction increases when humification proceeds. We also find decreasing yields of lignin-derived CuO-oxidation products and an increasing degree of oxidative decomposition (side-chain oxidation) in the lignin-derived structures as determined by acid/aldehyde ratios. The higher number of carboxyl groups due to ring cleavage and side-chain oxidation results in an increasing percentage of aromatic C extractable with alkaline solvents.

Assuming that lignin is the primary precursor of the aromatic-C components of humic acids in forest soils (Ertel *et al.*, 1988; Hayes *et al.*, 1989b), then the lignin structure is altered considerably during humification resulting in lignin-derived aromatic structures with a high degree of C substitution and carboxyl functionality. The complex picture obtained for the structural changes of lignin in forest soils leads to the conclusion that lignin undergoes several humification processes. The microbial degradation of lignin leads to the mineralization of lignin with concomitant ring cleavage and side-chain oxidation. According to Martin and Haider (1986) and Haider (1991) the processes of lignin mineralization are extracellular and lignin does not provide a carbon or energy source for microorganisms. Therefore a microbial resynthesis of lignin carbon does not occur. The more refractory lignin components are selectively preserved. There is evidence for a direct transformation of lignin, resulting in a decrease of the relative percentage of O-aryl C and a relative increase of C-substituted aromatic structures with depth.

The leaves and barks of several tree species are high in tannins (Benner *et al.*, 1990; Wilson and Hatcher, 1988). Tannins are heterogeneous compounds and provide problems for analysis by wet-chemical methods. Therefore, information on the structural changes they undergo during humification is scarce. Nonetheless, evidence is provided by dipolar dephasing ^{13}C NMR for the presence of tannins in the humic acid fractions extracted from forest soils. From these data it can be estimated that tannins contribute up to about 20% to the total aromatic-carbon content of forest soils. Certainly, the fate of tannins during humification needs further attention.

Polysaccharides are the major component of forest litter. The plant litter input of polysaccharides into forest soils is comprised of about 20–25% cellulose and 20–30% hemicelluloses (Kögel *et al.*, 1988a). The major structural difference is that cellulose is a crystalline polymer of glucose, whereas hemicelluloses are composed of various pentoses and hexoses. The amounts and types of monomers are different for different tree species (Fengel and Wegener, 1984). Plant carbohydrates provide the major carbon and energy source for microorganisms. Most of the C of plant carbohydrates is therefore mineralized within several months or years (Martin and Haider, 1986). The major part of the remaining C is present as hydrolyzable carbohydrates (Stott and Martin, 1990). In soils, the plant-derived polysaccharides are decomposed preferentially compared to lignin (Haider, 1986a). This is reflected by the complete loss of cellulose with depth. The cellulose content in forest soils decreases with depth from 20–25% in the litter layer to less than 3% in the A horizon (Hempfling *et al.*, 1987; Kögel-Knabner *et al.*, 1988). Similar results are obtained in laboratory experiments (Haider, 1986a,b, 1991; Ziegler, 1990). The concurrent increase of carbohydrates from microbial sources points to microbial resynthesis as the major humification process (Kögel-Knabner *et al.*, 1988, 1990; Baldock *et al.*, 1990). The polysaccharides comprise a high proportion of the O-alkyl carbon in the humified horizons (Oh, Ah). There is still a lack of data providing evidence for the processes which lead to a stabilization of these polysaccharides.

CPMAS ^{13}C NMR spectroscopy has shown that forest soil organic matter contains significant amounts (15–30%) of alkyl carbon (Hempfling *et al.*, 1987; Hempfling and Schulten, 1990; Kögel-Knabner *et al.*, 1988, 1990; Preston *et al.*, 1990). The major alkyl-C components of forest soil organic matter are extractable and bound lipids (Ziegler, 1989), the plant polyesters cutin and suberin, and other nonsaponifiable aliphatic components (Kögel-Knabner *et al.*, 1989, 1992a; Tegelaar *et al.*, 1989). The lipid fraction of forest soil organic matter most likely originates from microbial resynthesis (Stott and Martin, 1990). The non-lipid alkyl components constitute a major fraction of soil organic matter and associated humic substances. Cutin is decomposed or transformed in forest soils and does not accumulate with depth (Kögel-Knabner *et al.*, 1989). This has been confirmed in litter-bag experiments in the field as well as in laboratory decomposition experiments (Kögel-Knabner *et al.*, 1991; Ziegler and Zech, 1990).

Analysis by solid-state CPMAS ^{13}C NMR and Curie-point pyrolysis–gas chromatography–mass spectrometry indicated that the alkyl-carbon moieties are being altered significantly with increasing depth and decomposition in the soil profile. The NMR data show that the aliphatic structures in forest soil organic matter can be assigned to rigid carbon moieties and mobile carbon moieties (Kögel-Knabner et al., 1992a). Mobile carbons are found in compounds which are near their melting point or in gel-like structures. They have also been found in microbial residues isolated from soils (Baldock et al., 1990; Wilson et al., 1988; Kögel-Knabner et al., 1992b). With depth in the soil profiles investigated the mobile components are lost but the rigid aliphatic components appear to be selectively preserved. It was hypothesized that the mobile and rigid carbon types are possibly associated with different types of macromolecules (Kögel-Knabner et al., 1992b). The polyesters cutin and suberin from leaves, barks, and roots show a high proportion of mobile carbon structures, whereas the resistant nonsaponifiable aliphatic biomacromolecule which has been identified recently in the cuticles of several plants (Nip et al., 1986; Tegelaar et al., 1989, 1993) is composed almost exclusively of rigid carbon structures. The aliphatic materials in the humified soil horizons bear no resemblance to the resistant aliphatic nonsaponifiable biomacromolecules in fresh leaf cuticles. This lack of resemblance is probably due to the fact that selective preservation of resistant, nonsaponifiable plant macromolecules is not the dominant process leading to the accumulation of alkyl-carbon moieties in forest soil organic matter. Alternatively, it seems possible that the structural differences observed between the alkyl-carbon moieties in forest litter and humified soil horizons are due to a direct transformation of the material during humification resulting in increased cross-linking and therefore reduced mobility.

VI. Control of Organic Matter Transformations

Soil organic matter transformations represent mainly microbially mediated processes which are influenced by different categories of variables (Swift et al., 1979). Most important are the site-specific variables temperature, soil water regime, and available nutrients (e.g., N, P). The decomposition of plant debris is retarded at low temperature or at persistent water logging, and a net surplus of organic matter can accumulate under swamp conditions even in the tropics (Driessen and Dudal, 1989). There are only few field-related studies concerning the influence of site-specific parameters on humification. Since in advanced stages of humification aromaticity increases, Zech et al. (1989) and Zech and Haumaier (1989) studied the relationship between site-specific variables and the relative percentage of aromatic-carbon contents of the soil organic matter in eight soils of temperate, mediterranean, and tropical climates. They found that aromaticity (and thus probably intensity of humification) is significantly controlled by pH, C-to-N ratio, and the temperature regime. Using multiple regression analysis it could be shown that about 73% of the variation of aromaticity is due to changes of the temperature/precipitation ratio only.

Field experiments and laboratory studies (Ellenberg *et al.*, 1986; Wilhelmi and Rothe, 1990) showed that the rate of CO_2 production strongly depends on temperature. According to van't Hoff's law an increase in the mineralization of soil organic matter by a factor of about 2–3 can be expected when temperature rises by 10°C. In a laboratory study, raising the temperature from 5 to 22°C accelerated the decay of beech litter (Fig. 11) by a factor of 1.5 in aerobic microcosms under favorable moisture conditions (50–70% of maximum water-holding capacity). At 32°C the mineralization of organic matter was diminished 0.8-fold probably due to the fact that decomposing aerobic microorganisms develop optimally between 20 and 30°C (Wilhelmi and Rothe, 1990). Under water stress microbial activity is limited. Excess of water leads to anaerobic conditions, and oxygen deficiency has been shown to reduce beech litter decomposition by a factor of 1.2 to 2.3 at the same temperature regime (Fig. 11). Growth optimum of anaerobic microorganisms, e.g., the very common *Clostridium* species, ranges from 30 to 40°C (Schlegel, 1991). Therefore,

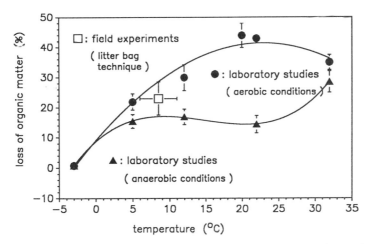

Figure 11   Temperature-dependent loss of organic matter from beech litter (mineralization) during the first year of (exclusively microbial) decomposition in the field (calculated from Heath *et al.*, 1966; Mommaerts-Billiet, 1971; Anderson, 1973; Edwards and Heath, 1975; Herlitzius and Herlitzius, 1977; Jörgensen, 1987) and under aerobic and anaerobic conditions in the laboratory.

the maximum of litter decay at low E_h was observed at 32°C, the highest temperature of the experiment. The data of litter-bag experiments in the field under naturally varying temperature and moisture conditions fit well into the results of controlled laboratory studies (Fig. 11).

In forest ecosystems the environmental variables (both abiotic and biotic) interact to such an extent that the separate effect of each is difficult to evaluate (Wallwork, 1976). The site conditions (climatic and soil factors) and the resulting quality of the litter material determine the composition of the decomposer community which in turn determines the amounts and distribution of carbon in soil. The humus types mull, moder, and mor are closely related to the composition of the soil biota (Schaefer and Schauermann, 1990). The extremely unfavorable (because cold, wet, and acid) humus type mor is characterized by a low biological activity (low zoomass, dominated by the mesofauna; fungi dominating over bacteria) leading to the formation of thick organic layers which overlie the mineral A horizon with a very sharp boundary. In the strongly contrasting mull humus type representing most favorable conditions both fungi and bacteria are active as well. The major portion of the high zoomass in mull soils is often constituted by earthworms which act as powerful driving variables in the redistribution of soil carbon (Anderson, 1988; Anderson *et al.*, 1989). Despite of these differences between mull, moder, and mor the results of our studies support the idea that the pathways of soil organic matter transformation are very similar and humic substances formation follows primarily the biopolymer degradation model—even under different site conditions. The mechanisms of control preferentially regulate the intensity of the organic matter transformation and less the general line of the pathways. In Fig. 12 the following regulations can be identified:

A. Quality and quantity of the above- and belowground primary resources are controlled by climatic and site factors; also by human activities (e.g., establishing monocultures).

B. The intensity of the microbial transformation depends on the quality and quantity of primary resources and climatic plus soil factors regulating the amount of mineralized end products (CO_2, H_2O, CH_4, etc.) as well as the amount of recalcitrant plant and microbial polymers. Conditions favorable for rapid mineralization are high temperature, optimal pH and oxygen contents, adequate nutrient supplies, no restriction of biodegradation due to high lignin contents, precipitation, sorption, or intercalation to the mineral soil matrix.

C. Less information is available about catalytic effects on transformation of primary and secondary resources into humic substances (for instance due to Mn, Fe, Al, Si). According to laboratory experiments

Bayreuth,1992

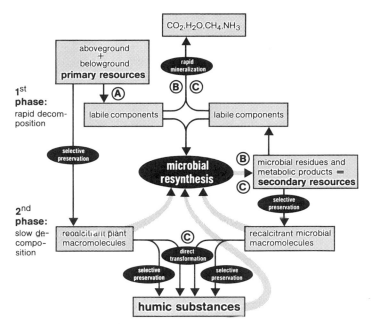

Figure 12 Processes of soil organic matter transformations and their control by climatic and soil factors. (A) Control of the quality and quantity of the primary resources, (B) control of the microbial activities, (C) possible catalytic effects on direct transformation and microbial activity.

these elements promote the direct transformation of low-molecular-weight organic precursors to humic macromolecules. Catalytic effects probably stimulate also microbial activity.

Although soil animals contribute much less than microorganisms to the total soil respiration, the macrofauna (particularly earthworms, isopods, and diplopods) plays a primary role in comminuting surface litter, incorporating the fragmented products into the mineral soil, thus bringing organic material in close association with mineral particles, and lastly converting both to organomineral complexes. Figure 2B (bottom) illustrates that beech litter–artificial soil mixtures which were incubated in the presence and absence of litter-consuming earthworms *Eisenia fetida* (SAV.) under controlled laboratory conditions (Ziegler and Zech, 1992) differ significantly from each other in their structural properties: The residual material obtained after decomposition with *E. fetida* consists mainly of stable aggregates where the mineral matter is intimately associated with organic debris (crumby structure), whereas the worm-free treatment represents a very loose side-by-side arrangement of mineral particles and coarse leaf fragments. The earthworms bound most parts of the total organic and mineral matter of beech litter and unstructured artificial soil in stable aggregates (organomineral complexes). This effect observed within a 446-day period is considerable since *E. fetida,* which can be classified into the ecological group of épigées in the sense of Bouché (1977), prefers organic-rich materials as the natural environment. Consequently, for the endogées or anécique species living in the mineral soil an even higher stabilizing efficiency should be expected under similar conditions.

In the A horizons of mull soils more than two-thirds of the total organic carbon is sequestered in organomineral complexes. In contrast, the less active moder and mor soils show only 50–20% organomineral associations (Fig. 13). The incorporation of relatively undecomposed organic matter from aboveground into the mineral soil (bioturbation) causes a much lower degree of humification in mull as compared to mor A horizons. Essentially, there is only a negligible input of particulate L, Of, or Oh

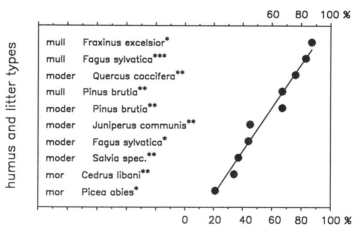

Figure 13   Relative proportions of organic carbon bound in organomineral complexes of A horizons of different mull, moder, and mor humus profiles that have developed in the temperate and mediterranean climate and under controlled laboratory conditions. *Temperate climate. **Mediterranean climate. ***Laboratory experiments.

material into the mineral soil in the case of mor, so that the organic matter of mor A horizons is scarcely renewed and strongly humified.

VII. Conclusions

The biodegradation and humification processes in soils can be studied by the structural characterization of forest soil organic matter in different horizons which are in turn characteristic of individual degrees of humification. Depending on the chemical nature of the individual biomacromolecules in plant and microbial residues several humification processes operate concurrently. The occurrence of one or the other humification process in soils is modulated by the chemical composition of the primary and secondary resources and environmental factors. For carbohydrates and proteins the major humification pathway seems to be microbial resynthesis, although some of the polysaccharides may survive due to protection by recalcitrant molecules. It is clear that there is a transformation of lignin structural units such that the aromatic units lose phenolic groups as depth in the soil profile increases. These results are not in agreement with models on humification in soils (Stevenson, 1982; Hedges, 1988; Hatcher and Spiker, 1988), which would lead to an increase in phenolic structures during humification of lignin or high contents of phenolic OH functionality in humic acids. The more recalcitrant lignin components are selectively preserved. Our knowledge on the biodegradation and humification of tannins is scarce and certainly needs further attention. The same holds for the humification pathway that the microbial melanins undergo in soils. Also, the humification of alkyl-C structures in forest soils has to be investigated further. In contrast to previous assumptions, the biodegradation and humification processes in forest soils lead to the accumulation of alkyl-carbon structures in the surface A horizon. The process of transformation of cutin/suberin and other aliphatic biomacromolecules of plant and microbial origin has still to be delineated. At the moment we cannot decide whether the cross-linking occurs via O or C linkages. Another major problem, which has not been addressed in the present work is the humification pathways leading to the formation of N-, P-, and S-containing humic compounds. New techniques like ^{31}P NMR spectroscopy show promising results also in these fields (Zech *et al.*, 1987a). For many components of the soil organic matter, it remains to be investigated how the stabilization against microbial degradation occurs. Nonetheless, a picture is emerging of humic substances as a complex mixture of different plant- and microbial-derived compounds, which exist in soils in a continuum of degradative stages.

Acknowledgments

We thank all colleagues and students who substantially contributed to our research on soil organic matter, in particular, G. Beudert, R. Bochter, G. Guggenberger, L. Haumaier, R. Hempfling, A. Miltner, R. Pöhhacker, and F. Ziegler. We also thank the visiting scientists R. Candler (University of Alaska, Fairbanks), M. B. David (University of Illinois, Urbana) F. Gil-Sotres (University of Santiago de Compostela, Spain), J. Hejzlar (Czechoslovak Academy of Sciences, Ceske Budejovice, Czechoslovakia), M.-B. Johansson (Swedish University of Agricultural Sciences, Uppsala, Sweden), T. Miano (Università di Bari, Italy), E. M. Perdue (Georgia Institute of Technology, Atlanta), and C. M. Preston (Pacific Forestry Center, Victoria, Canada). K. Haider (Bundesforschungsanstalt für Landwirtschaft, Braunschweig, FRG), P. G. Hatcher (The Pennsylvania State University, College Park, PA), R. L. Malcolm (U.S. Geological Survey, Denver), and N. Senesi (Università di Bari, Italy) provided helpful suggestions. The late Mrs. I. Ludwig and Mrs. C. Schreiber typed many manuscripts. Figures have been drawn by T. Engelbrecht. Last, but not least, we gratefully acknowledge the generous financial support by the Deutsche Forschungsgemeinschaft.

References

Anderson, J. M. (1973). The breakdown and decomposition of sweet chestnut (*Castanea sativa* Mill.) and beech (*Fagus sylvatica* L.) leaf litter in two deciduous woodland soils. I. Breakdown, leaching and decomposition. *Oecologia* **12**, 251–274.

Anderson, J. M. (1988). Invertebrate-mediated transport processes in soils. *Agric. Ecosyst. Environ.* **24**, 5–19.

Anderson, J. M., Flanagan, P. W., Caswell, E., Coleman, D. C., Cuevas, E., Freckman, D. W., Jones, J. A., Lavelle, P., and Vitousek, P. (1989). Biological processes regulating organic matter dynamics in tropical soils. *In* "Dynamics of Soil Organic Matter in Tropical Ecosystems" (D. C. Coleman, J. M. Oades, and G. Uehara, eds.), pp. 97–123. Univ. of Hawaii Press, Honolulu.

Angehrn-Bettinazzi, C., Lüscher, P., and Hertz, J. (1988). Thermogravimetry as a method for distinguishing various degrees of mineralization in macromorphologically-defined humus horizons. *Z. Pflanzenernähr. Bodenkd.* **152**, 177–183.

Babel, U. (1975). Micromorphology of soil organic matter. *In* "Soil Components" (J. E. Gieseking, ed.), Vol. 1, pp. 369–473. Springer-Verlag, Berlin.

Baldock, J. A., Oades, J. M., Vassallo, A. M., and Wilson, M. A. (1990). Solid-state CP/MAS ^{13}C N.M.R. analysis of bacterial and fungal cultures isolated from a soil incubated with glucose. *Aust. J. Soil Res.* **28**, 213–225.

Bell, A. A., and Wheeler, M. H. (1986). Biosynthesis and functions of fungal melanins. *Annu. Rev. Phytopathol.* **24**, 411–451.

Benner, R., Hatcher, P. G., and Hedges, J. I. (1990). Early diagenesis of mangrove leaves in a tropical estuary: Bulk chemical characterization using solid-state ^{13}C NMR and elemental analyses. *Geochim. Cosmochim. Acta* **54**, 2003–2013.

Beudert, G., Kögel-Knabner, I., and Zech, W. (1989). Micromorphological, wet-chemical and ^{13}C NMR spectroscopic characterization of density fractionated forest soils. *Sci. Total Environ.* **81/82**, 401–408.

Blume, H.-P. (1965). Die Charakterisierung von Humuskörpern durch Streu- und Humus-Stoffgruppenanalysen unter Berücksichtigung ihrer morphologischen Eigenschaften. *Z. Pflanzenernähr., Düng., Bodenkd.* **111**, 95–114.

Bouché, M. B. (1977). Stratégies lombriciennes. *Ecol. Bull.* **25,** 122–132.

Bracewell, J. M., and Robertson, G. W. (1987). Indications from analytical pyrolysis on the evolution of organic materials in the temperate environment. *J. Anal. Appl. Pyrol.* **11,** 355–366.

Driessen, P. M., and Dudal, R. (1989). "Lecture Notes on the Geography, Formation, Properties and Use of the Major Soils of the World." Agricultural University, Wageningen.

Edwards, C. A., and Heath, G. W. (1975). Studies in leaf litter breakdown. III. The influence of leaf age. *Pedobiologia* **15,** 348–354.

Eijsackers, H., and Zehnder, A. J. B. (1990). Litter decomposition: A Russian matriochka doll. *Biogeochemistry* **11,** 154–174.

Ellenberg, M., Mayer, R., and Schauermann, J. (1986). "Ökosystemforschung." Ulmer, Stuttgart.

Ellwardt, P.-C., Haider, K., and Ernst, L. (1981). Untersuchungen des mikrobiellen Ligninabbaus durch 13-C-NMR-Spektroskopie an spezifisch 13-C-angereichertem DHP-Lignin aus Coniferylalkohol. *Holzforschung* **35,** 103–109.

Ertel, J. R., and Hedges, J. I. (1984). The lignin component of humic substances: Distribution among soil and sedimentary humic, fulvic, and base-insoluble fractions. *Geochim. Cosmochim. Acta* **48,** 2065–2074.

Ertel, J. R., and Hedges, J. I. (1985). Sources of sedimentary humic substances: Vascular plant debris. *Geochim. Cosmochim. Acta* **49,** 2097–2107.

Ertel, J. R., Behmel, P., Christman, R. F., Flaig, W. J. A., Haider, K. M., Harvey, G. R., Hatcher, P. G., Hedges, J. I., Martin, J. P., Pfaender, F. K., and Schulten, H.-R. (1988). Genesis. *In* "Humic Substances and Their Role in the Environment" (F. H. Frimmel and R. F. Christman, eds.), pp. 105–112. Wiley, Chichester.

Fengel, D., and Wegener, G. (1984). "Wood: Chemistry, Ultrastructure, Reactions." de Gruyter, Berlin.

Gunnarsson, T., Sundin, P., and Tunlid, A. (1988). Importance of leaf litter fragmentation for bacterial growth. *Oikos* **52,** 303–308.

Haider, K. (1986a). Changes in substrate composition during the incubation of plant residues in soil. *In* "Microbial Communities in Soil" (V. Jensen, A. Kjoller, and L. H. Sorenson, eds.), pp. 133–147. Elsevier, London.

Haider, K. (1986b). The synthesis and degradation of humic substances in soil. *Trans. Congr. Int. Soc. Soil Sci., 13th,* Vol. VI, pp. 644–656.

Haider, K. (1991). Problems related to the humification processes in soils of the temperate climate. *Soil Biochem.* **7,** 55–94.

Haider, K., Kern, H. W., and Ernst, L. (1985). Intermediate steps of microbial lignin degradation as elucidated by ^{13}C NMR spectroscopy of specifically ^{13}C-enriched DHP-lignins. *Holzforschung* **39,** 23–32.

Harmon, M. E., Franklin, J. F., Swanson, F. J., Sollins, P., Gregory, S. V., Lattin, J. D., Anderson, N. H., Cline, S. P., Aumen, N. G., Sedell, J. R., Lienkaemper, G. W., Cromack, K., and Cummins, K. W. (1986). Ecology of coarse woody debris in temperate ecosystems. *Adv. Ecol. Res.* **15,** 133–302.

Hatcher, P. G., and Spiker, E. C. (1988). Selective degradation of plant biomolecules. *In* "Humic Substances and Their Role in the Environment" (F. H. Frimmel and R. F. Christman, eds.), pp. 59–74. Wiley, Chichester.

Hayes, M. H. B., MacCarthy, P., Malcolm, R. L., and Swift, R. S. (1989a). The search for structure: Setting the scene. *In* "Humic Substances II" (M. H. B. Hayes, P. MacCarthy, R. L. Malcolm, and R. S. Swift, eds.), pp. 3–31. Wiley, Chichester.

Hayes, M. H. B., MacCarthy, P., Malcolm, R. L., and Swift, R. S. (1989b). Structures of humic substances: The emergence of 'forms'. *In* "Humic Substances II" (M. H. B. Hayes, P. MacCarthy, R. L. Malcolm, and R. S. Swift, eds.), pp. 689–733. Wiley, Chichester.

Heath, G. W., Arnold, M. K., and Edwards, C. A. (1966). Studies in leaf litter breakdown. I. Breakdown rates of leaves of different species. *Pedobiologia* **6**, 1–12.

Hedges, J. I. (1988). Polymerization of humic substances in natural environments. *In* "Humic Substances and Their Role in the Environment" (F. H. Frimmel and R. F. Christman, eds.), pp. 45–58. Wiley, Chichester.

Hedges, J. I., Blanchette, R. A., Weliky, K., and Devol, A. H. (1988). Effects of fungal degradation on the CuO oxidation products of lignin: A controlled laboratory study. *Geochim. Cosmochim. Acta* **52**, 2717–2726.

Hempfling, R., and Schulten, H.-R. (1989). Selective preservation of biomolecules during humification of forest litter studied by pyrolysis-field ionization mass spectrometry. *Sci. Total Environ.* **81/82**, 31–40.

Hempfling, R., and Schulten, H.-R. (1990). Chemical characterization of the organic matter in forest soils by Curie point pyrolysis-GC/MS and pyrolysis-field ionization mass spectrometry. *Org. Geochem.* **15**, 131–145.

Hempfling, R., Ziegler, F., Zech, W., and Schulten, H.-R. (1987). Litter decomposition and humification in acidic forest soils studied by chemical degradation, IR and NMR spectroscopy and pyrolysis field ionization mass spectrometry. *Z. Pflanzenernähr. Bodenkd.* **150**, 179–186.

Hempfling, R., Zech, W., and Schulten, H.-R. (1988). Chemical composition of the organic matter in forest soils. 2. Moder profile. *Soil Sci.* **146**, 262–276.

Hempfling, R., Simmleit, N., and Schulten, H.-R. (1991). Characterization and chemodynamics of plant constituents during maturation, senescence and humus genesis in spruce ecosystems. *Biogeochemistry* **3**, 27–60.

Hepper, C. M. (1975). Extracellular polysaccharides of soil bacteria. *In* "Soil Microbiology" (N. Walker, ed.), pp. 93–110. Butterworth, London.

Herlitzius, R., and Herlitzius, H. (1977). Streuabbau in Laubwäldern. Untersuchungen in Kalk- und Sauerhumus-Buchenwäldern. *Oecologia* **30**, 147–171.

Jenkinson, D. S. (1988). Soil organic matter and its dynamics. *In* "Russell's Soil Conditions and Plant Growth" (A. Wild, ed.), pp. 564–607. Longman, Harlow, UK.

Jörgensen, R. G. (1987). Flüsse, Umsatz und Haushalt der postmortalen organischen Substanz und ihrer Stoffgruppen in Streudecke und Bodenkörper eines Buchenwald-Ökosystems auf Kalkgestein. *Goettinger Bodenkd. Ber.* **91**, 1–407.

Kassim, G., Martin, J. P., and Haider, K. (1981). Incorporation of a wide variety of organic substrate carbons into soil biomass as estimated by the fumigation procedure. *Soil Sci. Soc. Am. J.* **45**, 1106–1112.

Kögel, I. (1986). Estimation and decomposition pattern of the lignin component in forest soils. *Soil Biol. Biochem.* **18**, 589–594.

Kögel, I., Hempfling, R., Zech, W., Hatcher, P. G., and Schulten, H.-R. (1988a). Chemical composition of the organic matter in forest soils. I. Forest litter. *Soil Sci.* **146**, 124–136.

Kögel, I., Ziegler, F., and Zech, W. (1988b). Lignin signature of subalpine Rendzinas (Tangel- and Moderrendzina) in the Bavarian Alps. *Z. Pflanzenernähr. Bodenkd.* **151**, 15–20.

Kögel-Knabner, I., Zech, W., and Hatcher, P. G. (1988). Chemical composition of the organic matter in forest soils. III. The humus layer. *Z. Pflanzenernähr. Bodenkd.* **151**, 331–340.

Kögel-Knabner, I., Ziegler, F., Riederer, M., and Zech, W. (1989). Distribution and decomposition pattern of cutin and suberin in forest soils. *Z. Pflanzenernähr. Bodenkd.* **152**, 409–413.

Kögel-Knabner, I., Hatcher, P. G., and Zech, W. (1990). Decomposition and humification processes in forest soils: Implications from structural characterization of forest soil organic matter. *Trans. Int. Congr. Soil Sci., 14th*, Kyoto, Vol. 5 pp. 218–223.

Kögel-Knabner, I., Hatcher, P. G., and Zech, W. (1991). Chemical structural studies of forest soil humic acids: Aromatic carbon fraction. *Soil Sci. Soc. Am. J.* **55**, 241–247.

Kögel-Knabner, I., Hatcher, P. G., Tegelaar, E. W., and De Leeuw, J. W. (1992a). Aliphatic components of forest soil organic matter as determined by solid-state ^{13}C NMR and analytical pyrolysis. *Sci. Total Environ.* **113**, 89–106.

Kögel-Knabner, I., De Leeuw, J. W., and Hatcher, P. G. (1992b). Nature and distribution of alkyl carbon in forest soil profiles: Implications for the origin and humification of aliphatic biomacromolecules. *Sci. Total Environ.* **117/118**, 175–185.

Kolattukudy, P. E. (1981). Structure, biosynthesis, and biodegradation of cutin and suberin. *Annu. Rev. Plant Physiol.* **32**, 539–567.

Laatsch, W. (1957). "Dynamik der mitteleuropäischen Mineralböden." Verlag Steinkopff, Dresden.

Martin, J. P., and Haider, K. (1986). Influence of mineral colloids on turnover rates of soil organic carbon. *In* "Interactions of Soil Minerals with Natural Organics and Microbes" (P. M. Huang and M. Schnitzer, eds.), pp. 283–304. *Soil Sci. Soc. Am.*, Madison, WI.

McClaugherty, C. A., Aber, J. D., and Mellilo, J. M. (1984). Decomposition dynamics of fine roots in forested ecosystems. *Oikos* **42**, 378–386.

Meentemeyer, V., Box, E. O., and Thompson, R. (1982). World patterns and amounts of terrestrial plant litter production. *BioScience* **32**, 125–128.

Mommaerts-Blllet, F. (1971). Aspects dynamiques de la partition de la litière de feuilles. *Bull. Soc. R. Bot. Belg.* **104**, 181–195.

Nip, M., Tegelaar, E. W., De Leeuw, J. W., Schenck, P. A., and Holloway, P. J. (1986). A new non-saponifiable highly aliphatic and resistant biopolymer in plant cuticles. Evidence from pyrolysis and ^{13}C-NMR analysis of present-day and fossil plants. *Naturwissenschaften* **73**, 579–585.

Nip, M., De Leeuw, J. W., Holloway, P. J., Jensen, J. P. T., Sprenkels, J. C. M., De Poorter, M., and Sleeckx, J. J. M. (1987). Comparison of flash pyrolysis, differential scanning calorimetry, ^{13}C NMR and IR spectroscopy in the analysis of a highly aliphatic biopolymer from plant cuticles. *J. Anal. Appl. Pyrol.* **11**, 287–295.

Norwood, D. L. (1988). Critical comparison of structural implications from degradative and nondegradative approaches. *In* "Humic Substances and Their Role in the Environment" (F. H. Frimmel and R. F. Christman, eds.), pp. 133–148. Wiley, Chichester.

Oades, J. M. (1988). The retention of organic matter in soils. *Biogeochemistry* **5**, 35–70.

Oades, J. M. (1989). An introduction to organic matter in mineral soils. *In* "Minerals in Soil Environments" (J. B. Dixon and S. B. Weed, eds.), pp. 89–159. *Soil Sci. Soc. Am.*, Madison, WI.

Ohta, S., Suzuki, A., and Kumada, K. (1986). Experimental studies on the behavior of fine organic particles and water-soluble organic matter in mineral soil horizons. *Soil Sci. Plant Nutr.* **32**, 15–26.

Peberdy, J. F. (1990). Fungal cell walls—a review. *In* "Biochemistry of Cell Walls and Membranes in Fungi" (P. J. Kuhn, A. P. J. Trinci, M. J. Jung, M. W. Goosey, and L. G. Copping, eds.), pp. 5–30. Springer-Verlag, Berlin.

Preston, C. M. (1992). The application of NMR to organic matter inputs and processes in forest ecosystems of the pacific northwest. *Sci. Total Environ.* **113**, 107–120.

Preston, C. M., Schnitzer, M., and Ripmeester, J. A. (1989). A spectroscopic and chemical investigation on the de-ashing of humin. *Soil Sci. Soc. Am. J.* **53**, 1442–1447.

Preston, C. M., Sollins, P., and Sayer, B. G. (1990). Changes in organic components for fallen logs by ^{13}C nuclear magnetic resonance spectroscopy. *Can. J. For. Res.* **20**, 1382–1391.

Raich, J. W., and Nadelhoffer, K. J. (1989). Belowground carbon allocation in forest ecosystems: Global trends. *Ecology* **70**, 1346–1354.

Riederer, M., Matzke, K., Ziegler, F., and Kögel-Knabner, I. (1993). Occurrence, distribution and fate of the lipid plant biopolymers cutin and suberin in temperate forest soils. *Org. Geochem.* (in press).

Rogers, H. J., Perkins, H. R., and Ward, J. B. (1980). "Microbial Cell Walls and Membranes." Chapman & Hall, London.

Saiz-Jimenez, C., and De Leeuw, J. W. (1984). Pyrolysis gas chromatography mass spectrometry of isolated synthetic and degraded lignins. *Org. Geochem.* **56**, 417–422.

Saiz-Jimenez, C., and De Leeuw, J. W. (1986). Chemical characterization of soil organic matter fractions by analytical pyrolysis-gas chromatography-mass spectrometry. *J. Anal. Appl. Pyrol.* **9**, 99–119.

Sarkanen, K. V., and Ludwig, C. H. (1971). "Lignins." Wiley (Interscience), New York.

Schaefer, M., and Schauermann, J. (1990). The soil fauna of beech forests: Comparison between a mull and a moder soil. *Pedobiologia* **34**, 299–314.

Schlegel, H. G. (1991). "Allgemeine Mikrobiologie." Thieme, Stuttgart.

Schlesinger, W. H. (1977). Carbon balance in terrestrial detritus. *Annu. Rev. Ecol. Syst.* **8**, 51–81.

Schlesinger, W. H. (1990). Evidence from chronosequence studies for a low carbon-storage potential of soils. *Nature (London)* **348**, 232–234.

Schnitzer, M. (1990). Selected methods for the characterization of soil humic substances. *In* "Humic Substances in Soil and Crop Sciences: Selected Readings" (P. MacCarthy, C. E. Clapp, R. L. Malcolm, and P. R. Bloom, eds.), pp. 65–89. Soil Sci. Soc. Am, Madison, WI.

Schulten, H. R., and Schnitzer, M. (1991). Supercritical carbon dioxide extraction of long-chain aliphatics from two soils. *Soil Sci. Soc. Am. J.* **55**, 1603–1611.

Seelenfreund, D., Lapierre, C., and Vicuña, R. (1990). Production of soluble lignin-rich fragments (APPL) from wheat lignocellulose by Streptomyces viridosporus and their partial metabolism by natural bacterial isolates. *J. Biotechnol.* **13**, 145–158.

Stevenson, F. (1982). "Humus Chemistry." Wiley, New York.

Stott, D. E., and Martin, J. P. (1990). Synthesis and degradation of natural and synthetic humic material in soils. *In* "Humic Substances in Soil and Crop Sciences: Selected Readings" (P. MacCarthy, C. E. Clapp, R. L. Malcolm, and P. R. Bloom, eds.), pp. 37–63. Soil Sci. Soc. Am., Madison, WI.

Swift, M. J., Heal, O. W., and Anderson, J. M. (1979). "Decomposition in Terrestrial Ecosystems." Blackwell, Oxford.

Tegelaar, E. W., De Leeuw, J. W., and Saiz-Jimenez, C. (1989). Possible origin of aliphatic moieties in humic substances. *Sci. Total Environ.* **81/82**, 1–17.

Tegelaar, E. W., Hollman, G., van der Vegt, P., De Leeuw, J. W., and Holloway, P. J. (1993). Chemical characterization of the outer bark tissue of some angiosperm species: Recognition of an insoluble, non-hydrolyzable highly aliphatic biopolymer (suberan). *Org. Geochem.* (in press).

Theng, B. K. G., Tate, K. R., Sollins, P., Moris, N., Nadkarni, N., and Tate, R. L. (1989). Constituents of organic matter in temperate and tropical soils. *In* "Dynamics of Soil Organic Matter in Tropical Ecosystems" (D. C. Coleman, J. M. Oades, and G. Uehara, eds.), pp. 5–32. Univ. of Hawaii Press, Honolulu.

Voelkel, R. (1988). Hochauflösende Festkörper-^{13}C-NMR-Spektroskopie von Polymeren. *Angew. Chem.* **100**, 1525–1540.

Vogt, K. A., Grier, C. C., and Vogt, D. J. (1986). Production, turnover, and nutrient dynamics of above- and belowground detritus of world forests. *Adv. Ecol. Res.* **15**, 303–377.

Waksman, S. A. (1938). "Humus: Origin, Chemical Composition and Importance in Nature." Ballière, London.

Wallwork, J. A. (1976). "The Distribution and Diversity of Soil Fauna," pp. 200–242. Academic Press, London.

Wessels, J. G. H., and Sietsma, J. H. (1981). Fungal cell walls: A survey. *Encycl. Plant Physiol. New Ser.* **13B,** 294–352.

Wilhelmi, V., and Rothe, G. M. (1990). The effect of acid rain, soil temperature and humidity on C-mineralization rates in organic soil layers under spruce. *Plant Soil* **121,** 197–202.

Wilson, M. A. (1987). "NMR Techniques and Applications in Geochemistry and Soil Chemistry." Pergamon, Oxford.

Wilson, M. A., and Hatcher, P. G. (1988). Detection of tannins in modern and fossil barks and in plant residues by high-resolution solid-state ^{13}C nuclear magnetic resonance. *Org. Geochem.* **12,** 539–546.

Wilson, M. A., Heng, S., Goh, K. M., Pugmire, R. J., and Grant, D. M. (1983). Studies of litter and acid insoluble soil organic matter fractions using ^{13}C-cross polarization nuclear magnetic resonance spectroscopy with magic angle spinning. *J. Soil Sci.* **34,** 83–97.

Wilson, M. A., Batts, B. D., and Hatcher, P. G. (1988). Molecular composition and mobility or torbanite precursors: Implications for the structure of coal. *Energy & Fuels* **2,** 668–672.

Zech, W., and Haumaier, L. (1989). Zur Aromatizität der organischen Bodensubstanz. *Mitt. Dtsch. Bodenkd. Ges.* **59,** 501–504.

Zech, W., Kögel, I., Zucker, A., and Alt, H. G. (1985). CP-MAS-^{13}C-NMR-Spektren organischer Lagen einei Tangelrendzina. *Z. Pflanzenernähr. Bodenkd.* **148,** 481–488.

Zech, W., Alt, H. G., Haumaier, L., and Blasek, R. (1987a). Characterization of phosphorus fractions in mountain soils of the Bavarian Alps by ^{31}P NMR spectroscopy. *Z. Pflanzenernähr. Bodenkd.* **150,** 119–123.

Zech, W., Johansson, M.-B., Haumaier, L., and Malcolm, R. L. (1987b). CPMAS ^{13}C and IR spectra of spruce and pine litter and of the Klason lignin fraction at different stages of decomposition. *Z. Pflanzenernähr. Bodenkd.* **150,** 262–265.

Zech, W., Haumaier, L., and Kögel-Knabner, I. (1989). Changes in aromaticity and carbon distribution of soil organic matter due to pedogenesis. *Sci. Total Environ.* **81/82,** 179–186.

Zech, W., Hempfling, R., Haumaier, L., Schulten, H.-R., and Haider, K. (1990a). Humification in subalpine Rendzinas: Chemical analyses, IR and ^{13}C NMR spectroscopy and pyrolysis-field ionization mass spectrometry. *Geoderma* **47,** 123–138.

Zech, W., Ziegler, F., Miltner, A., Wiedemann, P., and Cepel, N. (1990b). Litter decomposition and humification in mediterranean and temperate forest soils under cedar (West Taurus, Turkey) and spruce (Fichtelgebirge, F. R. Germany). *Proc. Int. Cedar Symp.,* Antalya, pp. 933–942.

Zech, W., Ziegler, F., Kögel-Knabner, I., and Haumaier, L. (1992). Humic substances distribution and transformation in forest soils. *Sci. Total Environ.* **117/118,** 155–174.

Zelibor, J. L., Romankiw, L., Hatcher, P. G., and Colwell, R. R. (1988). Comparative analysis of the chemical composition of mixed and pure cultures of green algae and their decomposition residues by ^{13}C Nuclear Magnetic Resonance Spectroscopy. *Appl. Environ. Microbiol.* **54,** 1051–1060.

Ziegler, F. (1989). Changes of lipid content and lipid composition in forest humus layers derived from Norway spruce. *Soil Biol. Biochem.* **21,** 237–243.

Ziegler, F. (1990). Zum Einfluss von Regenwürmern (*Eisenia fetida;* Lumbricidae) und mineralischer Substanz auf die Zersetzung von Buchenstreu (*Fagus sylvatica*) und Gerstenstroh (*Hordeum vulgare*) im Modellversuch. *Bayreuther Bodenkd. Ber.* **13.**

Ziegler, F., and Zech, W. (1989). Distribution pattern of total lipids and lipid fractions in forest humus. *Z. Pflanzenernäehr. Bodenkd.* **152,** 287–290.

Ziegler, F., and Zech, W. (1990). Decomposition of beech litter cutin under laboratory conditions. *Z. Pflanzenernäehr. Bodenkd.* **153,** 373–374.

Ziegler, F., and Zech, W. (1992). Formation of water-stable aggregates through the action of earthworms. Implications from laboratory experiments. *Pedobiologia* **36,** 91–96.

Ziegler, F., Kögel, I., and Zech, W. (1986). Alteration of gymnosperm and angiosperm lignin during decomposition in forest humus layers. *Z. Pflanzenernähr. Bodenkd.* **149,** 323–331.

Ziegler, F., Kögel-Knabner, I., and Zech, W. (1992). Litter decomposition. *In* "Responses of forest ecosystem changes" (A. Teller, P. Mathy, and J. N. R. Jeffers, eds.), pp. 697–699. Elsevier Applied Science, London.

10

The Effect of Aggregation of Soils on Water, Gas, and Heat Transport

R. Horn

I. Introduction

One of the earliest publications on soil aggregation by Wollny in 1898, "Untersuchungen über den Einfluss der mechanischen Bearbeitung auf die Fruchtbarkeit des Bodens," described in great detail the positive effect of soil structure on root growth, on water availability, and on the gas transport in soils. The positive effects of soil structure on soil strength were mentioned also and it was concluded that the mechanisms involved in the interaction of soil structure, plant growth, and yield needed to be investigated in greater detail in the future. The positive effects of a favorable soil structure and negative effects of soil compaction on crop growth and yield have been repeatedly described (Scheffer and Schachtschabel, 1992; Blanck, 1929–1939; Dexter, 1988; Horn, 1989). Although why crops responded favorably to good soil structure was often speculated, the cause and effect of these relationships were rarely investigated experimentally. Rynasiewicz (1945) and Page and Willard (1947) demonstrated the effect of a good soil structure on plant growth and yield. They pointed to interactions among soil structure, water status, and soil aeration. Also Emmerson *et al.* (1978) described effects of soil structure on aeration, plant-available water, root penetration, and compressibility of soil and hence on plant growth and crop yield.

In earlier studies, the main focus was on macroscopic effects rather than on the physical properties of single structural elements of soils. Largely due to this lack of detailed information it was impossible to predict physical, chemical, or biological properties of soils or to validate corresponding models. Contradicting results were obtained when aggregated soils were investigated. The physicochemical properties of aggregates as the main compartment for any chemical reaction in ecosystems were not taken into consideration. This resulted in difficulties of predicting soil processes because experiments on root growth in structured soils suggested that the pattern of root development depended on certain properties of soil aggregates (Strassburger, 1969).

Soil structure is "the spatial heterogeneity of the different components of properties of soil" (Dexter, 1988). While genetic descriptions of hydromorphic soil properties (Blume, 1968) could have been used as an indicator for the necessity to differentiate between physical and chemical properties of structured soils at the macro (m, km, m^2, m^3), and the microscale (range of μm to cm), detailed measurements of soil physics have only seldomly been carried out to understand physicochemical processes in structured soils. Thus, there is an urgent need to apply physical and chemical methods, in order to understand and predict fluxes of water, heat, and gases in such systems, which are based on a detailed knowledge of soil structure. In the following the process of aggregate formation and the variation of physical and physicochemical properties are described at different scales in order to provide a better understanding of differences in physical properties of structured soils. Special emphasis is given to a lysimeter experiment in which almond trees were grown in large lysimeters at Avdat (Israel) during the period 1984 to 1988. The experiment was carried out to quantify relationships between plant growth and effects of soil aggregation processes including dynamic processes of pore size distribution and changes due to swelling and shrinking during irrigation of loess soil. Lysimeters (diameter, 3 m; depth, 1, 2, or 3 m; volume, 7, 14, or 21 m^3 of soil) were filled with homogenized loess, planted with a single almond tree, and watered only at the beginning of each growing season to field capacity (pF = 1,8). The effect of repeated watering and drying on soil aggregation was compared in 3-m-deep lysimeters with and without plants, which were rewatered to pF 1.8 when dried to pF 2.7.

II. Processes of Aggregate Formation

In soils containing more than 15% clay (particle size< 2 μm) the mineral particles (sand, silt, and clay) tend to form aggregates. Usually the process occurs when soils dry and swell, and it is further enhanced by biological activities (Hillel, 1980). Aggregates may show great variation in size from crumbs (diameter < 2mm) to polyhedres or subangular blocks of 0.005–0.02 m, or even to prisms or columns of more than 0.1 m.

During the first period of shrinkage, mineral particles are tied together by capillary forces which increase the number of points of contact and result in a higher bulk density (Horn *et al.*, 1989). The initial aggregates always have rectangular-shaped edges because, under these conditions, stress release would occur perpendicular to an initial crack and stress would remain parallel to the crack (strain-induced fracturing). However, due to the increased mechanical strength, the mobility of particles in the aggregates against vertical stress is reduced. According to the theory of Terzaghi (cited by Horn, 1989), nonrectangular shear plains are also created which after repeated swelling and shrinking processes result in fractures in which the value of the angle of internal friction determines the deviation from a 90° angle (shear-induced fracturing; Hartge and Horn, 1977; Hartge and Rathe, 1983).

In newly formed aggregates, the number of contact points depends on the range of moisture potential and on the distribution of particle sizes as well as on their mobility (i.e., state of dispersion, flocculation, and cementation). Soil shrinkage, including crack formation, increases bulk density of aggregates. The increase in bulk density with the initial watering and drying of the soil permits the aggregates to withstand structural collapse. The increase of the strength of single aggregates is further enhanced by a more pronounced particle rearrangement, if the soil is nearly saturated with water increasing the mobility of clay particles due to dispersion and greater menisci forces of water (Utomo and Dexter, 1981; Horn and Dexter, 1989). Drying causes enhanced cohesion by capillary forces. Consequently, in order to carry the same soil load the bulk density of aggregates and thus the number of contact points decreases (Fig. 1). With increasing intensity of drying of the moulded soil, its ability to perform reversible volume changes decreases. In wetter soils, the smaller proportion of residual to normal shrinkage (i.e., the greater reversibility during swelling) causes more intensive particle mobility and rearrangement in order to reach a state of minimum free energy (Horn, 1976). Although swelling may lead to partial expansion of contracted particles following rewetting of aggregated soils, a complete disaggregation is not possible if there is no additional input of kinetic energy, as has been demonstrated also by the puddling or kneading of rice soils (Horn, 1976). Thus, aggregate strength will depend on (i) capillary forces, (ii) intensity of shrinkage (normal/residual), (iii) number of swelling and shrinkage cycles, (iv) mineral particle mobility (i.e., rearrangement of particles in order to reach the status of lowest free energy), and (v) bounding energy between particles in/or between aggregates or in the bulk soil.

The effects of these parameters on aggregate formation have been tested by crushing tests in combination with measurements of the aggregate bulk density and mean aggregate diameter. The tensile strength of

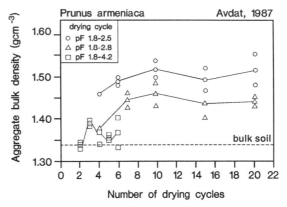

Figure 1 Tensile strength of newly formed aggregates as a function of bulk density, aggregate size (aspect ratio), and number of swelling and drying cycles. From Semmel *et al.,* 1989.

aggregates increased with the number of drying events as well as with the degree of dryness reached in each event. Also, the diameter of aggregates became smaller during repeated drying events (Horn and Dexter, 1989). This could be derived from data on aggregates obtained during the lysimeter experiment in Avdat (Israel). With increasing amounts of water available to the plants, aggregates became weaker. The frequency of wetting and drying events increased the tensile strength and reduced the size of aggregates (Fig. 2; Semmel *et al.,* 1989).

The strength of aggregates and their diameters were also affected even by single particles, which increased the total area of water menisci and reduced the diameter of the newly formed pores at the same time (Zhang, 1991). Aggregation was also enhanced by biological and chemical processes such as flocculation and cementation by organomineralic bondings (Dexter *et al.,* 1988). In the latter process, polysaccharides and organic substances have to be considered as well (Hempfling *et al.,* 1990). Aggregate stabilization by extracellular metabolic products of colonies of bacteria and by root exudates has been demonstrated. The effects appear to be due to reactions at the contact points of mineral particles (Cheshire, 1979; Goss and Reid, 1979; Martin, 1977; Tippkötter, 1988). Thus, soil structure formation by shrinkage and swelling results in well-known heterogenization of the pore size distribution in the bulk soil (macroscale) due to formation of coarser interaggregate pores and finer intraaggregate pores (microscale). This is obvious from the increased bulk density of single aggregates. In addition, substantial variations in physical and chemical properties of single aggregates occur (Horn and Taubner, 1989). For example, in humid climates the bulk density values of prisms in loess range between 1.45–1.65 g/cm^3 compared to 1.45–1.55 g/cm^3 for the

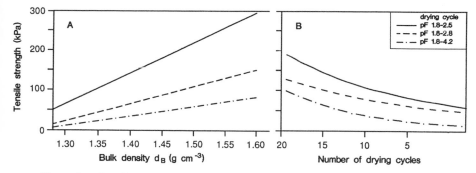

Figure 2 The effect of wetting and drying intensity on aggregate tensile strength. The higher the amount of plant-available water the weaker the aggregates even at higher bulk density. The more often aggregates had been wetted and redried the stronger they became at a smaller diameter. From R. Horn, 1989. Reprinted with permission of Kluwer Publ.

bulk soil. Values for subangular blocky aggregates in Pelosols range between 1.8–2.0 g/cm^3 in comparison to 1.45–1.55 g/cm^3 for the bulk soil (Horn, 1981). The standard deviations of bulk density determinations for aggregates or for their tensile strength may be explained by such small-scale effects.

Aggregates always exhibit a higher mechanical strength than the bulk soil. This can be derived from measurements of the penetration resistance (Horn *et al.*, 1987). Within individual aggregates an increase of strength of the outer skin of aggregates has been described by Becher (1991). Moreover, the surface of aggregates may have minor areas of increased weakness besides very dense and strong ones. The latter can be taken as an indication of further microaggregate cracking (plains of weakness), partial rooting, the existence of earthworm channels, and variation in grain size distribution.

During wetting and drying, capillary forces also cause particle transport and differences in the particle size distribution within single aggregates. Due to hydraulic gradients induced by the wetting and drying front, convex menisci may pull single particles or even microaggregates to the outer edges. Because of the very limited reversibility of particle mobilization during consecutive and long-lasting swelling processes (see above), the outer skin of aggregates gets a clay content higher than that of the inner part. The amount of coarser particles like silt and sand would increase in the center of the aggregates (Horn, 1987). During all these processes soil structure and aggregate properties would be only in a quasi-steady state, because they are always exposed to climatic and other effects. Both anthropogenically or naturally formed aggregates could be further deformed or altered by freezing and thawing events as well as by biological activity and changes in the chemical composition of the soil.

For example, if strong aggregates (plates, subangular blocks) are frozen, ice lense formation results in a pealing off of the outer aggregate skin forming smaller and/or less-dense units. The process is often called "soil curing." In contrast, denser aggregates may also be formed, when ice pressure exceeds internal soil strength and when soil compaction occurs because of the expansion of the water volume during freezing (Horn, 1985; Schababerle, 1990).

III. Hydraulic Aspects

A. Water Retention Curve

Aggregation due to swelling and shrinking is affected by hydrologic and hydraulic properties of the soil. With an increase in the number of drying cycles the total porosity first decreases. Later it may increase again (Horn and Dexter, 1989). The volume of fine pores (i.e., the volumetric water content at pF > 4,2) is enhanced by decreasing drying intensity (Fig. 3). In addition, the amount of water available to plants (i.e., water content at pF of 1,8–4,2) is reduced with more intensive soil drying. Only at more negative water potentials is the air entry value exceeded depending how wet the soil had been kept (Horn *et al.*, 1989). The latter effect is determined by the correspondingly steep slope of the pF/water content curve at pF < 1.8.

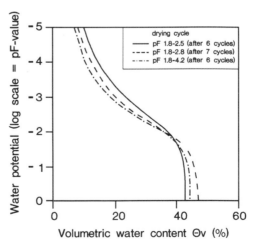

Figure 3 Effect of drying intensity on changes in the water retention curve. Fine pore volume was enhanced more by less-severe drying.

B. Darcy Law: Hydraulic Conductivity

Given a laminar flow and a homogeneous pore system, the water flux in soils can be described and quantified by Darcy's law. Generally, the values of the hydraulic gradient vary only by half an order of magnitude depending on water potential, grain, and pore size distribution (Hartge and Horn, 1977). If the soil–plant interaction is also taken into account, values of hydraulic gradients of up to 9 kPa $\cdot$ m^{-1} can be calculated when water content differences are expressed as pF at a given time and assuming a constant water content/pF curve (Bohne, 1988). The values of the hydraulic conductivity of soils range between 10^{-4} and 10^{-13} m $\cdot$ s^{-1} depending on water potential, texture, and structure. Under saturated conditions, hydraulic conductivities range between 10^{-4} and 10^{-5} m $\cdot$ s^{-1} in a sandy soil and between 10^{-6} and 10^{-9} m $\cdot$ s^{-1} in a clay. Hydraulic conductivity is affected by structure and texture. It is high when the soil is highly porous, fractured, or aggregated and low when it is tightly compacted and dense. The hydraulic conductivity depends not only on the pore volume but also on the continuity of conducting pores. In structured soils with very large cracks the hydraulic conductivity for the bulk soil increases while flow velocity is strongly reduced inside the aggregates due to shrinkage (Table I). The hydraulic conductivity may decrease by 4 orders of magnitude in single aggregates compared to the bulk soil unless the aggregates contain more sand than silt and clay, in which case there would be no difference compared to the bulk soil. The effects of structure on hydraulic conductivity persist under unsaturated conditions. Also, changes in structure directly affect the degree of variation in the hydraulic conductivity (Fig. 4). At less-negative values of water potential, the unsaturated hydraulic conductivity of single aggregates decreases with the compaction of the structural elements (prisms less than polyhedrons or subangular blocks) compared to fluxes in bulk soils. Only in weak aggregates are differences and ranges smaller (Gunzelmann and Horn, 1985). After exceeding the crossover potential values at very negative potentials (Hillel, 1980), higher values of hydraulic

Table I Saturated Hydraulic Conductivity kf (m s^{-1}) for Structured Bulk Soil Samples and Single Aggregates

Structure	Texture	kf (m s^{-1})	
		Bulk soil	Aggregates
Subangular-blocky	Loamy clay	1.1×10^{-4}	$3.5 \times 10^{-8} \pm 2.2 \times 10^{-8}$
Blocky	Loamy clay	1.6×10^{-5}	$4.8 \times 10^{-8} \pm 2.7 \times 10^{-8}$
Prismatic-blocky	Loamy clay	2.6×10^{-7}	$6.0 \times 10^{-8} \pm 2.5 \times 10^{-8}$
Prismatic	Loamy clay	3.8×10^{-5}	$3.4 \times 10^{-8} \pm 2.4 \times 10^{-5}$

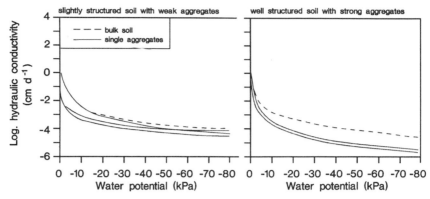

Figure 4 Hydraulic conductivity (cm · d⁻¹) as a function of soil water potential (kPa) in soil with weak aggregates and soil with strong aggregates

conductivity are obtained in aggregates as compared to the bulk soil. This heterogenization of the flow paths in aggregates compared to the bulk soil is further enhanced, since the outer skin of aggregates contains more clay than the center and the latter has more coarse pores than the outer part (Horn, 1987). Consequently, water and air flow of single aggregates is reduced further, which can also be linked from the increasing tortuosity of the pore system at different positions in the aggregate.

C. Darcy Law: Hydraulic Gradient

With respect to the water transport out of single aggregates and the effect of pore continuity, a retarded change in water potentials between the different positions in the aggregates and a delayed water flux from the center to the surface has been demonstrated (Türk *et al.*, 1991). Under laboratory conditions as well as under field situations in lysimeter experiments (0.1 m³) it could be shown that the water potential decrease over time due to evaporation was more rapid in the homogenized loess material compared to single aggregates embedded in homogenized loess at the same depth of 20 cm. Denser aggregates resulted in a more-delayed decrease of potential (Fig. 5). The tensiometer cup size in those experiments was always 1 mm in diameter and the cups were installed both in the homogenized bulk soil and in single aggregates (for technique, see Gunzelmann and Horn, 1985).

Even when a single aggregate is placed in a completely homogenized loess soil material at the same bulk density the increases in water potential inside the single aggregate and in the bulk soil are not identical. Differences increase with drying of the soil. Therefore, it can be concluded that aggregate formation always induces a multidimensional water flux even if the aggregates are rather soft and coarse.

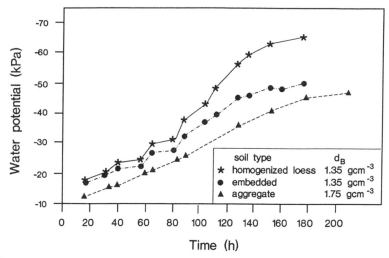

Figure 5 Effect of bulk density and pore continuity of the homogenized soil and of single aggregates embedded in the homogenized loess on changes in soil water (kPa) potential with time. Although the bulk density of the homogenized soil and single aggregates are the same, the homogenized soil is drier than the aggregates. Increasing aggregate bulk density results in a further slowing of the dessication.

D. Modeling Water Fluxes

Modeling the effect of hydraulic properties on water transport in structured soils with an unknown history of aggregate formation revealed that in a vertisol derived from clay ("Amaltheenton") the interaggregate water flux could be up to 8500 times higher than that in the intraaggregate pore system. Furthermore, the directions of water fluxes also differed (Gunzelmann, 1989). Thus, the effects of interaggregate pores become more significant in structured soils (mobile water). On the other hand, the water transport in intraaggregate pores becomes vanishingly small (immobile water; see Beven and Germann, 1982). Even in sandy soils a one-dimensional flux would be the exception because of microcrack formation and water-repellant effects at particle surfaces. The well-described process of fingering of the water flow in sandy soils may be based on the same principles.

IV. Thermal Aspects

Due to changes in the number of particles per aggregate volume and variations in the particle arrangement, the number of particle contact points and the higher water saturation at a given soil water potential should change the thermal properties of soils. Measured temperature gradients in homogenized or in hydraulically well-controlled reaggre-gated loess soil, lysimeters were modeled by a Fourier series and were fitted assuming a one-dimensional nonstationary heat transport in soils (according to Horton *et al.*, 1983). Aggregation by swelling and shrinking would increase the thermal conductivity and the heat conductivity, which both depend on water content (Figs. 6a and 6b), since heat flow depends not only on the continuity of contact points (conductance) but also on the continuity of water-filled pores (convection and diffusion).

V. Aspects of Soil Aeration

In general, soil aeration is governed by two processes (a) transport of oxygen from the atmosphere into the soil (atmospheric air contains 20.5 vol% O_2, soil air 10–20%), and (b) consumption of oxygen by bio-logical respiration or by chemical reactions.

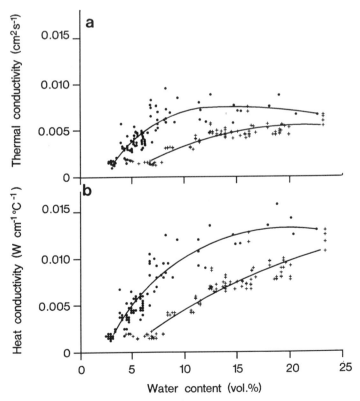

Figure 6 Apparent thermal conductivity (a) and heat conductivity (b) in disturbed (+) and in structured (·) loess as a function of water content.

Gas transport in the soil occurs both as a viscous flow along a pressure gradient and as diffusional flow with a concentration gradient in air-filled pores. Gas transport phenomena in soils are affected by pore size distribution, pore continuity, and water saturation (Scheffer and Schachtschabel, 1992). Currie (1965) dealt with the problem of a bimodal pore system on gas-exchange processes in soils and pointed out that soils with highly developed natural aggregates should have distinct zones of aggregate (i.e., intraaggregated) pores separated by a more continuous system of interaggregate pores. Gas transport in soil profiles will occur preferentially through this interaggregate pore system formed by macropores.

Besides diffusion, diurnal pressure and temperature changes allow an exchange of soil macropore air by mass flow processes. However, the transported gas volume is rather small and exchanged gas volumes are mainly in the top 10 cm of the soil (Glinski and Stepniewski, 1985). Oxygen diffusion to sinks (respiration by soil microorganisms) takes place in the intraaggregate pores and is induced by the concentration gradient resulting from respiration.

When the oxygen demand within soil aggregates is high and O_2 diffusion is limited by partial or even complete water saturation (oxygen diffusion will be reduced by a factor of 300,000 in water saturated pores) and low pore continuity, anoxic sites may develop even if the interaggregate pore space contains sufficient oxygen. These conditions often have been described in the field with respect to denitrification and root growth conditions (Flühler *et al.*, 1976; Smith, 1980; Tiedje *et al.*, 1984). Thus, in well-structured soils research on aeration should focus on oxygen transport within aggregates and on oxygen consumption. Greenwood and Goodman (1967) were the first to determine the oxygen distribution within single aggregates saturated with KCl. Using platinum electrodes they measured anoxic zones in aggregates of only 8-mm diameter. Sextone *et al.* (1985) used shielded microelectrodes with a tip diameter of 30–50 μm to determine the oxygen distribution within water-saturated aggregates. Both authors stated that O_2 gradients inside single aggregates appeared to be steeper in artificial or disturbed aggregates.

A. Effect of Texture on Gas Transport

Studies of the effects of soil texture on gas transport and on the composition of soil air in artificial aggregates reveal a strong correlation between the pore size distribution and the air entry value. The air entry value is the soil water potential at which gas diffusion to the aggregate center increases because water menisci are removed from continuous pores (Fig. 7).

In prisms of sandy-loamy texture the increase in O_2 partial pressure occurred at a soil water potential of about -15 kPa while polyhedrons

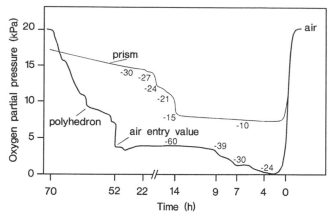

Figure 7 Oxygen partial pressure changes in a prism (loamy sand) and in a polyhedron (loamy clay) as a function of soil water potential (kPa) (numbers at distinct positions of the curves).

with loamy-clay texture became aerated at more negative water potentials of < − 60 kPa (Zausig et al., 1990). According to Fick's law, gas flow depends on the diffusion coefficient and concentration gradient. Assuming an oxygen content of the air surrounding the aggregate of 20%, O_2 transport within the aggregate would require additional time. If chemical and microbial oxygen demand inside the aggregate exceeds oxygen supply, zones of low oxygen partial pressure or even anoxic microsites may develop. Stepniewski et al. (1991) described a method where O_2-sensitive microelectrodes were pushed through soil aggregates at a constant speed of 0.00166 mm · s^{-1}. By this procedure continuous radial profiles of oxygen partial pressure could be measured. The method was used to compare the internal oxygen status of artificially formed spherical aggregates (diameter of 24 mm) of six different soil samples at soil water potentials ranging from − 1 to − 6 kPa (Zausig et al., 1993). It was found that the intensity of anoxia and the diameter of anoxic centers would be controlled not only by microbial and chemical oxygen demand but also by parameters such as aggregate hydraulic conductivity and pore size distribution, i.e., by the soil texture (Fig. 8).

In fine-textured aggregates of an A horizon of a Vertisol derived from Amaltheenton (site Tröbersdorf, northeast Bavaria) oxygen diffusion was severely restricted at a soil water potential of − 4 kPa while an increase in sand resulted in an oxygen content increase at even higher (less negative) soil water potential. A pronounced reduction of the O_2 partial pressure with increasing distance from the aggregate surface is found in aggregates containing clay because of the very pronounced tortuosity of the pore system. Only within 50 h after saturating spherical aggregates

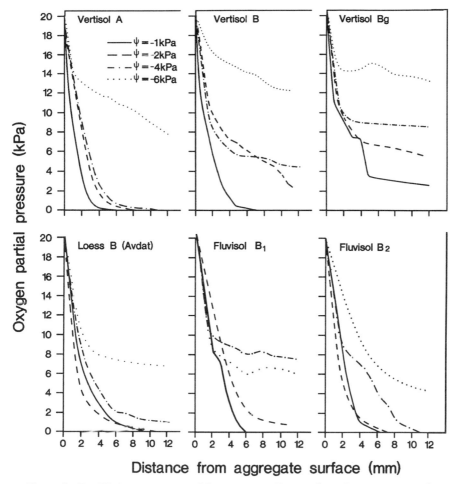

Figure 8 Equilibrium oxygen partial pressure vs distance from the aggregate surface for soil aggregates from varied soil horizons at −1, −2, −4, and −6 kPa soil water potential

(diameter 24 mm) of homogenized soil material of the Vertisol A horizon to −1 kPa soil water potential was an equilibrium state reached with an anoxic zone of 16 mm diameter. In sandy loamy aggregates anoxia was less pronounced.

B. The Effect of Soil Structure on Aeration

Aggregated soils always include secondary large interaggregate pores (coarse pores >50 μm) and small intraaggregate pores (finer pores <0 μm). This results in a heterogenization of the pore system and of the texture within the aggregates which strongly affects transport phenomena. The smaller the biological activity and the smaller the degree of organization of soil particles the smaller the intraaggregate pores which in turn restrict gas diffusion. A large reduction in O_2 partial pressure or even anoxia within aggregates under *in situ* conditions results from restricted pore space (diameter and continuity) as well as a source of reduced carbon, provided O_2 consumption by microbes is the main factor causing anoxia. Thus, natural soil aggregates should have an anoxic or less-aerated center if the soil water potential is < -60 kPa. This has been demonstrated for naturally developed aggregates of the A horizon of a Vertisol derived from Amaltheenton at the Tröbersdorf site (Zausig and Horn, 1992).

Spatial heterogenity in soils also occurs as a consequence of a nonhomogeneous distribution of microbial populations and organic substances. Several authors describe an accumulation of organic substances at the surface of soil aggregates due to (a) the flux of dissolved organic carbon in the coarse pores, (b) root growth and subsequent decay processes, and (c) the excretion of organic material by animals (Allison, 1968; Hattori, 1988; Alef and Kleiner, 1986; Christensen *et al.*, 1990; Augustin, 1992). In particular, aggregates of the subsoil horizons fulfill these conditions. They are predominantly coated with dissolved organic carbon in the outer skin (pore walls) which is then translocated down to deeper layers by rain water infiltrating the macropore system. Eventually, the material is deposited on pore walls, i.e., on aggregate surfaces. For example, naturally formed aggregates from the B and Bg horizons of the Vertisol derived from Amaltheenton (site, Tröbersdorf) exhibit internal oxygen partial pressures that were smaller than 8 kPa only when water potential was about -1 to -2 kPa. A sharp decrease of pO_2 was observed only within the first 2 mm from the aggregate surface, while the inner part maintained a constant level of pO_2 of 8 to 10 kPa (Zausig and Horn, 1992; Fig. 8). Thus, in subsurface horizons oxygen consumption by microbial respiration will happen mainly on the surfaces of aggregates by aerobic microorganisms (Hattori, 1988). Biogenic aggregate formation in the A horizons leads to intensive mixing of organic material with mineral soil particles. Thus, in surface horizons microorganisms are evenly distributed over the entire volume of loose and porous aggregates. Aerobic microbial activity would then cause oxygen depletion within the aggregate. Anoxic aggregate centers would develop where facultative anaerobic microorganisms become dominating (Horn *et al.*, 1993; Fig. 9).

Significant amounts of organic substances may induce large decreases in redox potential soon after saturation with water (Fig. 10). Thus, in humic A horizons redox potentials should drop rapidly upon wetting while in subsurface horizons with low contents of organic substances only

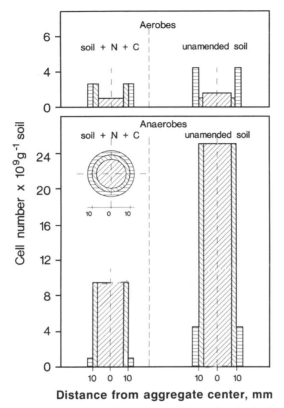

Figure 9   Distribution pattern of aerobes and anaerobes in single aggregates.

slow small redox potential changes occur. When, for example, artificial aggregates were prepared from soil material of a Bg horizon (0.3% org. C), almost the entire change in oxygen partial pressure was measured after 1 week of saturation at a temperature of 20°C. However, adding peat soil extract or sucrose solution (10%) during aggregate production caused the redox potential to decrease within only 2 days and created an anoxic zone 16 mm thick (Fig. 11; Zausig and Horn, 1991).

The intensity and speed of changes of redox potential may depend on more than the content of organic matter. The chemistry of the mineral soil components appears to be important as well. Soils containing clay show less-intensive changes of redox potential than silty soils. In aggregates with sandy texture, the largest decreases of redox potential occurred. Thus, fine-textured soils seem to contain more substances that function as a redox buffer, whereas the quartz fraction of sandy soils is

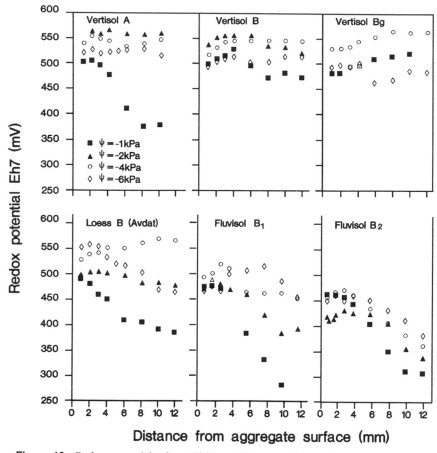

Figure 10 Redox potential values (Eh7) at different positions of spherical aggregates (diameter 24 mm) at -1, -2, and -3 kPa soil water potential

more or less inert and does not affect chemical processes. One of the first substances used as an electron acceptor by anaerobic microorganisms is nitrate-N. Only 12 h after watering aggregates from N- and C-enriched soil material to -0.5 kPa soil water potential denitrification was observed even in aggregates of 2 mm diameter (Fig. 12). Larger aggregates had bigger anoxic volumes and thus the amount of denitrified N increased (Horn et al., 1993). Also, the type of microorganisms differed between the inner and the outer part of the aggregates. In the outer skin aerobes did exist, while in the center denitrifiers dominated.

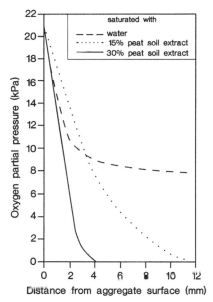

Figure 11 Oxygen partial pressure vs distance from aggregate surface in spherical aggregates of the Bg horizon of a Vertisol at -4 kPa.

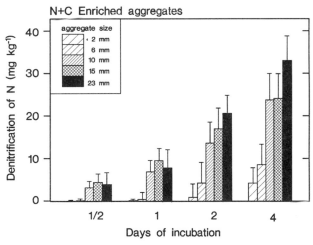

Figure 12 Denitrification rate in single aggregates as a function of time of incubation and aggregate diameter for N- and C-enriched artificial aggregates.

VI. Consequences of Aggregation on Plant Water Use

A. Water Fluxes in Lysimeters

The importance of newly formed aggregates for water transport depends on the intensity of aggregate formation, texture, maximum drying intensity, and number of swelling and drying cycles. Aggregate formation is also affected by water uptake of the plant (intensity, quantity) which depends on factors such as root length density and water uptake efficiency. Interactions between roots and soil were investigated in a lysimeter experiment which was carried out in the Negev (Israel) under well-defined arid climatic conditions. The lysimeters (depth 1, 2, or 3 m and diameter 3 m) were all filled with completely homogenized loess and watered once a year to field capacity at the beginning of each growing season for 4 years. The plant and root growth as well as the changes in the hydraulic properties of the soils were registered at the end of each growing season. In addition water content and soil water potential changes were determined continuously in the lysimeters. Soil water potential became more negative as expected during the growing season in the lysimeters. This was more pronounced in the upper soil layers due to the higher water uptake by roots and evaporation. It was shown that the change in soil water potential and the shape of the time-dependent decrease in water content was related to both the initial total amount of plant-available water in the lysimeters and to the plant age (Fig. 13). It was obvious that the water uptake was completed earlier with increasing age of the trees and with decreasing amount of plant-available water. As the total amount of plant-available water (pF 1.8–4.2) could not be transpired by the almond trees in 2-m- and especially in 3-m-deep lysimeters during the first growing season, the steepness of the water uptake rate (decline in the plant-available water with time) increased each year. The remaining amount of water at the end of each growing season decreased to zero with increasing age of the trees. Almond trees were capable of extracting water from soil below pF 4.2 probably due to osmotic adjustment in the root.

Such intense drying is expected to affect soil structure formation on a macroscale. This can be demonstrated by: (a) the increasing amount of large air-filled pores, (b) the slight increase in saturated hydraulic conductivity, and (c) the steeper decline of the corresponding unsaturated hydraulic conductivity. Less water had to be added at the beginning of each subsequent growing season to bring the soil to field capacity. Aggregate formation was also obvious from the more pronounced steepness of the soil water potential decline in the center of the lysimeters during the growing season and from the more rapid water infiltration in the center of the lysimeters (Fig. 14).

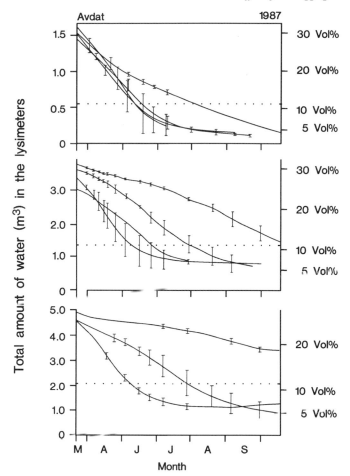

Figure 13 Changes in the total amount of water with time and treatment (1-, 2-, and 2-m-deep lysimeters filled with homogenized loess). The dotted line marks the amount that is not available to the plant.

The effect of reaggregation on the pore size distribution varied with the number of growing seasons and initial amount of plant-available water. A smaller amount of water had to be added to the 1-m-deep lysimeters at the beginning of the second growing season compared to the first year (1800–1580 liter). The 2-m-deep lysimeters were refilled with approximately 0.1 m³ less water. Thus, aggregate formation initially induced a reduction of the amount of plant-available water mainly by reduction of the amount of intermediate pores (ϕ, 50–10 μm). Only after hydraulic and strength properties reached equilibrium was the amount of

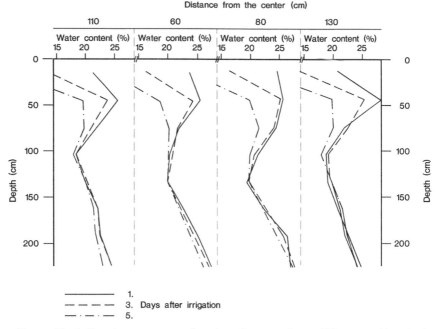

Figure 14 Infiltration pattern as a function of aggregation at different positions in the lysimeter over 5 days (lysimeter experiment Avdat/Israel). The more aggregated that the soil is, the deeper the water infiltrates.

plant-available water and the number of large air-filled pores increased. At this stage, aggregate bulk density and the bulk soil reached a steady state, the so-called "normal bulk density".

B. Interrelation between Plant Growth and Water Uptake

Plants may overcome the limitation of water availability in aggregated soils through increasing the hydraulic gradient at the root surface by decreasing plant water potential and increasing the root length density. Bohne and Hartge (1990) stated that the hydraulic gradients in the rhizosphere of barley or spinach could reach 9 (kPa m^{-1}) and thus increase water uptake by plant roots. Root growth increased with water uptake and root length density decreased the radius of soil cylinders per root under the assumption of an isotropic root growth (Fig. 15). In the lysimeter experiment root growth reached a final length density which depended on the hydraulic properties of the soil. During the 4 years of the lysimeter experiment in Israel a value of approximately $2 \cdot 10^3$ m $\cdot$ m^{-3} was obtained. However, the total amount of roots, which were excavated at the end of each growing season, did not represent the situation during the growing season. In addition to fine root losses during

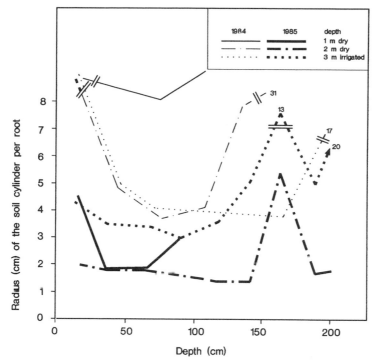

Figure 15　Changes in the radius of soil cylinders per root depending on the plant-available water for differently aggregated soils (loess, almond tree, Avdat/Israel).

harvesting and root washing, root losses occurred during the growing season. These roots decayed at a rapid rate under the warm climatic conditions. Fine root losses during washing may have amounted to 20% of total fine root length. In order to calculate a complete water balance including the seasonal changes of the hydraulic properties and the altered root length density in the second year of tree growth, a finite element model (FEM) was formulated and validated by the measured changes in water content or water potential. The FEM calculations showed that in the 3-m-deep lysimeters during the first 2 years the calculated changes in the water content or in water potential coincided with the measured ones (Fig. 16). The basis for the finite element model was the differentation of soil physical and chemical properties which can be defined for distinct soil volumes by triangulation techniques in all directions. Thus, hydraulic properties are defined in the *x, y, z* directions and the root length density and root water potential volumes as a function of time and treatment are available, then the transformed Laplace equation can be solved. The detailed description of the FEM as well as of the input

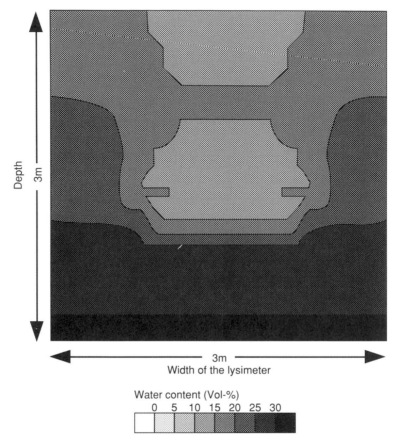

Figure 16 Calculated water content in the 3-m-deep lysimeter after 39 weeks of reaggregation. The degree of dryness differs at the various positions in the lysimeter at the same depth depending on the rooting density and corresponding water uptake efficiency (FEM technique).

parameters were described elsewhere (Richards *et al.*, 1993). In general, if well-documented changes in root growth over time and space as well as the corresponding alterations in the hydraulic properties are available, then the physical water flux model based on finite element techniques may be used to incorporate anisotropic soil properties for the pore size distribution as well as for the hydraulic conductivity in the vertical and horizontal direction. In addition, such models are also applicable for the quantification of the total amount of water loss or for the nutrient uptake.

ABA concentration in the xylem sap is an indicator of shortage of plant-available water in the soil (Wartinger *et al.*, 1990; see also this volume, Chapter 7). It was unclear if the maximum soil drying would also affect the ABA production. If the soil lysimeters (0.1 m³) planted with almond trees were watered and dried repeatedly to certain soil water potential values no ABA concentration increase in the xylem sap could be determined. If however, rewatering was delayed and the final soil water potential value was more negative as compared to the initial minimum value then the ABA concentration in plants generally increased proportionaly to the actual soil water potential (i.e., amount of plant-available water; Fig. 17). Thus, it could be postulated that there was no relation between the "non-plant-available" water (pores < 0.2 μm) and stomatal conductance. However, the smallest actual amount of available water determines the production of ABA, if the soil is further dried out.

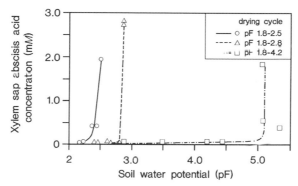

Figure 17　Abscissic acid concentration (ABA) increase in the xylem sap of almond trees due to a single more intensive drying as compared to the maximum previously defined dryness stage (i.e., pF value). Several cycles of wetting and drying to specific soil water potentials had occurred in advance.

VII. Conclusions

The results verify the idea that processes of aggregation do affect

- the availability of water for plant uptake, and
- the accessibility of water-filled pores for roots.
  This is due to the formation of finer pores in the outer skin of the soil aggregates where water is available only at more negative soil water potentials. This results in larger anaerobic soil volumes in aggregates as well as in an altered physicochemical behavior.
  A summary table of the aggregation processes and functions shows the various effects (Table II).

Table II Processes of Aggregation and Changes in Functions

Stage	Process	Aggregate type	Consequences	Hydraulic effects	Aeration effects	Thermal effects
I. Mechanical effects						
Early	Rectangular cracks	Singular, Coherent Prism	Increasing number of contact points Increasing aggregate bulk density Formation of inter- and intraaggregate pores	Decreased intraaggregate pore volume and saturated hydraulic conductivity Increased macropore flux	Reduced intraaggregate aeration Increased differentiation of aerobic and anaerobic zones	Increased spec. heat capacity Increased thermal conductivity
Increased number of drying cycles or drying intensity	Crack formation by shear forces Oblique shear planes	Polyhedron Subangular block	Rearrangement of particles inside aggregate Clay skins Accumulation of coarser particles in aggregate center Increased intraaggregate tortuosity	Decreased hyraulic conductivity as compared to bulk soil Smaller decline of the k/ψ curve of aggregates	Differentiation of gaseous composition Steeper decline of pO_2 at a given water potential inside the aggregate	Theoretical increase of conductivity in outer skin
Final	Reaching smallest free entropy and natural aggregate density	Prism Polyhedron Subangular block Spheroid	Reduced aggregate bulk density Increased aggregate strength	Increased plant available water Steeper slope of k/ψ curve	Improved aeration	Reduced thermal conductivity
II. Biological effects						
	Biological activity	Crumb	Homogenization by mixing processes reduced bulk density Increased strength	Increased saturated hydraulic conductivity Steeper slope of the k/ψ curve	Increased aeration Increased oxygen consume Decreased redoxpotential	Reduced thermal conductivity

The modeling of processes involving the effects of sequences of water uptake, plant growth, and aggregation on water fluxes is possible using a finite element analyses. However, this requires that the input parameters and especially those for the plant (root length, transpiration, etc.) are available.

Acknowledgments

Many of the results determined by the author and his co-workers were obtained in experiments financially supported by the German Research Foundation in the SFB 137. The author thanks especially Dr. T. Baumgarth and J. Zausig for their instructive collaboration during the preparation of this text. The author is thankful for this financial support.

References

Alef, K., and Kleiner, D. (1986). Arginine amonification, a simple method to estimate activity potentials in soils. *Soil Biol. Biochem.* **18,** 34–39.

Allison, R. E. (1968). Soil aggregation—some facts and fallacies as seen by a microbiologist. *Soil Sci.* **106,** 136–143.

Augustin, S. (1992). Mikrobielle Stofftransformationen in Bodenaggregaten, 152 S. Doctoral dissertation, Universität Göttingen.

Becher, H. H. (1991). Festigkeitverteilungen in Einzelaggregaten. *Mitt. Dtsch. Bodenkd. Ges.* **66,** 99–101.

Beven, K., and Germann, P. (1982). Macropores and water flow in soils. *Water Resour. Res.* **18,** 1311–1325.

Blanck, E. (1929-1939). "Handbuch der Bodenlehre," Vol. 7. Springer-Verlag, Heidelberg.

Blume, H. P. (1968). "Stauwasserböden," Ulmer, Stuttgart.

Bohne, H. (1988). Utilization of water by rye seedlings under conditions of restricted rooting. *Proc. ISTRO Conf.*, pp. 13–18. Edinburgh.

Cheshire, M. V. (1979). "Nature and Origins of Carbohydrates in Soils." Academic Press, London.

Christensen, S., Simkins, S., and Tiedje, J. M. (1990). Spatial variation in denitrification: Dependence of activity centers on the soil environment. *Soil. Sci. Soc. Am. J.* **54,** 1608–1613.

Currie, J. A. (1965). Diffusion within soil microstructure. A structural parameter for soils. *Soil Sci.* **16,** 278–289.

Dexter, A. R. (1988). Advances in characterisation of soil structure. *Soil Tillage Res.* **11,** 199–239.

Dexter, A. R., Horn, R., and Kemper, W. (1988). Two mechanisms of age hardening. *J. Soil Sci.* **39,** 163–175.

Emmerson, W. W., Bond, R. D., and Dexter, A. R., eds. (1978). "Modification of Soil Structure." Wiley, Chichester.

Flühler, H., Stolzy, L. H., and Ardakani, M. S. (1976). A statistical approach to define soil aeration in respect to denitrification. *Soil Sci.* **122,** 115–123.

Glinski, J., and Stepniewski, W. (1985). "Soil Aeration and Its Role for Plants." CRC Press, Boca Raton, FL.

Goss, M., and Reid, F. B. (1979). Influence of perennial ryegrass roots on aggregate stability. *Annu. Rep.—Agric. Res. Counc. Letcombe Lab.* pp. 24–25.

Greenwood, D. J., and Goodman, D. (1967). Direct measurement of the distribution of oxygen in soil aggregates and in columns of fine soil crumbs. *J. Soil Sci.* **18**, 182–196.

Gunzelmann, M. (1989). Quantifizierung und Simulation des Wasserhaushaltes von Einzelaggregaten und strukturierten Gesamtböden unter besonderer Berücksichtigung der Wasserspannungs-/Wasserleitfähigkeits-Beziehung von Einzelaggregaten. *Bayreuther Bodenkd. Ber.* **11**, 1–178.

Gunzelmann, M., and Horn, R. (1985). Wasserhaushaltsuntersuchungen in natürlich gelagerten Bodenaggregaten. *Mitt. Dtsch. Bodenkd. Ges.* **42**, 239–245.

Hartge, K. H., and Horn, R. (1977). Spannungen und Spannungsverteilungen als Entstehungsbedingungen von Aggregaten. *Mitt. Dtsch. Bodenkd. Ges.* **25**, 23–33.

Hartge, K. H., and Rathe, I. (1983). Schrumpf- und Scherrisse-Labormessungen. *Geoderma* **31**, 325–336.

Hattori, T. (1988). Soil aggregates as microhabitats of microorganisms. *Rep. Inst. Agric. Res., Tohoku Univ.* **37**, 23–36.

Hempfling, R., Schulten, H. R., and Horn, R. (1990). Relevance of humus composition for the physical/mechanical stability of agricultural soils: A study by direct pyrolysis—mass spectrometry. *J. Anal. Appl. Pyrol.* **17**, 275–281.

Hillel, D. (1980). "Fundamentals of Soil Physics." Academic Press, London.

Horn, R. (1976). Festigkeitsänderungen infolge von Aggregierungs-prozessen eines mesozoischen Tones. Doctoral Dissertation, Technische Universität, Hannover.

Horn, R. (1981). Die Bedeutung der Aggregierung von Böden für die mechanische Belastbarkeit. *Schriften. Tech. Univ. Berlin* FB14; 10, 1–200.

Horn, R. (1985). Der Einfluss der Frostgare auf bodenphysikalische Kenngrössen. *Z. Kulturtech. Flurbereinig.* **26**, 42–51.

Horn, R. (1987). The role of structure for nutrient sorptivity of soils. *Z. Pflanzenernaer. Bodenkd.* **150**, 13–16.

Horn, R. (1989). Aggregate characterisation as compared to soil bulk properties. *Soil Tillage Res.* **19**, 268–289.

Horn, R., and Dexter, A. R. (1989). Dynamics of soil aggregation in an irrigated desert loess. *Soil Tillage Res.* **13**, 252–266.

Horn, R., and Taubner, H. (1989). Effect of aggregation on potassium flux in a structured soil. *Z. Pflanzenernaehr. Bodenkd.* **152**, 99–104.

Horn, R., Stork, J., and Dexter, A. R. (1987). Untersuchungen über den Einfluß des Bodengefüges für den Eindringwiderstand in Böden. *Z. Pflanzenernaehr. Bodenkd.* **150**, 342–347.

Horn, R., Taubner, H., and Hantschel, R. (1989). Effect of structure on water transport, proton buffering and nutrient release. *Ecol. Stud.* **77**, 323–340.

Horn, R., Stepniewski, W., Wlodarczyk, T., Walensik, G., and Eckhardt, E. F. M. (1993). Denitrification rate and microbial distribution within homogenous soil aggregates. *Soil Sci. Soc. Am. J.* (in press).

Horton, R. P. J., Wierenga, P., and Nielsen, D. R. (1983). Evaluation of theoretically predicted thermal conductivities of soils under field and laboratory conditions. *Soil Sci. Soc. Am. J.* **41**, 460–466.

Martin, J. K. (1977). Factors influencing the loss of organic carbon from wheat roots. *Soil Biol. Biochem.* **9**, 1–7.

Page, J. B., and Willard, C. J. (1947). Cropping systems and soil properties. *Soil Sci. Soc. Am. Proc.* **11**, 81–88.

Richards, B. G., Horn, R., and Baumgartl, T. (1993). FEM technique for the prediction of water transport processes in the SPAC. In preparation.

Rynasiewicz, J. (1945). Soil aggregation and cotton yield. *Soil Sci.* **60**, 387–396.

Schababerle, P. (1990). Stofftransport und Gefügeänderungen beim partiellen Gefriercn von Ton-Barrieren. Schriftenreihe Angew, Geol. Karlsruhe **7**, 1–214.

Scheffer, F., and Schachtschabel, P. (1992). Lehrbuch der Bodenkunde, 13th ed., p. 491. Enke Verlag, Stuttgart.

Semmel, H., Horn, R., Hell, U., Dexter, A. R., and Schulze, E.-D. (1989). The dynamic of aggregate formation and the effect on soil physical properties. *Soil Technol.* **3**, 113–129.

Sextone, A. J., Revsbech, N. P., Parkin, T. B., and Tiedje, J. M. (1985). Direct measurement of oxygen profiles and denitrification rates in soil aggregates. *Soil Sci. Soc. Am. J.* **49**, 645–651.

Smith, K. A. (1980). A model of the extend of anaerobic zones in aggregated soils, and its potential application to estimates of denitrification. *J. Soil Sci.* **31**, 263–277.

Stepniewski, W., Zausig, J., Niggemann, S., and Horn, R. (1991). A dynamic method to determine the O_2-partial pressure distribution within soil aggregates. *Z. Pflanzenernäehr. Bodenkd.* **154**, 59–61.

Strassburger, E. (1969). "Lehrbuch der Botanik," Fischer, Stuttgart.

Tiedje, J. M., Sextone, A. J., Parkin, T. B., Revsbech, N. P., and Shelton, D. R. (1984). Anaerobic processes in soil. *Plant Soil* **76**, 197–212.

Tippkötter, R. (1988). Aspekte der Aggregierung. Habilitationsschrift, Universität Hannover.

Türk, T., Mahr, A., and Horn, R. (1991). Tensiometrische Untersuchungen an Aggregaten in homogenisiertem Löss *Z. Pflanzenernaehr. Bodenkd.* **154**, 361–368.

Utomo, W. H., and Dexter, A. R. (1981). Age hardening of agricultural top soils. *J. Soil Sci.* **32**, 335–350.

Wartinger, A., Heilmeier, H., Hartung, H., and Schulze, E.-D. (1990). Daily and seasonal courses of leaf conductance and abscisic acid in the xylem sap of almond trees (*Prunus dulcis* (Miller) D. A. Webb) under desert conditions. *New Phytol.* **116**, 581–587.

Wollny, E. (1898). Untersuchungen über den Einfluss der mechanischen Bearbeitung auf die Fruchtbarkeit des Bodens. *Forsch. Geb. Agrik. Phys.* **20**, 231–290.

Zausig, J., and Horn, R. (1991). Der Belüftungszustand eines Pelosol Gleyes als Funktion des Bodenwasserhaushaltes. *Mitt. Dtsch. Bodenkd. Ges.* **66**, 55–58.

Zausig, J., and Horn, R. (1992). Soil water relations and aeration status of single soil aggregates, taken from a gleyic vertisol. *Z. Pflanzenernaehr. Bodenkd.* **155**, 237–245.

Zausig, J., Hell, U., and Horn, R. (1990). Eine Methode zur Ermittlung der wasserspannungsabhängigen Änderung des Sauerstoffpartial–druckes und der Sauerstoffdiffusion in einzelnen Bodenaggregaten. *Z. Plfanzenernaehr. Bodenkd.* **153**, 5-10.

Zausig, J., Stepniewski, W., and Horn, R. (1993). Oxygen concentration and redox potential gradients in different model soil aggregates at a range of low moisture tensions. *Soil Sci. Soc. Am. J.* (in press).

Zhang, H. (1991). Der Einfluss der organischen Substanz auf die mechanischen Eigenschaften von Böden. Doctoral Dissertation, Universität Hannover.

IV

Flux Control at the Population and Ecosystem Level

11

Structure and Biomass Transfer in Food Webs: Stability, Fluctuations, and Network Control

H. Zwölfer

I. Introduction

Food webs are an essential element of any ecosystem as they mediate the transfer of energy and biomass from the producer to the consumer levels. Since food webs connect plants, animals, and microorganisms within ecological communities, they are appropriate for describing the organization of ecological systems and the trophic interrelationships of organisms. However, food webs of large ecosystems are usually so complex and contain components which are so difficult to assess that a precise inventory of the species and a quantitative representation of food relationships are not feasible. For this reason there is a growing interest in the analysis of tritrophic plant–insect systems, i.e., of food webs which consist of plants and phytophagous and entomophagous insects. Examples are the studies by Price *et al.* (1980), Price and Clancy (1986), Ehler (1992), and Tscharntke (1992). For our study of the structure and function of food webs we have chosen a series of complexes of phytophagous and entomophagous insects associated with the flower heads and stem galls of the host plant taxon Cardueae (family Asteraceae). These plant–insect complexes constitute ecological microsystems, which have been studied for several dozen related plant species. They represent "evolutionary replicates" of a particular type of food web. Cardueae–insect food webs can be investigated along different geographical transects and under different ecological and climatological conditions, thus offering a chance to compare ecological variants. They also allow the comparison of autochthonous and allochthonous systems, as a considerable number of European Cardueae insects has been used as biocontrol agents against weedy Cardueae species in North America (Harris, 1991).

Interest in the energetics of food webs started with the classical study of Lindeman (1942). Since this study much information on the flux and transfer rates of energy along food chains (Phillipson, 1966; Odum, 1983; Ellenberg *et al.*, 1986) has become available. Problems related to the topographic structure of food webs and the consequences with regard to niche space and food web stability have gained increasing attention (Cohen, 1978; Pimm, 1982, 1984, 1988; Pimm *et al.*, 1991). In the context of this book, which investigates principles and mechanisms guiding the flow of energy and matter in ecological systems, this contribution focuses on the question of to what extent and by which control mechanisms the flux of energy and biomass in the investigated food webs is regulated. A necessary corollary is a discussion of the organization, predictability, and variability of the food webs studied.

II. Cardueae–Insect Food Webs

The following chapter provides an inventory of insect species associated with Cardueae host plants (Zwölfer, 1965, 1988, 1990). Information on food webs in flower heads is available for the following Cardueae genera: *Arctium* (4 spp.), *Staehelina* (1 sp.), *Carduus* (9 spp.), *Cirsium* (16 spp.), *Galactites* (1 sp.), *Onopordum* (6 spp.) (all in the subtribe Carduinae), and *Centaurea* (17 spp.), *Microlonchus* (1 sp.), *Carthamus* (2 spp.) (all in the subtribe Centaureinae). In addition we investigated the insects associated with the Carlineae genera: *Xeranthemum* (2 spp.), *Carlina* (3 spp.) and the Echinopeae genus *Echinops* (2 spp.). In the following I only deal with Cardueae food webs, as they are more diversified than those of the other two Asteraceae tribes. Our most intensively studied system, the *Cirsium arvense–Urophora cardui* food web, involves stem galls and not flower heads. However, this system can be derived from an ancestral flower head food web (Zwölfer and Arnold-Rinehart, 1992) and can be seen as an ecological analogue of *Urophora* food webs in Cardueae flower heads.

A. Ecological Context and Structure

1. Two System Levels: Macro- and Microhabitats   The food webs investigated are formed by endophytic larvae of holometabolus insects, which occupy flower heads or plant galls, i.e., discrete microhabitats. These microhabitats are recurrent and relatively short-lived subunits (merocenoses sensu Tischler, 1949) of the vegetation. They form the environment and resources of eggs, larvae, and pupae of the primary and secondary consumers of the food web. The carrying capacity of these microhabitats is measurable (Zwölfer, 1979) and strongly limited. As the larvae of almost all flower head and gall inhabitants are entirely confined to their respective microhabitats, interactions at the consumer levels comprise not only predator–prey relationships and cannibalism, but also competition for food and space. The conditions for the development, survival, and interaction of the insect community within the microhabitat "flower head" are determined by the behavior of their adult stages, which live in the macrohabitat, i.e., the sites occupied by the populations of the host plants. Here they use their repertoire of sensory and discriminatory capacities to collect information on resources such as food, mates, and oviposition sites. Mating and oviposition are the key behavior patterns, setting the initial conditions for the processes occurring within the microhabitats. Whereas the dominant activity in the microhabitat is feeding, i.e., the acquisition and accumulation of energy and biomass, the insect's activities in the macrohabitat are primarily characterized by "informational processes." The adults of phytophagous insects and their parasitoids gain information about environmental variables which they use in decision-making procedures (Zwölfer, 1985a).

2. Organization and Complexity Figures 1A–1C represent examples of a typical organization of tritrophic food webs in Cardueae flower heads (Table I). The arrow at the left side of the flow charts symbolizes the input of energy and assimilates into the system, i.e., the amount sequestered by the herbivores. The output of the system consists of adult herbivores and parasitoids (right side of the chart) and heat due to the respiration of the system (indicated by "heat sink" symbols; Odum, 1983). Another output of the flower head is achenes. (The impact of the food web on the achene production of the host plant is discussed in Section II.B.1).

Each graph of Fig. 1 shows the pathways of energy flow for a particular field population (= a sample of flower heads collected at a given date at one of our observation sites). We did not calculate flux rates but assessed the amount of energy (in Joules) temporarily stored in different subunits of the food web (gall/callus tissue, mature larvae of herbivores and parasitoids, adults of herbivores). The energy values shown in the graphs have been calculated for samples of 100 flower heads. The flow charts use the symbols of the "energy circuit language" discussed by Odum (1983). Lines and arrows represent pathways for the flow of energy and materials through the food web. The "tank symbols" (compartments of storage of energy and material) at the upper left corner of the charts indicate galls and/or plant callus, i.e., tissues, the growth of which has been induced by specialized herbivores belonging to the family Tephritidae (Diptera) (*Urophora stylata* (Fig 1A, 2A), U. *cuspidata* (Fig 1B) or U. *jaceana* (Fig.

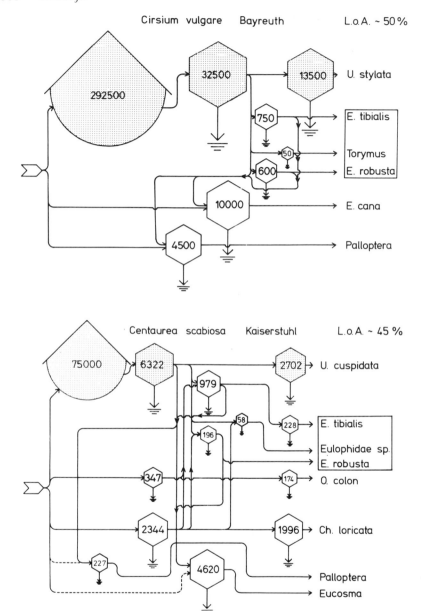

Figure 1 Foodwebs in flower head populations of *Cirsium vulgare* (A), *Centaurea scabiosa* (B), and *Centaurea jacea* (C). Arrows indicate the direction of the energy flow. Symbols (Odum, 1983) indicate a compartment of storage ("tank") and consumers which transform and store energy. Energy stored in the *Urophora* galls and in mature larvae and adults of herbivore and parasitoid species has been calculated in Joules for a population of 100 flower heads. Further explanations are given in the text (modified from Zwölfer, 1985a).

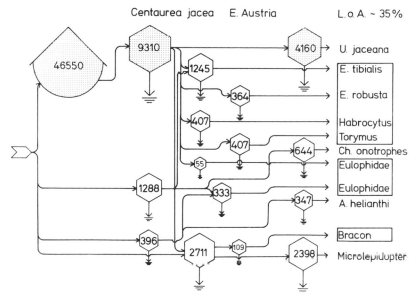

Figure 1   (*continued*)

Table I   Number of Host Plant Species for Which Different Trophic Strategies
Could be Recorded in the Herbivore Guilds in Flower Heads

Tribe and genus of host plant	Strategy 1, 2, and 3 occur	Strategy 2 and 3 occur, 1 needs confirmation	Only strategy 3 and/or 2
Cardueae			
Arctium	3	1	—
Carduus	6	1	2
Cirsium	12	2	1
Silybum	1	—	—
Galactites	—	—	1
Onopordum	3	1	—
Centaurea	13	2	3
Microlonchus	1	—	—
Serratula	—	1	—
Carthamus	3	—	—
Total	42	8	7
Carlineae			
Carlina	—	—	5
Xeranthemum	—	—	2
Staehelina	—	—	1
Echinopeae			
Echinops	—	—	3
Total	0	0	11

1C). The figures in the tank symbols refer to that part of the energy and matter stored in nontrophic, protective tissues, i.e., they represent energy used in structuring the microhabitat by the formation of sclerotized tissues of the gall complexes (Arnold-Rinehart, 1989). Energy and matter of the trophic tissues of the galls or of callus tissue are consumed by the larvae of the gall formers and thus enter the food web. The energy content of insect larvae (and for some species also of adults) is indicated by the size of the "consumer symbol" (= units that transform energy quality, store energy, and feed back autocatalytically to ensure inflow; Odum, 1983). The "heat sink" symbols (Odum, 1983) connected to the consumer symbols represent the positions of the food web where respiration transforms chemical energy into heat.

A total of 20–50% of the energy and matter present in the mature larvae of primary consumers (in Fig. 1, *Urophora* spp.) reaches the stage of the adult herbivores (which, together with adult parasitoids and the protective gall structures, form the final, net production of the system). The difference in energy is used for metabolic processes (pupation) or passes to the third trophic level which is dominated by larvae of parasitoids (larvae of the hymenopterans *Eurytoma tibialis*, *Eurytoma robusta*, *Torymus* sp., *Habrocytus* sp. and Eulophidae gen. sp.). The parasitoid *E. robusta* which is closely associated with *Urophora* galls is capable of operating at two trophic levels. If not enough biomass of host larvae is available, it consumes the trophic layer of *Urophora* galls (Zwölfer, 1979).

Gall inducers (Weis *et al.,* 1988) such as *Urophora* larvae (Figs. 2C and 2D) consume the content of the enriched cells of the trophic gall tissue (Arnold-Rinehart, 1989). Production efficiency (= percentage of assimilated energy of gall tissue which is incorporated into new herbivore biomass) has not been measured for *Urophora*, but data available for an ecologically similar gall-forming tephritid (*Eurosta solidaginis* on *Solidago* spp.; Stinner and Abrahamson, 1979) suggest that gall formers have a production efficiency of about 40%. Production efficiency of achene and receptacle feeders varies from 20 to 25% (Zwölfer, unpublished) and production efficiency of parasitoids in Cardueae flower heads (% biomass of host larvae transformed into biomass of parasitoid larvae) varies from 29 to 43% (Zwölfer, 1979; Romstöck, 1982; Michaelis, 1984).

Most species of the herbivore guild living in Cardueae flower heads do not induce galls or callus growth but feed on achenes and receptacle tissues. Examples are the tephritids *Chaetostomella onotrophes* Loew and *Acanthiophilus helianthi* Rossi (both Fig. 1B) and *Orellia colon* Meig. and *Chaetorellia loricata* Rond. (both Fig. 1C).

A third group of herbivores comprises species which can switch their diet and become important predators in the Cardueae food web. Examples in Fig. 1 are the larvae of several moth species (*Eucosma cana*

Haw., *Eucosma* sp. (Lepidoptera: Tortricidae) and other Microlepidoptera (Pyralidae (*Homeoesoma* spp.) and Gelechiidae (*Metzneria* spp.)). Among the saprophagous inhabitants of Cardueae flower heads, *Palloptera* larvae (Diptera: Pallopteridae) (Figs. 1A and 1C) feed not only on decaying tissues but also on any insect larva with which they come into contact. Figure 1 shows that insect complexes in Cardueae flower heads are characterized by numerous interactions between species occupying the second and third level of the food web. A common measure of such food web complexity is the degree of "connectance," that is, the actual, divided by the possible number of interspecific interactions (Pimm, 1982). In the food webs in Cardueae flower heads which we have investigated connectance ranges from 0.25 to 0.8, with a mean of 0.52 and a 95% confidence interval of ± 0.18. If these values are compared to the data of Cohen (1978) and Pimm (1982), connectance in our Cardueae food webs is clearly higher than the published average (Michaelis, 1984; H. Zwölfer, unpublished). This is also the case in the three food webs represented in Fig. 1. Pimm (1982, Fig. 5.1) calculated a calibration curve which shows the mean connectance for a given number of species in 26 terrestrial and aquatic food webs. This curve would predict a connectance of 0.5 for the *Cirsium vulgare* food web in Fig. 1A (actual connectance = 0.87), a connectance of 0.31 for the *Centaurea jacea* food web in Fig. 1C (actual connectance = 0.40) and a connectance of 0.42 for the *Centaurea scabiosa* food web in Fig. 1B (actual connectance = 0.50). The comparatively high values of connectance in food webs in Cardueae flower heads occur because certain phytophagous species (strategy 3 in Section II.A.3) can operate at more than one trophic level, and also because many parasitoids in these systems are niche specific and not taxon specific, i.e. they attack different herbivore host species in the same flower head population (Capek and Zwölfer, 1990).

3. Guild Structure and Energetic Aspects   In the flower heads of the majority of Cardueae species (Table I) the herbivore guild includes species with three different trophic strategies (Zwölfer, 1987):

Strategy 1 ("early aggregated attack").   These herbivores oviposit into immature flower heads. Usually their larvae occur gregariously within individual heads. They form either structural galls into which an additional flow of assimilates is induced (e.g., the members of the tephritid genus *Urophora* or of the cynipid genus *Isocolus*), or they exploit callus tissues (e.g., the weevil genera *Rhinocyllus* and *Bangasternus* and some members of the tephritid genus *Tephritis*). The species with this strategy tend to be host specific and often have evolved biotypes (Zwölfer and Romstöck-Völkl, 1991).

Strategy 2 ("achene and receptacle feeders").   Members of this group

attack a more advanced developmental stage of the flower head. They occur singly or are moderately aggregated and feed on mature tissue of the achenes and receptacle. In contrast with the first group, these species (mainly tephritids and members of the weevil genus *Larinus*) are not capable of inducing gall or callus tissues.

Strategy 3 ("operation at two trophic levels").   Species belonging to this group (e.g., members of the gelechiid genus *Metzneria,* the pyralid genus *Homoeosoma,* the tortricid genera *Eucosma* and *Epiblema,* and the anobiid genus *Lasioderma*) are active during the maturation phase of flower heads. If they come into contact with other herbivores, they switch to cannibalism or carnivory. In most cases only a single individual occupies a flower head. Members of some genera (e.g., *Homoeosoma, Eucosma*) are able to leave one flower head and enter another.

The extent to which the herbivore guild drains assimilates and energy from the flower heads of their host plants varies greatly with their trophic strategies. Figure 3 shows the average amount of energy stored in various compartments (protective gall tissue, mature larvae of herbivores (strategies 1,2,3) and mature larvae of parasitoids) and in the output of the

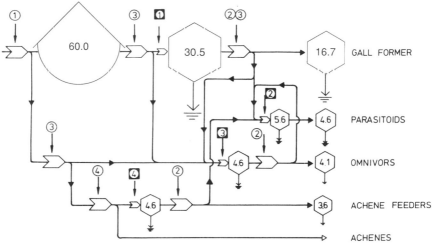

Figure 3 Summary of the energy content in the compartments of foodwebs in *Cirsium* and *Centaurea* heads. (Average values for 12 populations (*Cirsium vulgare, C. helenoides, Centaurea scabiosa, C. jacea*).) Symbols as in Fig. 1. Insects have been classified as herbivores with the trophic strategy I, parasitoids, and herbivores with the trophic strategies II and III. Figures in circles indicate points where the direction of the energy flow is controlled by members of the foodweb. Further explanations are given in the text (modified from Zwölfer, 1985a).

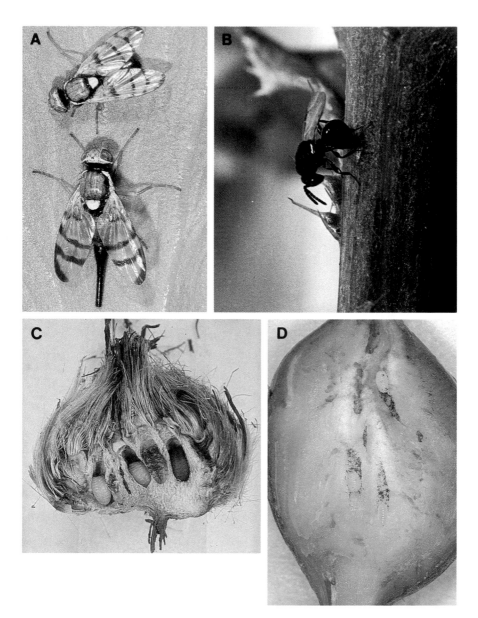

Figure 2　(A) Male and female of *Urophora stylata*, a tephritid species which induces galls in the flower heads of the spear thistle *Cirsium vulgare*. (B) A female of the chalcid wasp *Eurytoma serratulae* which probes a stem of creeping thistle, *Cirsium arvense*, for young larvae of *Urophora cardui*, a tephritid which induces stem galls in this host plant. (C) Cross section through a mature multilocular gall of the tephritid *Urophora solstitialis* in the flower head of the nodding thistle, *Carduus nutans*. The figure shows the lignified part of the gall with four chambers, occupied by a parasitized *Urophora* larvae (at left) and two mature, hibernating third-instar larvae of *Urophora* (center and at right). (D) Immature stem gall of *Urophora cardui* in *C. arvense*. The figure shows three immature larvae, which entered from above (right part of the gall) and feed on the surrounding trophic tissue.

system (adults of herbivores (strategies 1,2,3) and parasitoids). It is based on 12 analyzed food webs for the heads of two *Cirsium* spp. and two *Centaurea* spp. The figures in the Odum symbols represent the percentages of the energy content of the compartments. They have been calculated on the basis of the combined energy content of the protective gall tissue and mature herbivore larvae, which has been set to 100%. Figure 3 indicates that the highest amount of energy in the food web is invested in the protective structures of the galls (60%) and that on the average mature larvae of herbivore species which employ strategy 1 (early aggregated attack) store much more energy in their biomass (30.5%) than mature parasitoid larvae (5.6%) or herbivore larvae using strategy 2 (4.9%) or strategy 3 (4.6%). The dominant position of the strategy of gall formers is also shown in Table II which analyzes the *C. vulgare–U. stylata* food webs studied by Michaelis (1984) and H. Zwölfer (unpublished) near Bayreuth. It completes Fig. 3 as it gives the energy content of the food web and plant compartments of the flower head (achenes, receptacle, bracts).

The energy content of adult insects, i.e., the output of the systems described in Fig. 3 (adults gall formers (16.7%), adult parasitoids (4.6%), adults omnivores (4.1%) and adult of achene feeders (3.6%)) amounts to a total of 29%. Thus, an average of 63.6% of the energy content of mature insect larvae (herbivores and parasitoids) represents the net production of adult insects, whereas an average of 36.4% is used for metabolic processes (energy loss due to pupation and respiration). If the fate of the larval biomass of the gall formers is analyzed, its net production is only 55%, i.e., energy losses due to respiration and parasitism amount to 45%. In the analyzed samples parasitism of the other two groups of

Table II Example of the Energy Content of the Compartments in an Average *Cirsium vulgare* Flower Head with *Urophora stylata* Galls (with 3 to 6 cells)

Compartment	Energy
Mature flower head	25000 J/head = 100%
Achenes	531 J/head = 2.1%
Tissues of receptacle and bracts	13469 J/head = 53.9%
Gall + *U. stylata* larvae	11000 J/head = 44% (40–51%)
Protective gall tissues	10162 J/head = 40.65% (37–48%)
U. stylata + parasitoid larvae	838 J/head = 3.35% (2.4–4.2%)
U. stylata larvae	582 J/head = 2.33%
Parasitoid larvae	256 J/head = 1.02%

Data from Michaelis (1984); percentage in brackets, unpublished data by H. Zwölfer.

herbivores was low. Therefore the values given for the energy content of the output of adults of these groups may not be very representative.

The most important aspect of Fig. 3 is the fact that herbivores which induce galls or callus tissue sequester on the average nine times more energy and materials than the phytophagous species which do not stimulate tissue growth. However, an important part of this amount of assimilates and energy is invested in protective gall tissue, i.e., it is stored in a nontrophic compartment of the food web (Table II). As the majority of parasitoids also attack the herbivores using strategy 1, this group obviously occupies a key position in these Cardueae food webs.

Urophora galls create an additional sink for assimilates in the Cardueae flower heads (Harris, 1980; Zwölfer, 1985a). As the galls drain and accumulate matter and energy from other inflorescences of the plant, flower heads with *Urophora* galls have distinctly more biomass and a higher energy content than unattacked heads (Michaelis, 1984; Zwölfer, 1985a). *Tephritis conura* Loew, the dominant herbivore in heads of *Cirsium helenoides* and a typical representative of the strategy of early aggregated attack feeds on callus which it induces in the receptacle of the flower head. Romstöck (1987) demonstrated by calorimetric measurements that the growth of callus results merely from a reallocation of assimilates within the individual flower heads, i.e., *T. conura* does not convey additional energy into the attacked flower heads.

B. Degree of Resource Utilization

An important aspect of the stability of food webs is the proportion of the available resources (energy and materials) which is utilized by primary and secondary consumers. In the following sections we discuss first resource utilization by herbivores (which is measured by the proportion of attacked flower heads and destroyed achenes) and then that by parasitoids. Resource utilization in the *C. arvense–U. cardui* system is discussed in a separate section.

1. Resource Utilization by Herbivores

Zwölfer (1985b) analyzed the relationship between the attack on flower heads and the guild size of Cardueae herbivores. In most cases there is a significant correlation between the number of herbivore species and the percentage of flower heads attacked (Fig. 4). This is particularly high in *Centaurea solstitialis*, in which up to 12 herbivore species can coexist in a flower head population (Sobhian and Zwölfer, 1985) and in which the herbivore guild size explains 85% ($r^2 = 0.847$***) of the variation in the utilization of flower heads. The coefficients of determination of *C. scabiosa* ($r^2 = 0.562$ ***), *Centaurea diffusa* ($r^2 = 0.387$***), *C. arvense* ($r^2 = 0.334$***), *Carduus nutans* ($r^2 = 0.311$***), *C. jacea* ($r^2 = 0.275$***), and *Centaurea maculosa*

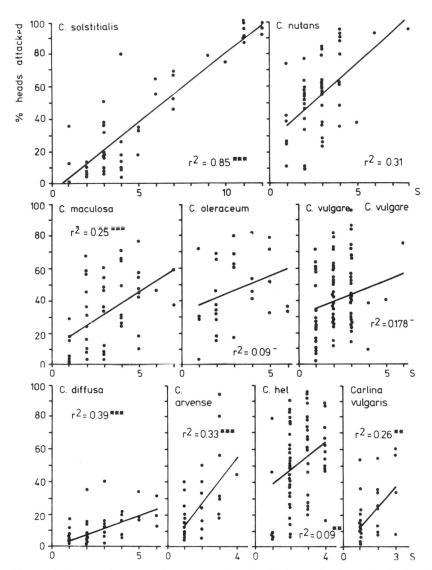

Figure 4 Scatter diagrams of resource utilization (% flower heads attacked) plotted against herbivore species richness (S, α diversity). Each dot represents a flower head population of 50–200 heads (modified from Zwölfer, 1985b).

(r^2 = 0.248***) are also highly significant. Correlations between guild size and attack on flower heads are less pronounced or statistically insignificant in *C. helenoides* (r^2 = 0.092**), *Cirsium oleraceum* (r^2 = 0.087[ns]), and *C. vulgare* (r^2 = 0.178[ns]) as these guilds are dominated by a single herbivore species (*T. conura* or *U. stylata*, respectively).

The average values of attack rates on flower heads for the examples shown in Fig. 4 range from 8.6% (*C. diffusa*) to 54.6% (*C. nutans*) and have a mean of 34.2% ± 9.49% (95% confidence interval). Attack rates are higher in Cardueae species with large flower heads (e.g., *C. nutans*, head diameter 25–30 mm) than in species with small heads (e.g., *C. diffusa*, head diameter 5–7 mm), but the extent to which achenes in attacked heads are consumed is higher in small heads.

Figure 4 represents the utilization of flower heads by herbivore guilds and Table III gives data on the impact of single herbivore species. Here I use information from Michaelis (1984), Romstöck (1987), Angermann (1987), Sturm (1988), and unpublished personal data to calculate rates of seed loss due to consumption by herbivores. Average attack rates on flower heads range from 2.8 to 46.9% (mean = 25.3%) and rates of seed loss range from 2 to 40% (mean and standard error = 13.03 ± 2.18%). *Tephritis conura* populations exploiting heads of *C. helenoides*, which occur in simple and particularly stable food webs, have an overproportionally high consumption rate (Romstöck, 1987). If their values are excluded from Table III, the average rate of seed loss is 11.6 ± 1.74%. These values which originate from Cardueae populations in Central Europe are lower than those obtained by Petney (1988) from eight Cardueae species (12 populations) in Jordan. If the small sample of *Onopordum alexandrinum* with an estimated percentage achene loss of 92.7% is excluded, his data suggest an average seed loss of 33.7 ± 7.29%. A comparison with our data is difficult, because the tephritid fauna in Jordan is depauperate compared with Europe while weevils are more abundant than in Europe (Petney and Zwölfer, 1985). The low average rate of consumption leads to the question of which mechanisms limit and stabilize the level of resource utilization of the herbivore guild in flower heads (Section III).

Müller (1984) investigated the insect fauna exploiting the roots of several *Centaurea* species. He found that 29% (*C. diffusa*, n = 550), 25% (*C. arenaria*, n = 531), 33% (*C. maculosa*, n = 3071), and 33% (*C. vallesiaca*, n = 577) of the roots were attacked by phytophagous insects. With an average of 30% these values are comparable to those obtained from dissections of flower heads (Table III).

Compared to the infestation of Cardueae flower heads and roots Freese (1991, 1992, 1993a) found a much higher attack rate in stems of *Cirsium* and *Carduus* spp. In five *Cirsium* and two *Carduus* species the number of

Table III   Examples of Resource Utilization of Herbivores in Cardueae
Flower Heads

Herbivore	Host plant	N Heads dissected	% Heads attacked	% Loss of achenes	Source
Urophora vulgare	*Cirsium stylata*	5500	22%	7.7%	(1)
U. vulgare	*C. stylata*	1756	38%	13.3%	(2)
U. vulgare	*C. stylata*	920	39.4%	13.8%	(3)
U. vulgare	*C. stylata*	1633	41.8%	14.6%	(4)
U. vulgare	*C. stylata*	5529	24.7%	8.7%	(5)
U. vulgare	*C. stylata*	940	23.9%	8.4%	(6)
Terellia serratulae	*C. stylata*	5500	8%	3.6%	(1)
Tephritis conura	*Cirsium helenoides*	3600	46.9%	35–40%	(7)
T. conura	*Cirsium oleraceum*	2200	29.5%	24.5%	(8)
T. conura	*C. oleraceum*	2000	24%	19.9%	(9)
T. conura	*Cirsium acaule*	1700	11%	?	(9)
T. conura	*Cirsium palustre*	3000	15%	?	(9)
T. conura	*Cirsium erisithales*	700	43%	?	(9)
Xyphosia miliaria	*C. palustre*	350	25%	22.5%	(10)
X. miliaria	*Cirsium arvense*	530	24.2%	19.4%	(11)
X. miliaria	*C. arvense*	3500	20.3%	16.3%	(13)
Orellia ruficauda	*C. arvense*	3500	2.8%	>2%	(13)
Tephritis bardanae	*Arctium* spp.	2925	22.5%	10.4%	(12)
Orellia tussilag.	spp.	662	16.9%	3.4%	(12)
Urophora affinis	*Centaurea maculosa*	6430	20%	3.8%	(14)
Guild of 8 herbivores	*C. maculosa*	6430	33.1%	4.7%	(14)

Sources. (1) Michaelis (1984); (2) Bayreuth, 1986; Zwölfer, unpublished data; (3) Bayreuth, 1987; Zwölfer, unpublished data; (4) Bayreuth, 1988; Zwölfer, unpublished data; (5) Zwölfer (1972); (6) Redfern (1968); (7) Romstöck (1982); (8) Eschenbacher (1982); (9) Romstöck (1987); (10) Arnold (1985); (11) Angermann (1984); (12) Sturm (1988); (13) Angermann (1987); (14) Zwölfer (1978).

% loss of achenes = estimates for the whole flower head population on the basis of loss of achenes in heads with average numbers of herbivore larvae.

stems occupied by larvae of phytophagous insects ranged from 68.1% in *C. arvense* to 100% in *C. nutans* with a mean of 89.4% (SE ± 3.8%).

2. Resource Utilization by Parasitoids   To assess parasitization rates, a large series of dissections of flower head samples were made. They all showed that the parasitoid guilds of the flower head food webs usually consume only moderate proportions of herbivore larvae and pupae. For the parasitoid complex of *U. stylata* in *C. vulgare*, Redfern (1968), H. Zwölfer (unpublished), and Angermann (1987) estimate parasitization rates of 20.4% (*n* = 1420 host larvae), 37.4% (*n* = 357 host larvae), 21% (*n* = 1000 host larvae), and 24% (*n* = 1100 host larvae), respectively. P Sturm (unpublished) obtained rates varying from 9.7 to 22.9% for parasitoid attack on the tephritid *Tephritis bardanae* in Arctium heads. J. Arnold (unpublished) found that on average 22.2% of the *Xyphosia miliaria* larvae but up to 50% of the *Tephritis cometa* larvae in *Cirsium palustre* heads were parasitized.

Additional data are given in the scatter plot in Fig. 5 which compares parasitization rates in 80 populations of different phytophagous inhabitants of Cardueae heads (*T. conura* on *C. oleraceum* and *C. helenoides; U. stylata* on *C. vulgare; X. miliaria* and *Orellia ruficauda* on *C. arvense; Urophora affinis* on *C. diffusa*) with the percentage of flower heads occupied by phytophagous host larvae. For the parasitoids (mainly species of the Chalcoidea genera *Eurytoma, Pteromalus,* and *Torymus*) the mean and standard deviation of host insect utilization is 21.93% (± 19.9%) and the median, 16.3 ± 2.12%. Thus, parasitization rates vary greatly, but on the average resource utilization of the parasitoids seems lower than that of their phytophagous hosts (mean and SD = 38.2% ± 21.9%; median = 39.0 ± 2.45%). This difference disappears, however, if we take into consideration that the parasitods consume more or less the whole content

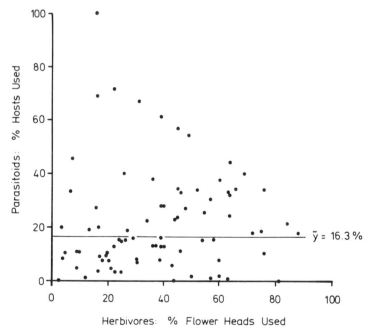

Figure 5   Resource utilization by parasitoids (% hosts attacked) and herbivores (% flower heads attacked) in 80 populations of Cardueae (*Cirsium vulgare, C. oleraceum, C. helenoides, C. arvense, Centaurea maculosa, C. diffusa*). Each dot represents a flower head population (samples from 50–200 heads) for which the attack rate by herbivores and their parasitoids has been established by dissections.

of the host larva or pupa, whereas the Cardueae flower head insects usually only destroy a part of the resource (Table III). The highest parasitization values in Fig. 5 (rates from 60 to 100%) are found at relatively low host densities (less than 40% of flower heads occupied by host larvae) which could suggest a tendency for an inverse density dependence. A statistical analysis of the data in Fig. 5 shows, however, no significant correlation ($r = 0.033$, ns) between the degree of resource exploitation of the parasitoids and that of their host insects. Thus, on the whole the investigated parasitoid faunas do not show a trend to concentrate their attack at high or low host densities.

3. Resource Utilization in the Urophora cardui System   A remarkable example of the underexploitation of a host plant is presented by the tephritid, *U. cardui*, which is a highly specialized herbivore, forming well-visible galls on the shoots of *C. arvense* (Fig. 2D). The densities of field populations of *U. cardui* studied in France, Austria, and Germany were assessed by a time-sampling method and the subsequent calculation of the log-transformed number of galls which could be collected in 100 min (log density index, LDI). An average carrying capacity of 6 galls/ *C. arvense* ramet was estimated from cage tests (Zwölfer, unpublished; maximal exploitation rates = 15 to 20 galls on single ramets) and the maximal gall densities found under field conditions (averages of 4 to 6 galls/ramet). By counting the average number of galls/ramet for populations with known values of LDI, the calibriation function, %RU = 0.00428*EXP(2.917*LDI), was calculated to assess the percent resource utilization (%RU) of *U. cardui*. A frequency distribution of this parameter (Fig. 6A) shows that the great majority of *U. cardui* populations investigated by us used less than 10% of the available host plant resource (median = 6.3% resource utilization). A similar situation has been found in the extended study of Schlumprecht (1990) on *U. cardui* populations in the Upper Rhine Valley and northern Bavaria. As our density values refer only to the relatively small proportion of thistle stands where *U. cardui* galls are present, the actual exploitation rate of *U. cardui* is much lower still. The same is the case in North America, where *U. cardui* has been introduced as a biocontrol agent against *C. arvense* (Peschken and Harris, 1975). In contrast with other introduced *Urophora* spp. (Julien, 1987; Harris, 1991) which exploit 80–100% of their host stands, *U. cardui* in Canada occupies only about 1% of the available *C. arvense* stands (Dr. P. Harris, personal communication Regina, Canada).

The parasitoid complex in European populations of *U. cardui* (Figs. 6B and 6C) comprises the highly specialized endoparasitoid *Eurytoma serratulae* (Fig. 2B) and a group of three less specialized ectoparasitoid species (*E. robusta, Pteromalus elevatus, Torymus chloromerus*). The endopar-

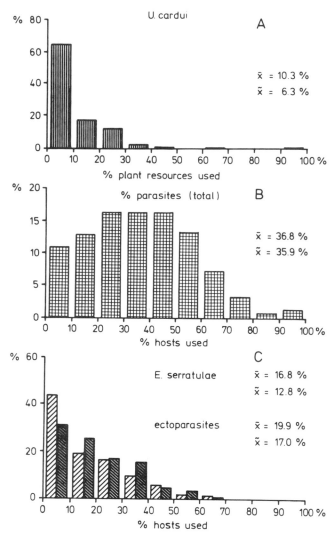

Figure 6 (A) Frequency distribution of the plant resource utilization for 366 populations of *U. cardui* (origin: Germany, France, Italy, Austria). (100% resource utilization is
reached with an average of 6 galls/ramet.) (B) Distribution of total parasitization rates of
433 European *U. cardui* populations. (C) Distribution of parasitization rates of the koinobiont endoparasitoid *Eurytoma serratulae* and a group of three idiobiont ectoparasitoids
(*Eurytoma robusta, Pteromalus elevatus,* and *Torymus chloromerus*).

asitoid can be classified as koinobiont (Askew and Shaw, 1986), as it attacks an early larval host instar and allows the host to complete its larval feeding period before it is killed and consumed. The ectoparasitoids, which attack *U. cardui* larva in a more advanced stage, are idiobionts (Askew and Shaw, 1986) as they consume their host immediately. The total parasitism of *U. cardui* reaches an average of 36.8% and is distinctly higher than that of other European *Urophora* spp. Between *E. serratulae* and the ectoparasitoids there are distinct competitive interactions at the level of the individual galls (Section III.B.2). But there is no significant interaction at the level of entire *U. cardui* populations, as the parasitization rates of the 433 investigated samples of *E. serratulae* and ectoparasitoids showed only a very weak negative correlation ($r = -0.093, b = -0.097,$ $p = 0.054$).

C. Local and Regional Aspects of Food Web Stability

The dynamics of the investigated Cardueae insect food webs vary with the spatial and temporal scales of the investigation. The results of single populations analyzed at one locality (local scale) usually differ from food web studies extended to cover the localities of a region or several regions (regional scale). At a local scale the dynamics of many of the investigated food webs may change greatly, but at a regional scale the structure and the components of most food webs in Cardueae flower heads become highly predictable. The basic structures (Section II.A.2) remain stable despite considerable density fluctuations in single seasons and localities. The stability (= predictability) of the single elements of the investigated food webs varies with their position in the trophic hierarchy. Keystone species, such as host plants or dominant phytophagous insects, are usually more predictable than species at higher trophic levels.

In the following sections we focus on the aspect of the temporal persistence of local food webs, i.e., on their "durational stability" sensu Southwood (1976) and on density fluctuations (temporal variability) of their elements.

1. Plant Populations The stability of host plant populations depends on their life history strategy. Populations of annual Cardueae, such as *Centaurea cyanus,* are less stable than biennial Cardueae such as *C. nutans, C. vulgare, C. maculosa,* or *C. diffusa* and these are less stable than perennial Cardueae such as *C. helenoides, C. oleraceum* or *C. scabiosa.* Personal long-term observations in the Swiss Jura provide evidence that in undisturbed habitats, stands of *C. oleraceum, Cirsium erisithales, Cirsium eriophorum, C. scabiosa,* and *C. jacea* may survive for at least 20 years.

Another important factor is the type of habitat of the host plant. The perennial *C. arvense* forms stable populations in certain types of habitats, such as pastures and riverine forests. A long-term survey of this thistle species, which was started in 1962 in the Swiss Jura and 1972 in the Upper Rhine Valley and continued at intervals of several years until 1991, strongly suggests that in neglected pasture land, along river banks, and even in some stable ruderal habitats, local *C. arvense* populations can survive for more than 25 years. On the other hand, ramets of *C. arvense* populations could often be tracked only for a few years in unstable habitats such as cultivated fields, gardens, road sides, or waste land (Zwölfer, 1979; Angermann, 1987).

2. Primary Consumers: Stability and Fluctuations   At the second trophic level (phytophagous insects) population stability can be measured by the length of persistence (= durational stability) and by the amplitude of population fluctuations. Species for which single populations could be tracked for more than 8 years and where local extinction rates were found to be below 10% are classified below as highly persistent (high durational stability), whereas species with local extinction rates of 50% or more have low durational stability. Density fluctuations have been assessed by calculating the ratios of population maxima to population minima (Strong *et al.*, 1984) or by the correlation of population densities between following years (Fig. 7). The dynamic behavior of most of the investigated phytophagous Cardueae insects can be described by one of the following three patterns:

i. High durational stability and low fluctuations. The populations of *T. conura* on *C. helenoides* studied by Romstöck (1987; Romstöck-Völkl, 1990a,b) (average N_{max}/N_{min} ratio = 3.0; maximal $N_{max}N_{min}$ ratio = 6.0; r^2 of density t2 vs t1 = 0.50) and of *U. cuspidata* and *O. colon* in flower heads of *C. scabiosa* growing at undisturbed sites (Völkl *et al.*, 1993) ($N_{max}N_{min}$ ratio = 1.8 and 5.7) are examples of phytophagous insects with an extraordinary high stability (compare the above data with Figs. 5.1 and 5.2 in Strong *et al.*, 1984). It is noteworthy that these stable populations were found on perennial host plants growing in relatively undisturbed habitats (small patches of grassland in mountain sites, abandoned pasture land on calcareous soils).

ii. Medium durational stability and moderate to high fluctuations. This group contains phytophagous Cardueae insects with populations which in our observation area could be followed for 5 to 8 years. Their $N_{max}N_{min}$ ratios vary between 10 and 100. This range corresponds with the majority of published data (see Fig. 5.1 in Strong *et al.*, 1984).

Medium values of stability are found in many phytophagous insects in the flower heads of biennial Cardueae (e.g. *X. miliaria* on *C. palustre*, *U. stylata* on *C. vulgare*, *U. solstitialis* and *Rhinocyllus conicus* on *C.*

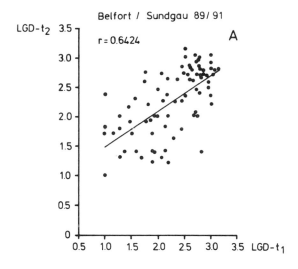

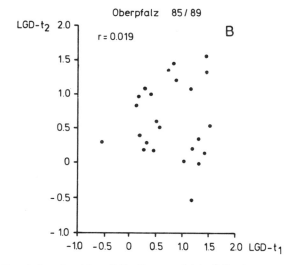

Figure 7 Population densities of *Urophora cardui* in following generations (T1, T2). Each dot represents a population. (A) Belfort–Sundgau area. Scale, log (*n* galls/100 min). Correlation between densities in subsequent years, $r = 0.6424$ ($P < 0.001$). (B) Oberpfalz (data from Schlumprecht, 1990). Scale, log (*U. cardui* larvae/m^2). Correlation not significant.

nutans, T. bardanae on *Arctium tomentosum,* (Zwölfer, unpublished)). Phytophagous populations for which Angermann (1987) observed annual turnover rates below 50% (Table IV) can be included here. Among the phytophagous fauna of perennial host plants, populations with medium durational stability and moderate or high fluctuations often occur in relatively unstable or regularly disturbed habitats. Völkl *et al.* (1993) give examples for phytophagous insects in flower heads of *C. scabiosa* growing in mown and sheep-grazed sites. We include here those *U. cardui* populations in which the density of the preceding generations explained from 15 to 40% of the variation of the subsequent population densities (Fig. 7). These populations in the Belfort area and in the Upper Rhine Valley have $N_{max}N_{min}$ ratios from 20 to 80. They occur in habitats (riverine forests, river banks) where the host plant, *C. arvense,* forms a dense network of small and relatively undisturbed patches.

iii. Low durational stability. A number of phytophagous Cardueae insects were found only sporadically in our observation area. They have a low durational stability (records are sometimes only available for one season) and they occur often only at low densities. Examples are observations made on host plant species which are marginal in the insect's host range (e.g., records of the tephritid *Tephritis cometa,* a tephritid which is mainly associated with *C. arvense,* from heads of *C. palustre*) or at localities which are situated at the margin of the insect's distribution area (e.g., records of the predominately Mediterranean weevil *R. conicus* from *C. nutans* in Upper Frankonia). On the other hand a low local durational stability may be caused by a metapopulation structure, as is discussed in the following section.

3. Primary Consumers: Population Structures   Most of the phytophagous species investigated occur in fragmented habitats and exploit plant populations with a patchy distribution. This discontinuous distribution pattern of resources, together with specific differences in the dispersal behavior of adult herbivors, results in different types of population structures. These can be characterized as (i) redistribution systems, (ii) meta-population systems, and (iii) source–sink systems.

i. An example of a redistribution system occurs in *T. conura* as adults leave the host habitats in late summer and hibernate at sites which remain unknown. Recolonization of the host plant populations in the following spring causes an annual redistribution of the populations. This is combined with density-dependent, postcolonization dispersal (Romstöck-Völkl, 1990a). A similar situation occurs in *Rhinocyllus* and *Larinus* populations which also leave the host plants as adults to hibernate (Zwölfer, 1979; Zwölfer and Harris, 1984).

ii. Metapopulations (Levins, 1969; Gilpin and Hanski, 1991) are sets of locally unstable subpopulations which interact via individuals moving among populations and which secure, in this way, the regional persistence of the species. A characteristic of metapopulations is the combination of a low durational stability of single populations and a high turnover rate, i.e., a high frequency of local extinctions and establishments of new populations by dispersers from existing local populations. Angermann (1987) found this pattern in his study of food webs in heads of *C. arvense* and *C. vulgare*. He emphasizes that the patchy distribution of Cardueae dependent on disturbance-generated habitats favors a metapopulation structure among the specialized herbivores. Examples of turnover rates from Angermann's study are given in Table IV.

An example of the development of an experimental metapopulation system for *U. cardui* is given by Zwölfer (1979). Releases of 60 females and 60 males at Delemont (where *U. cardui* does not naturally occur) resulted, in 1970, in a population of 546 galls (parasitism = 17.8%). This pioneer population (population A) gave rise, in 1971, to a second subpopulation (B) which was established in a thistle patch at a distance of 40 m. Population A developed 120 galls (parasitism = 95.5%) in 1971 and 80 galls (parasitism = 100%) in 1972 and as a consequence became extinct in 1973. Subpopulation B had 226 galls (parasitism = 65.4%) in 1971, 80 galls (parasitism = 60.2%) in 1972, 25 galls (parasitism = 56.8%) in 1973, and 15 galls (parasitism = 89.7%) in 1974. It became extinct in 1975, but in 1973 migrants of this subpopulation colonized a thistle patch at a distance of 50 m from subpopulation B. This subpopulation C had 30 galls in 1973

Table IV Annual Turnover Rates of Phytophagous Insects in Flower Heads of *Cirsium arvense*

Insect species	Host plant	Turnover rate	(N populations)
Xyphosia miliaria	*Cirsium arvense* (f)	0.08	25
Urophora stylata	*Cirsium vulgare*	0.17	18
X. miliaria	*C. arvense* (m)	0.20	15
Eucosma cana	*C. vulgare*	0.21	19
E. cana	*C. arvense* (m)	0.36	11
Orellia ruficauda	*C. arvense* (f)	0.37	19
E. cana	*C. arvense* (f)	0.48	23
Terellia serratulae	*C. vulgare*	0.64	11
Tephritis cometa	*C. arvense* (f)	0.82	11
Larinus sp.	*C. arvense* (f)	0.90	10

Angermann's (1987) data for 1984 and 1985. Turnover rates were calculated as average of the sum of local extinction plus colonization events per monitored host plant population).

(parasitism = 18.7%) and 15 galls (parasitism = 83.3%) in 1974. It became extinct in 1975. This experimentally established population system had some typical features of metapopulations such as low durational stability of subpopulations, formation of daughter populations, increasing rates of parasitism during the life time of the single subpopulations, and asynchronous population dynamics of the single subpopulations. In contrast to the metapopulation concept of Levins (1968, 1969) the field populations in this experiment as well as in a series of additional colonization experiments in the Swiss Jura (Fig. 14 in Zwölfer, 1979) did not achieve long-term persistence as the extinction rates were far higher than the colonization rates (Fig. 8).

iii. In population systems with a source–sink structure, habitat patches or regions support stable populations with growth rates balanced by emigration (= sources) while other less-favorable habitat patches are occupied by populations whose local growth rates make them dependent on immigration from source populations (Hanski and Gilpin, 1991; Harrison, 1991). We found this pattern during a long-term survey of *U. cardui* populations in the Belfort–Sundgau area (Zwölfer, unpublished). These consist of a central network of mostly

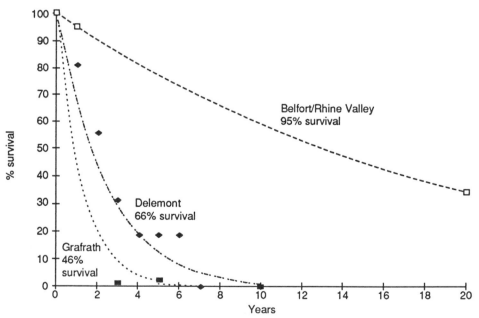

Figure 8 Survival rates of individual *Urophora cardui* populations in the Belfort–Sundgau area (annual survival = 95%), in the Delemont area (experimental populations, annual survival 66%), and in the Grafrath area (annual survival = 46%).

large populations in those parts of the area where altitudes are below 350 m above sea level and where habitats (mainly forests and river banks) are relatively undisturbed. Here, *U. cardui* remained stable from 1968 to 1991 (Figs. 7A and 8). Northwest and southeast of the central area, at altitudes from 350 to 700 m, small and relatively isolated, marginal *U. cardui* populations, which have high extinction rates, occur. Flight mill experiments with *U. cardui* (Remund and Zwölfer, 1993) have demonstrated that adults of this species can disperse over distances up to 10 km or more. This fact, the topography of the area, and the position and different dynamics of the populations allow us to conclude that the *U. cardui* populations of the Belfort–Sundgau area form a source–sink system in the sense of Harrison (1991). Schlumprecht's (1990) life table studies of *U. cardui* populations in the Upper Rhine Valley and his data, which suggest a density-dependent dispersal behavior in this fly, corroborate this conclusion.

4. Secondary Consumers An analysis of the parasitoid complexes of *T. conura* (Romstöck, 1987), *U. stylata* (Michaelis, 1984); Angermann, 1987), *U. cardui* (Schlumprecht, 1990; Zwölfer, 1979), *X. miliaria, O. ruficauda*, and other species (Angermann, 1987) has shown a relatively high degree of persistence in the populations of many of the parasitoid species. Despite the fact that single parasitoid species exhibit characteristic differences in their mean densities, parasitoids of Cardueae insects form guilds with a predictable composition of species. The amplitude of their density fluctuations is usually somewhat higher than that of their host populations. As an example Fig. 9 provides an overview of the average densities/gall of *U. cardui* and its two main parasitoids, *E. serratulae* and *E. robusta*, in the Upper Rhine Valley. In this food web it was not possible to predict the densities of parasitoid populations on the basis of the densities of the previous parasitoid generation, whereas under the optimal conditions of the Belfort–Sundgau area the densities of the host *U. cardui* could be predicted (Fig. 7A).

Even parasitoids associated with host species forming metapopulations are able to track their host populations, but there are differences in the span of time needed to locate newly established colonies. For the *T. conura* food web Romstöck (1987) and Romstöck-Völkl (1990a) could show that the parasitoid *Pteromalus caudiger* recolonizes host populations within 1 year after initially having been removed from the site. In *Eurytoma* sp. nr *tibialis* this process takes much more time. Because of its high dispersal capacity *Pteromalus* uses a higher proportion of host populations and shows lower population fluctuations than *Eurytoma*. *Eurytoma*, on the other hand, parasitizes on the average more *T. conura* larvae (mean =

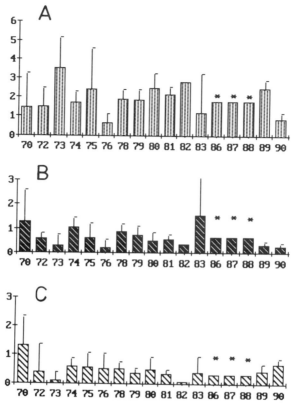

Figure 9   Average densities per gall and 95% confidence intervals of unparasitized *Urophora cardui* pupae and larvae (A), of *Eurytoma serratulae* (B), and of *E. robusta* (C) in the Upper Rhine Valley and the Sundgau. *For 1986–1988 a 3-year average (values from Schlumprecht, 1990) is given.

15.4%, maximum 56%) than *Pteromalus* (mean = 8.5%, maximum 28%). Similarly, in the *U. cardui* food web the ectoparasitoids, particularly *E. robusta,* are usually quicker in locating newly established host populations than the monophagous parasitoid *E. serratulae* (Schlumprecht, 1989; Eber, 1993).

5. *Range Extensions and Breakdown of the Urophora cardui Food Web*   A survey for *U. cardui* galls, which began in 1968 and was continued until now (H. Zwölfer, unpublished), showed that the European distribution of *U. cardui* differs from that of other *Urophora* spp. such as *U. solstitialis* (Möller-Joop, 1989), *U. stylata* (Michaelis, 1984), or *U. affinis* (Zwölfer, 1978). Whereas the distribution areas of these species cover large and continuous parts of the distribution of the respective host plants, *U. cardui*, which has the ubiquitous *C. arvense* as host, shows large gaps in its distribution pattern (Fig. 10). It occurs, for example, only at few places in Switzerland (Merz, 1991) and it is absent from large regions in central Germany (Zwölfer, unpublished). Moreover, borders of the *U. cardui* distribution are not stable. In the region of Bayreuth (northeastern Bavaria) *U. cardui* extended its range from 1977 to 1990 from between 20 to 40 km to the north (Fig. 11) (Eber and Brandl, 1993). Distinct range extensions also occurred for *U. cardui* from 1969 to 1990 in the Belfort area (Zwölfer, unpublished) and during the last decade in Schleswig Holstein (Pschorn-Walcher (Kiel), personal communication). Jansson (1992) gives a detailed documentation of the invasion of *U. cardui* in southern Finland, where it was first reported in 1981. He found that from 1986 to 1991 the mean annual range extension of *U. cardui* was 6.9 ± 4.2 km. It is remarkable that in all these cases newly established populations of *U. cardui* had acquired their parasitoids after a relatively short time and that even the monophagous *E. serratulae* was able to follow the range extensions of its host. Schlumprecht (1989), who investigated the dispersal behavior of *U. cardui* and *E. serratulae* over a period of 4 years, found that both the fly and the parasitoid have about the same average dispersal ability. In his observation area, average range extensions were 1.35 km/year for *U. cardui* and 1.26 km/year for *E. serratulae*. In contrast with the less-specialized ectoparasitoids (*E. robusta*, *P. elevatus*, *T. chloromerus*) *E. serratulae* reaches newly established populations of *U. cardui* usually with a delay of only 1 to several years (Fig. 12; Schlumprecht, 1989; Eber, 1993). The presence or absence of *E. serratulae* can therefore be used to estimate whether *U. cardui* populations are newly established, i.e., 1–2 years old, or older.

Figure 10 shows two areas where *U. cardui* populations declined and finally became extinct. These are the experimental populations in the Swiss Jura (Section II.C.3) and the populations in the area northwest of Grafrath (southern Bavaria). Here a storm had destroyed a forest area of more than 10 km^2 and subsequent afforestations allowed a population build-up of *C. arvense* and offered favorable conditions for the development of *U. cardui* for several decades. Gall densities (which were measured as the number of galls which could be collected in 100 min) were high in 1967 and 1968 (> 1000 galls/100 min) and still considerable in 1976 (744 galls/100 min). In 1979 (8 galls/100 min) and 1981 (16 galls/100 min) there was a drastic decline in the number of host plant patches and in the number and density of *U. cardui* populations and in 1990, despite a careful survey, no more *U. cardui* populations could be found. As the former clearings and young forest plantations had developed into

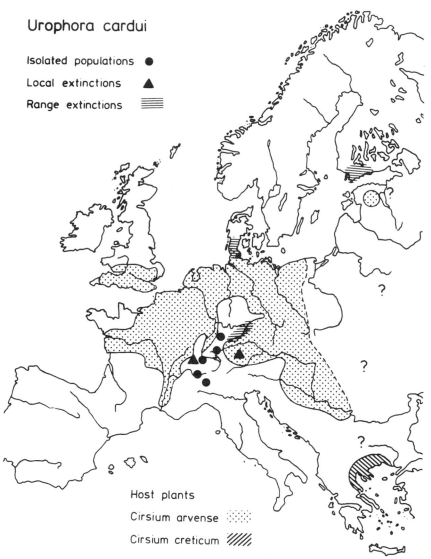

Figure 10 Distribution area of *Urophora cardui* in western and central Europe based on personal records and a literature review (records from southern Finland from Jansson, 1992).

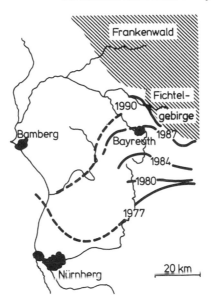

Figure 11 Range extension of *Urophora cardui* in the Bayreuth area. The northern borders of the distribution area were drawn by connecting the northernmost records from 1977 to 1990. The distribution border of 1987 has been estimated from data obtained by Eber (1993) and Schlumprecht (1989). Interrupted lines indicate that the exact border is not known. The mountain area of the Fichtelgebirge and the Frankenwald, which forms a natural distributional barrier for *U. cardui* is hatched.

semi-mature, closed spruce forests, the density of thistle patches had apparently become too low to support a viable metapopulation system of *U. cardui*.

III. Network Control of Food Webs

The foregoing sections dealt with structures, functions, and stability of food webs formed by Cardueae–insect systems. This section focuses on control mechanisms involved in the maintenance of food webs and discusses the processes which regulate the densities of their components.

A. Levels and Types of Control

The basis of the investigated food webs is Cardueae flower heads or galls, i.e., discrete microhabitats (Section II.A.1); these units form the primary level on which control factors operate. The second level comprises the attacked plants of a local host population. In the case of metapopulations

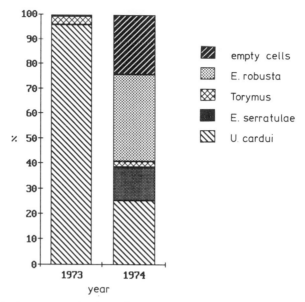

Figure 12   Average production of *Urophora cardui* galls of eight experimental field populations in 1973 (first generation, offspring field-relased adults) and 1974 (second generation) (Ludwigsburg).

(Section II.C.3) networks of local subpopulations or patches of plant populations constitute a third level where control processes can come into play. Control mechanisms may operate in the microhabitat as well as in the macrohabitat (Section II.A.1). In the microhabitat (i.e., in the individual flower head system or gall) possible regulatory factors such as availability of food resources, competition, parasitism, predation, and plant–insect interactions operate on the immature stages of primary and secondary consumers. Possible control mechanisms at the level of the macrohabitat (single host plant populations and landscapes with sets of host plant patches) are host selection, communication, and dispersal behavior of adult insects and adult mortality.

B. Availability of Resources and Competition

As the degree of resource utilization and the intensity of competition are mutually interdependent, both factors are discussed in this section.

1. Underexploitation of Resources   In any food web shortness of food, space, and other resources is an ultimate factor which limits the growth of its components. Continued overexploitation, for example the extreme rates of parasitization as observed in our experimental *U. cardui* populations (Section II.C.3), will eventually lead to a complete breakdown of the food web. In Section II.B (Table III and Figs. 4–6) we have shown that, apart from some exceptional cases, herbivores as well as parasitoids use only a relatively small fraction of their available food resources, if entire host plant or host insect populations are analyzed. Whereas the carrying capacity of the system is almost never reached at the level of the macrohabitat, it is a distinct limiting factor at the level of the microhabitat. In contrast to grazing animals or hunting predators, which can abandon overexploited spaces and seek new resources, the insects of the investigated food webs are contained in discrete plant structures from which, in the case of overexploitation, escape is not possible. The degree to which these resource units can be used depends entirely on the host selection and oviposition of adult herbivores and parasitoids. As at this level synchronization and spatial coincidence with the resource and information on the resource state is never perfect, an "economically" entirely perfect distribution of eggs cannot be achieved. Moreover, as flower head or gall inhabitants cannot make corrections during the larval stage, a complete utilization of the available resources is not possible at the population level. This means that a certain fraction of the host plant structures as well as of the host insect population always has a chance to escape the herbivores or parasitoids. The failure to exploit resources completely helps to avoid overexploitation and thus contributes to the stability of the investigated food webs.

2. Interspecific Competitive Interactions   At the level of the microhabitat, space and food are limiting factors. Therefore we expected interspecific competition to occur in cases of a simultaneous attack on the same flower head by different herbivore species. Indeed, these microhabitats are one of the few ecological systems where direct competition can be demonstrated (e.g., Zwölfer, 1979) for *Larinus sturnus* vs *L. jaceae; U. solstitialis* and *L. sturnus* vs *R. conicus*). However, the risk of direct interspecific competition is reduced by the use of different trophic strategies (Section II.A.3), different habitat preferences (Völkl *et al.*, 1993), asynchronous oviposition (Romstöck, 1987), and the ability of host-searching females to discriminate between occupied and unoccupied heads (Angermann, 1986). Michaelis (1984) showed that in the case of a direct contact in a flower head of *C. vulgare* the gall-forming tephritid *U. stylata* is superior in competition to the achene-feeding tephritid *Terellia serratulae.* Angermann (1987) confirmed this observation. Angermann (1986, 1987) also describes an important indirect and asymmetric competition between the tephritids *X. miliaria* and *O. ruficauda.* Both attack heads of *C. arvense,* but *X. miliaria* oviposits earlier and is therefore competitively superior. A comparable situation has been found by Freese (1991) in the stems of

C. arvense in which the distinct negative correlation between the weevils *Apion carduorum* and *Ceutorhynchus litura* suggests a direct or indirect competitive relationship.

Interactions between tephritids or weevils, on the one hand, and lepidopterous larvae, on the other, are common in Cardueae flower heads. Interactions with moth larvae constitute a form of predation rather than cases of interspecific competition, as the microlepidopterous larvae in Cardueae flower heads are capable of switching in the case of contact with other herbivores from phytophagy to carnivorous behavior (Section II.A.3).

There is little evidence for direct interspecific competition between secondary consumers, i.e., parasitoid spp., as species exploiting the same host usually show enough niche differentiation to avoid direct contact in a host larva and multiparasitism (development of two different parasitoid species in the same host) is not possible in parasitoid species associated with hosts on Cardueae. In the *T. conura* food web, Romstöck (1987) investigated the relationship between the two chalcid parasitoids *Eurytoma* sp. nr *tibialis* and *P. caudiger*. She found an additive parasitization effect, which speaks against an exploitative or aggressive competition among the parasitoids. Complex competitive interactions can be found in the parasitoid guild associated with the stem galls of *U. cardui* (Zwölfer, 1979). The flow chart in Fig. 13 shows that the outcome of a contact between the larvae of the ectoparasitoid *E. robusta* and the endoparasitoid *E. serratulae* depends on the age of the *E. serratulae* larva. If the latter is still immature and has not induced a sclerotization of the *U. cardui* larva [Fig. 13, node (7)] *E. robusta* (which then becomes a secondary parasitoid) consumes it together with the remainder of the primary host. Otherwise the *E. serratulae* larva is protected within the sclerotized host cuticle. If the *E. robusta* larva contacts a mature *U. cardui* larva within the gall (Fig. 13, node (4)) it develops as a primary parasitoid, otherwise (Fig. 13, node (5)) it may switch to phytophagy and feed on the trophic tissue of the gall, if the latter is well developed. The balance between *E. serratulae* and *E. robusta* is further influenced by the size of the gall. Zwölfer and Arnold-Rinehart (1993) show that *E. robusta* is more efficient in exploiting galls with a small diameter and few larval chambers, whereas *E. serratulae* has an advantage in galls with many larval chambers, as it oviposits before the growing gall tissues form a refuge for some of the *U. cardui* larvae (Fig. 2B). In small galls *E. serratulae* suffers considerable losses by secondary parasitism through *E. robusta*.

3. Avoidance of Intraspecific Competition   As has been discussed in Section II.A.3 herbivores with the strategy of early aggregated attack are either gall inducers (*Urophora* spp., *Isocolus* spp.), which convey additional

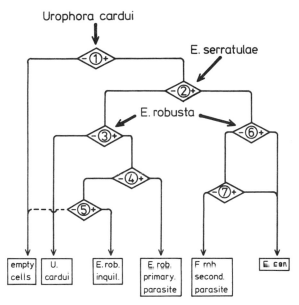

Figure 13 Flow chart of the development of the *Urophora cardui* foodweb. Figures indicate alternatives ($-$, no; $+$, yes). (1) Formation of trophic gall tissue; (2) oviposition by the endoparasitoid *Eurytoma serratulae;* (3) and (6) oviposition by the ectoparasitoid *E. robusta;* (4) *Urophora* larva grown up to the 3rd instar; (5) *E. robusta* larva finds enough trophic gall tissue; (7) *E. serratulae* larva in the prepupal stage, *Urophora* cuticle sclerotized.

assimilates to their feeding sites (Harris, 1980; Michaelis, 1984; Zwölfer, 1985a) or they cause callus growth (*T. conura* (Romstöck, 1987), *Rhinocyllus* (Shorthouse and Lalonde, 1984)) which also increases the food base in a flower head. Herbivores with early aggregated attack therefore run little risk of intraspecific competition for food and the number of *Tephritis* or *Urophora* larvae occupying a flower head or gall does not affect the weight or survival chances of the larvae (Romstöck, 1987; H. Zwölfer, unpublished). Moreover, larger galls provide more protection against parasitoids (Schlumprecht, 1990; Zwölfer and Arnold-Rinehart, 1993). These observations explain why we could not find oviposition deterrent pheromones in the tephritids using the strategy of early aggregated attack.

The situation is different in herbivores belonging to the group of achene and receptacle feeders (second trophic strategy, Section II.A.3). Angermann (1986) showed that females of the tephritids *X. miliaria* and *O. ruficauda,* whose larvae usually live as solitary achene feeders in *C. arvense* heads, mark the flower heads after oviposition which deters additional oviposition. *Tephritis bardanae,* an achene feeder in the heads of *A. tomentosum,* reduces high larval densities by larval interference competition. Sturm (1988) found that the average size of egg batches deposited by a female is 3.7 (SE ± 0.21). Due to multiple ovipositions in the Bayreuth area densities of eggs/head vary at different localities from 7.5 (SE ± 0.41) to 4.2 (SE ± 0.19). Figure 14 shows, however, that as early as the first larval instar, average larval numbers per *Arctium* flower head are adjusted to values of between 3 and 4, i.e., to densities which correspond with the size of a single batch of eggs. It is remarkable that a number of 3–4 *T. bardanae* larvae per *A. tomentosum* head use only 12–15 achenes from an average of 44. (Sturm, 1988). Thus, *T. bardanae* regulates its average larval densities to a level far below the theoretical carrying capacity of the flower head. (Sturm (1988) could show that in exceptional cases up to 12 *T. bardanae* larvae can develop in a single head.)

Freese (1991, 1992) describes a similar regulation of surplus densities by larval interference competition in experiments with the longhorn beetle *Agapanthia villosoviridescens.* By concentrating and confining adults on caged *C. palustre* he induced females to deposit up to 18 eggs on the branches of a single plant. Nevertheless only one larva per thistle stem survived.

In many tephritid species attacking Cardueae host plants aggressiveness and territorial behavior of adults contributes to spread oviposition and to avoid overexploitation of flower heads. Examples are *T. conura*

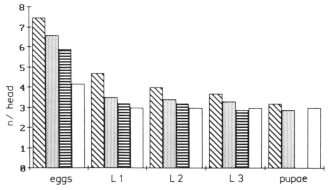

Figure 14 Development of the densities of eggs, larvae, and pupae of *Tephritis bardanae* in the heads of *Arctium tomentosum* at four localities near Bayreuth (Sturm, 1988). Whereas numbers of eggs per flower head vary greatly (difference between the four localities is significant at $P < 0.001$), larval densities from the second instar on are adjusted to an uniform level of an average of 3–4 larvae/head.

(Romstöck, 1987), *T. bardanae* (Straw, 1991), and *Urophora* spp. (Michaelis, 1984; Peschken and Harris, 1975; Harris, 1989).

C. Predator–Prey Relationships

Predator–prey relationships (which include parasitoid–host relationships) are often considered to be the key mechanism in regulating the population densities of phytophagous insects (Strong *et al.*, 1984). In the following sections I analyze the role of parasitoids in the investigated Cardueae food webs.

1. Tests for Density Dependence   In population dynamics density dependence may be found on a temporal as well as on a spatial scale (Hassell, 1986). It can occur between generations (temporal density dependence) as well as within one generation (spatial density dependence by nonrandom search of predators (Hassell, 1979, 1986; Morrison and Strong, 1980; Walde and Murdoch, 1988)). In cases where data from a series of successive life tables are available, the method developed by Varley *et al.* (1973) and discussed by Southwood (1978) allows us to infer temporal density dependence from a plot of *k*-values (a logarithmic measure of mortality) against a logarithmic density measure. Thus we can decide whether a parasitoid or some other mortality factor is directly density dependent, inversely density dependent, delayed density dependent, or density independent. This approach was followed in the studies of Michaelis (1984) (mortality factors affecting *U. stylata* on *C. vulgare*), Romstöck (1987) (mortality of *T. conura* on *C. helenoides* due to parasitoids), and Schlumprecht (1990) (mortality factors affecting *U. cardui* on *C. arvense*). Michaelis (1984) found, using tests for spatial density dependence, that the parasitoids (*E. tibialis, E. robusta, T. chloromerus, P. elevatus*) and the predators (*Palloptera* sp., *Eucosma cana*) are inversely density dependent. The test for temporal density dependence gave a slight trend toward a delayed density dependence in *E. tibialis* and the ectoparasitoids. Romstöck (1987) reports an inverse density dependence at the spatial and at the temporal scale for the parasitoid *Eurytoma* sp. near *tibialis* and spatial density independence and a slight delayed temporal density dependence for *P. caudiger*. Schlumprecht (1990) found in some observation areas and in 10–50% of the analyzed generations, a slight direct spatial density dependence in *E. serratulae* but density independence on a temporal scale. *Eurytoma robusta* and *T. chloromerus* were density independent on a spatial scale and possibly delayed density dependent on a temporal scale. Angermann (1987) who investigated the parasitization of the tephritids *X. miliaria* and *O. ruficauda* in *C. arvense* heads and of *U. stylata* in *C. vulgare* heads over 2 years found no signs of a spatial or temporal density dependence. In summary, none of the investigated parasitoid species operates as a distinct density-dependent mortality factor and there is also no evidence that single predator species can control Cardueae herbivores in a density-dependent way.

2. Population Trends in Urophora cardui: *A Path Analysis*   From 1989 to 1991, 40 *U. cardui* populations were monitored in the Belfort-Sundgau region (Section II.C.3). Variables measured included (1) the altitude above sea level of the population, (2) average diameter of galls/population, (3) cell number/gall/population, (4) empty cells/gall/population, (5) *E. robusta* larvae/gall/ population, (6) *E. serratulae* larvae/gall/population, (7) surviving *U. cardui* larvae/gall/population, and (8) gall densities (measured as LDI, see Section II.B.4). A path analysis (Sokal and Rohlf, 1981) was made to estimate the impact of values of these variables in the previous year (= T1) on the values in the following year (= T2). Figure 15 summarizes the results: If "gall density T2" is taken as a criterion variable, then only "gall density T1" and the survival rate of *U. cardui* at T1, but none of the other variables, are statistically significant predictors. A total of 29% of the variation of the gall density T1 can be explained by the variable "altitude" (see Section II.C.3). For the criterion variable "surviving *U. cardui* T1" there is one positive path (cells/galls) and three negative ones (empty cells and parasitoids) all being highly significant. The residual variable U_2 affects the criterion variable surviving *U. cardui* T1 only slightly, as in the microhabitat "gall" the mortality factors can be precisely measured. On the other hand the residual variables U_1 and U_3 (which affect criterion variables exposed to the influences of the macrohabitat) have high values and each account for more than 50% of the unexplained variation. It is interesting that the gall density in the previous year is a much stronger predictor of the gall density in the following year than the rates of surviving *U. cardui* larvae and that the variables "cells/gall" and "empty cells/gall" are much stronger predictors for the rate of surviving *U. cardui* than the two parasitoids. The variable "empty cells" represents chambers of the gall without larval development of *U. cardui* or of the parasitoids. Such cells often contain empty eggs of *E. robusta*.

An important aspect of the *U. cardui* food webs in the Belfort–Sundgau region is the fact that survival rates of *U. cardui* are only significant predictors of gall density T2 if they are combined with the variable gall density T1 and even then they raise the multiple coefficient of determination only by 3.5% (from $R^2 = 0.3245$ to $R^2 = 0.3688$, Fig. 15). In regressions where the survival rate of *U. cardui* is the single independent variable, no significant correlations exist between gall densities and the survival rates of *U. cardui* of the same generation (T1, $r = -0.1367$, ns; T2, $r = -0.1394$, ns) or subsequent generations (T1 > T2, $r = 0.131$,

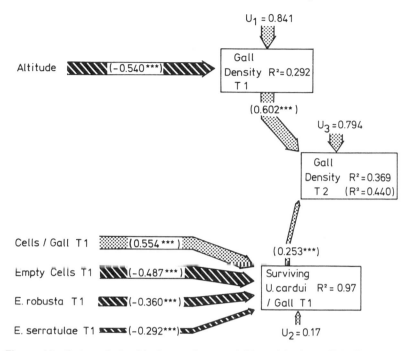

Figure 15 Path analysis with the predictor variables "altitude," cells/gall, empty cells/gall, *Eurytoma robusta* larvae/gall, and *E. serratulae*/gall and the criterion variables gall densities in the previous year (T1), surviving *U. cardui* larvae/gall in the previous year (T1), and gall densities in the following year (T2). U_1, U_2, U_3 are residual variables. Figures are path coefficients which are significant at ***$P < 0.001$ and ** $P < 0.01$ (**). The R^2 values of the criterion variables result from the predictor variables shown in the graph ($R_2 = 0.440$ for gall density T2 is obtained in a multiple regression with altitude, gall density T1, and *U. cardui* survival T1 as independent variables). For further explanations see the text.

ns). There is so far no evidence in our data sets to indicate that at the level of the microhabitat (i.e., in the galls) *U. cardui* populations in the high density area suffer from less parasitism and have better larval survival chances than other *U. cardui* populations. Our path analysis comes to the same conclusion as Schlumprecht's (1990) life table studies which indicated that parasitoids and other mortality factors operating within the galls have no or little influence on the population dynamics of *U. cardui*.

3. Densities in Allochthonous Herbivore Populations In several insect species which attack Cardueae species, autochthonous population densities in Europe can be compared with population densities of species used intentionally as biocontrol agents or accidentally introduced into North

America. An example is given in Fig. 16 which compares the biomass of *R. conicus* (fresh weight of mature larvae (mg) per gram of flower head dry weight) in *C. nutans* heads in the Upper Rhine Valley (locality, Mulhouse, 1971; data from Zwölfer, 1980) with that in Montana (Galatin Valley, 1977: data from Rees, 1978), where a *R. conicus* population originating from Mulhouse has been introduced as a biocontrol agent of *C. nutans*. The autochthonous *R. conicus* populations are integrated into a rich food web with competitors (*Urophora* sp., *Larinus* sp.), predators (*Homoeosoma* spp.), and parasitoids (*Bracon* spp., *Habrocytus*, *Pterandrophysalis*). In contrast, the allochthonous *R. conicus* populations are the only consumers in an extremely simplified food chain. If losses due to parasitoids (hatched areas in Fig. 16) are corrected for, the Mulhouse food web produced 17.5 mg *R. conicus* ($= 1.6 \pm 2.5$) larvae/g head dry wt, compared with 306 mg ($= 28 \pm 12$) larvae/g head dry wt in Montana. This net production of the introduced biocontrol agent is much higher than the combined net production of primary consumers in the complex Mulhouse food

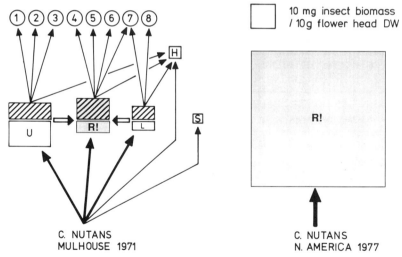

Figure 16   The food web in heads of *Carduus nutans* populations analyzed in 1971 near Mulhouse (France, Ht Rhin, Zwölfer, 1980) and in the Galatin Valley (Montana; Rees, 1978). The size of the squares represents the biomass of living mature larvae of primary consumers (in mg) per flower head dry weight. Hatched areas, biomass of primary consumers (mg) consumed by secondary consumers. U, *Urophora solstitalis* with the parasitoids (1) *Eurytoma tibialis*, (2) *E. robusta*, (3) *Torymus* sp.; R!, *Rhinocyllus conicus* with the parasitoids (4) *Pterandrophysalis levantina*, (5) *Habrocytus* sp., (6) *Bracon minuator*, (7) *Bracon urinator*; L, *Larinus sturnus* with the parasitoids (7) *B. urinator* and (8) *Tetrastichus crassicornis*; H, *Homoeosoma* spp.; S, additional phytophagous species of minor importance. Further details are given in the text.

web (60–70 mg if larvae consumed by parasitoids and predators are included). Its impact on the competitive capacity of the host plant, *C. nutans*, is high enough to have reduced the density of *C. nutans* populations in Canada, the United States, and New Zealand below the economic threshold (Zwölfer and Harris, 1984; Kok and Pienkowski, 1985; Julien, 1987).

Urophora affinis and *Urophora quadrifasciata*, two tephritids introduced from Europe to North America as biocontrol agents against the knapweeds *C. maculosa* and *C. diffusa*, have not as yet achieved commercially satisfactory control, but as in the case of *R. conicus* they destroy a higher proportion of achenes alone than the entire phytophagous guilds in Europe. In *C. diffusa* heads, the *U. affinis* density/head in Canada is 18.8 times higher than that in Europe, and in *C. maculosa* heads, densities of *U. affinis* and *U. quadrifasciata* are 3.3 and 13.7 times higher, respectively (H. Zwölfer, unpublished; and P. Harris, personal communication, Regina, Canada, 1992). In *U. stylata* (introduced to Canada against *C. vulgare*), *O. ruficauda* (on *C. arvense*, accidentally introduced to North America), and *Metzneria lapella* (on *Arctium minus*, accidentally introduced to North America (Hawthorn and Hayne, 1978)) the allochthonous populations in North America also have 5 to 10 times higher population densities than those in the autochthonous European food webs (H. Zwölfer, unpublished data). A remarkable exception is *U. cardui* on *C. arvense*, as the degree of resource utilization in populations introduced to North America is as low as that in our European observation area (Section II.B.4, and Harris, 1989; Peschken *et al.*, 1982).

Figure 17 compares the relationship betweeen species numbers in the herbivorous guilds of autochthonous European (Zwölfer, 1985b) and allochthonous Canadian (Harris, 1990) populations of *C. maculosa*. It shows that in each area the percentage of attacked flower heads increases with increasing guild size, but the introduced herbivores in Canada attack about three times more flower heads than those in the European food webs. A single introduced herbivore species (*U. affinis*) infests on average a higher proportion of hosts than the entire autochonous European food web.

Our data show that there are striking differences between average population densities of native phytophagous Cardueae insects and native food webs on the one hand, and average densities of introduced insects or guilds on allochthonous host plant populations, on the other hand. At first sight these data seem to contradict the conclusions of Sections III.C.1 and III.C.2, which assume a very low potential of single parasitoids to control the densities of the herbivorous fauna of Cardueae. The different results can, however, be reconciled by the facts that (i) in the allochthonous food webs the combined impact of the guild of parasitoids, preda-

Centaurea maculosa

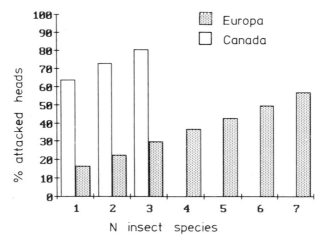

Figure 17   Phytophagous guilds in *Centaurea maculosa* heads: Resource utilization and guild sizes in Europe (autochthonous food webs) and Canada (introduced guilds; Harris, 1990)).

tors, and most competitors is absent and that (ii) the host plants often have considerably higher population densities.

D. Plant–Insect Interactions

In the Cardueae–insect systems which we have investigated plant–insect interactions play a dominant role as regards the functions and often also the dynamics of the food webs. In the following sections a summary of the more important aspects is given.

1. Synchronization and Coincidence   Most endophytic Cardueae insects and parasitoids can deposit their eggs only during a narrow "temporal window," i.e., they are dependent on a particular phase in the development of the resource (e.g., *T. conura* (Romstöck, 1987) or *U. stylata* (Michaelis, 1984; Redfern and Cameron, 1989)). But the phenologies of the Cardueae hosts and of primary and secondary consumers are not perfectly synchronized. Host plants are not a precisely predictable resource for specialized Cardueae herbivores due to meteorological factors and often to anthropogenous disturbances (e.g., *C. vulgare* for *U. stylata* (Michaelis, 1984), *C. arvense* for *U. cardui* (Schlumprecht, 1990), *C. helenoides* for *T. conura* (Romstöck, 1987)). The same is true for the relationship between phytophygous hosts and their parasitoids (e.g., *U. cardui* and *Eurytoma* spp. (Zwölfer, 1979; Schlumprecht, 1990)). The stochastic unpredictability of the coincidence of the appropriate developmental phases of the resources and consumers reduces the degree of resource utilization. It creates "temporal refuges" for the host plant or host insect (e.g., partial asynchrony between heads of *C. arvense* and *Xyphosia* or *Orellia* (Angerman, 1984), seasonal asynchrony between heads of *Carduus pycnocephalus* and *R. conicus* (Goeden and Ricker, 1985)) and can protect food webs against overexploitation by single components. It can also facilitate the coexistence of species belonging to the same guild (e.g., *Eurytoma serratulae* and *E. robusta* (Zwölfer, 1979)).

2. *Plant and Habitat Parameters and Oviposition Preferences* Besides its phenology other parameters of the host plant and its habitat may also determine whether the plant is accepted for oviposition. Examples are the heights of plant stems (*Urtica–Dasineura* (Bringezu, 1987)), leaf size (e.g., *Urtica–Trioza* (Bringezu, 1987)), the position of flower heads (e.g., *C. helenoides–T. conura* (Romstöck, 1987), *C. arvense–X. miliaria* (Angermann, 1984)), or the sex of the host plant (e.g., *O. ruficauda* oviposits only in female flower heads; *X. miliaria* prefers male flower heads (Angerman, 1984, 1986)). The density of certain herbivores decreases with increasing density and/or size of host plant patches (e.g., *U. cardui* on *C. arvense* (*Schlumprecht*), *X. miliaria* on *C. arvense* (Angermann, 1984), *Aglais urticae* on *Urtica dioica* (Bringezu, 1987)). Shaded host plant stands may be preferred (e.g., *U. cardui* on *C. arvense* (Zwölfer, unpublished), *T. bardanae* and *Cerajocera ceratocera* on *Arctium* (Straw, 1991)). The size of Cardueae flower heads can play a role in influencing the spatial patterns of parasitism and in providing host refuges (Price, 1988; Redfern and Cameron, 1989; Romstöck-Völkl, 1990b; Schlumprecht, 1990). Differences in such habitat and host preferences reduce interspecific competition among guild members.

Plant architecture is an important factor in resource partitioning in four species of aphids associated with *C. arvense* (Völkl, 1989) and the coexistence and dynamics of aphid–parasitoid complexes on Cardueae host plants (Völkl, 1990, 1991, 1992; Weisser, 1991).

3. *Plant-induced Mortality of Herbivores* Abscission or abortion of flower heads containing endophytic insect larvae can be a mortality factor of insects associated with *Cirsium* species. Angermann (1984, 1986) found that 10–15% of the flower heads of *C. arvense* were lost by abscission and that these heads contained a higher proportion of *O. ruficauda* and *X. miliaria* larvae than heads which remained on the plant. Romstöck (1987) describes an increasing probability of abortion and/or abscission in capitula of *C. helenoides* which occupied lower positions on the inflorescences. The risk of mortality in *T. conura* is low, as this tephritid prefers flower heads in upper positions. In stem miners of *C. arvense*, callus formation may kill larvae of the weevil *C. litura* and the agromyzid *Melanagromyza* sp. (Freese, 1991). In the weevil *C. litura* this can occur if there are too few larvae in a stem of the host plant *C. arvense* to consume enough of the growing callus tissues (Zwölfer and Harris, 1966).

4. Plant Galls   Galls on Cardueae hosts are the result of a manipulation of the growth pattern of the host plant by chemical and mechanical stimuli exerted by the early larval stages of the gall former (Arnold-Rinehart, 1989). They provide enriched food, shelter, hibernation and pupation sites, protection against most predators, and partial protection against parasitoids (Sections II.A.2 and III.B.3; Schlumprecht, 1990; Zwölfer and Arnold-Rinehart, 1993). On the other hand, a number of risks are involved in this trophic strategy. Galls in Cardueae flower heads constitute a target for certain "niche-specific" parasitoids (Capek and Zwölfer, 1990). Gall formers are also dependent on the proper responses of the host plant, which in turn may depend on environmental conditions. The success of *U. cardui* can be greatly diminished by the phenomenon of empty cells. Often these empty cells contain eggs of the ectoparasitoid *E. robusta* but neither *Urophora* nor parasitoid larvae. We also observed empty cells in cage experiments where parasitoids were precluded. The number of empty cells/gall is negatively correlated with the diameter of the gall. The proportion of empty cells significantly increases in years with dry summers (e.g. in 1976 and 1990) and in regions where the precipitation in July (the main period of gall growth) was low (Fig. 18). These observations strongly suggest that the main cause of empty cells is an inability of the young *Urophora* larva to induce an appropriate trophic tissue layer in its chamber, a risk which occurs particularly in host plants under water stress. Figure 18 also shows that with increasing precipitation in July there is a significant increase in the average gall diameter and in the proportion of surviving (i.e., unparasitized) *U. cardui* larvae. Moreover, the process of gall formation of *U. cardui* seems to involve a high waste of eggs: In a large series of oviposition experiments we (Freese and Zwölfer, unpublished observations) obtained an average of 11.3 eggs/oviposition whereas under field conditions an *U. cardui* gall contains only an average of 3 larval chambers.

5. Nutritional Quality of the Host Plant   It is well known that the nutritional quality of plant material consumed influences growth and development of insect herbivores (Scriber and Slansky, 1981). As a model system for the analysis of the interaction of plant quality in Cardueae and herbivore fitness Stadler (1989, 1990, and unpublished) investigated the relationship between *Centaurea jacea* and the aphid *Uroleucon jaceae*. This system is particularly complex, as aphids are *r* strategists which are highly adapted to an optimal exploitation of their host and to rapid bionomic

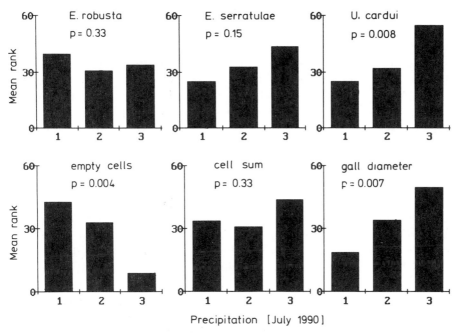

Figure 18   Influence of precipitation in July (1990) on parameters of *Urophora cardui* galls. Abscissa: 1, 10–25 mm; 2, 26–70 mm; 3, > 70 mm. Ordinate: mean ranks (Kruskal–Wallis one-way analysis). n = 64 populations originating from different regions in Germany.

adjustments to physiological changes in the host plant. In the *Centaurea–Uroleucon* food chain, the stage of development and the nutritional quality of the host determines, together with day length, the reproductive tactics of *Uroleucon,* i.e., the amount of female investment in gonads or soma tissues and the size, number, and development of embryos. The highest reproductive rate occurs during the stage of shoot formation of *Centaurea.* The nutritional quality of *Centaurea* also has a pronounced influence on certain of the behavior patterns of *Uroleucon.* A decrease in plant quality increases the time span before the "escape response" is shown. The "dropping response" of *Uroleucon* could only be observed on high-quality hosts and it is directly related to the physiological state of the aphid. In the *Centaurea–Uroleucon* system there is a feedback loop, as the consumption of phloem sap by *Uroleucon* modifies several plant parameters and plant quality. It reduces plant biomass (fresh weight and dry weight), number and size of leaves, and the concentration of amino acids.

E. Dispersal and Other Stabilizing Behavior Patterns

I have shown that in Cardueae food webs the degree of resource utilization is usually far below the carrying capacity of the resource (Section II.B) and also that predator–prey or host–parasitoid relationships are in most cases not sufficiently density dependent and strong to explain this underutilization (Section III.C). One stabilizing component in the Cardueae food webs which has been investigated is the patchy distribution and compartmental nature of the resources and the endophytic habits of the consumers (Section III.B.1). In the following sections I discuss the interaction of this distribution pattern of the resources with certain behavior patterns of the consumers.

1. Density Control in **Tephritis conura** *on* **Cirsium helenoides**   The following summarizes results obtained by Romstöck (1987), Romstöck-Völkl (1990a), and Romstöck-Völkl and Wissel (1989). Manipulations of field populations of *T. conura* showed that adult flies redistributed themselves in such a way that the densities in 1 year were comparable to these of the previous year. Neither by using an experimental removal nor by using a concentration of *T. conura* pupae was it possible to modify the popoulation densities in the field. Despite high densities in some plant populations the carrying capacity of the flower heads was only reached in 5% of populations. During the larval development within the flower heads, density-independent mortality factors (parasitism, predation) reduced populations by an average of 40%. Adult dispersal and winter mortality (90% of emerged adults) were the key factors (sensu Varley *et al.,* 1973) determining population changes. Colonization by overwintering adults was influenced by the necessity to synchronize the oviposition period with the varying phenologies of bud development in the single stands. At high experimental densities adult behavior (adult dispersal, ability to discriminate against buds with larvae, aggressive interactions between searching adults) exhibited the potential of density-dependent regulation.

2. Density Control in **Urophora cardui** *on* **Cirsium arvense**   As discussed in Section II.C.5 the population dynamics and population structure of *U. cardui* differ in our observation regions (Fig. 10), where stable situations (Belfort–Sundgau, Upper Rhine Valley, Oberpfalz), as well as range extensions (Bayreuth), area (Fig. 11), and population breakdowns (Grafrath area, experimental colonies at Delemont) were observed. It is not yet possible to evaluate the mechanisms of density control of *U. cardui* in the Bayreuth area with respect to the recent range extension of *U. cardui* (Eber and Brandl, 1993). But enough data are available to analyze the situation in the Belfort–Sundgau area and the Upper Rhine Valley, i.e., in regions where *U. cardui* populations have persisted since the start of my survey in 1968/1969. Schlumprecht (1990) carried out a series of *k*-factor analyses which showed that the only consistently effective density-dependent factor in the population dynamics of *U. cardui* was the frequency of oviposition events in the following generation, i.e., a process determined by the emigration and immigration behavior of adults. Data from Peschken and Harris (1975) and recent oviposition experiments (Freese and Zwölfer, unpublished) suggest that at high female densities the effect of this "density-dependent adult dispersal" (Schlumprecht, 1990) may be reinforced by a density-dependent waste of eggs due to an ovipositional overload. Schlumprecht's (1990) correlations between the *k*-values (a logarithmic measure) of "adult dispersal" and the log of population densities were always positive and highly significant. They explain from 20 to 56% of the variation in the *U. cardui* population densities and have slopes between 0.47 and 0.73. The path analysis (Fig. 15) of the factors influencing the gall densities in the Belfort–Sundgau populations (Section III.C.2) leads to the same conclusion, i.e., that local gall densities have a strong tendency to remain relatively stable at values distinctly below the carrying capacity of the host populations (Fig. 6A). The mortality factors operating within the gall (the microhabitat) are not density dependent and are poor predictors of the densities of the next gall generations. Thus, a combination of patchy population structure and adult behavior, recognizable in the annual redistribution of oviposition events (and perhaps reinforced by a density-dependent ovipositional overloading of hosts), also regulates the Belfort–Sundgau populations.

Field and cage experiments (Zwölfer and Peschken, 1993) provide evidence that the complex male–female communication system of *U. cardui* plays an additional stabilizing role. The presence of males influences the distribution patterns of galls which significantly ($P < 0.0001$) concentrate at or near thistles occupied by males. On thistles with previous male contact [and presumably marked with a secretion produced by male rectal glands, 4-methyl-3Z,5-hexadienoic acid (Frenzel *et al.*, 1990)] the average probability of gall formation was increased by a factor of 2 ($P < 0.0001$). If oviposition occured, the probability that three and more galls were induced at the same plant increased significantly in cage experiments ($P < 0.0001$). This tendency of females to concentrate oviposition at sites occupied by males obviously reinforces the coherence of local populations at the expense of new colonization events.

3. Density Manipulation Experiments   We experimentally increased the density of a number of insect species on Cardueae host plants held in cages (measuring either $2 \times 2 \times 2$ m or $0.5 \times 0.5 \times 0.8$ m) or in the field. Table V summarizes the results. With endophytic flower head insects (tephritids) larval densities (proportion of attacked flower heads) were increased compared to the average field situation by confining the

Table V   Density Manipulation Experiments with Arthropods on Cardueae

Species	Host plant	Type of experiment	Result
Tephritids			
Urophora cardui	*Cirsium arvense*	Cages (1)	20- to 50-fold increase
Tephritis conura	*Cirsium helenoides*	Cages (2)	Slight increase
T. conura	*C. helenoides*	Field (2)	No increase
Xyphosia miliaria	*C. arvense*	Cages (3)	2- to 4-fold increase
X. miliaria	*Cirsium palustre*	Cages (4)	2- to 3-fold increase
Chrysomelids			
Cassida rubiginosa			
larvae	*C. arvense*	Cages (5)	No increase
adults	*C. arvense*	Cages (5)	No increase
larvae	*C. arvense*	Field (5)	No increase
adults	*C. arvense*	Field (5)	No increase
Theridiidae			
Theridion impressum	*Cirsium vulgare*	Field (6)	No increase

(1) H. Zwölfer, unpublished, (2) Romstöck (1987), (3) Angermann (1987), (4) J. Arnold, unpublished, (5) Warnek, unpublished, (6) Scheidler (1989).

flies in cages. The effect was most pronounced in *U. cardui* and only slight in *T. conura*, a species which usually occurs in high densities in the field (Section II.B.2). The results suggest that in these species there is a strong tendency for adults to leave crowded sites, but that an increased oviposition rate can be induced if dispersal of the adults is prevented. With the ectophytic larvae or adults of the tortoise beetle *Cassida rubiginosa* or with the spider *Theridion impressum* an experimental increase in density was not achieved. Here the oviposition behavior and the dispersal of the larvae (*Cassida*) or aggressiveness (*Theridion*, Scheidler, 1989) reinforced the effect of adult dispersal.

IV. Discussion and Conclusions

This section begins with an overview of the results relating to the structure and stability of the investigated food webs. The second part summarizes different aspects of density control and of control mechanisms.

A. Structure, Resource, Utilization, and Stability

In Section II Cardueae–insect food webs have been described as often locally unstable but globally stable systems which exhibit a predictable structure and which are composed of keystone species and satellite elements. The most important attributes of the investigated food webs can be summarized as follows:

i. Transfer of energy and materials in the food webs occurs essentially within discrete microhabitats (flower heads, plant galls) by insect larvae (Section II.A.1).

ii. The species constellation of these food webs is determined by processes which take place in the macrohabitat (Section II.A.1) where important components are the dispersal behavior, search and selection of resources (mates and hosts), territoriality, and oviposition by adult herbivores and parasitoids, i.e., activities related to information processing and decision making.

iii. The investigated food webs have distinct and recurrent basic patterns of organization (Section II.A.2; Fig. 1) which differ among the host plant species and may exhibit regional modifications and annual fluctuations. With a mean connectance of 0.52 (confidence interval $= \pm 0.18$; i.e., with 52% of the possible connections between species realized) the complexity of the investigated food webs (Section II.A.2) is remarkably high (compare Pimm, 1982).

iv. As a rule the herbivore guilds associated with Cardueae flower heads present three different trophic strategies (Section II.A.3). The strategy early aggregated attack with gall or callus induction occupies a key position, as it requires on the average nine times more plant resources than the strategies achene and receptacle feeders and operation at two trophic levels (Fig. 3).

v. Herbivores of the investigated food webs attacked on the average from 25 to 45% of the flowerheads (Section II.B.1; Figs. 4 and 5) and consumed on the average 10–15% of the achenes (Table III). Parasitoids used on the average from 10 to 30% of their hosts (Section II.B.2; Fig. 5). An exception is the *U. cardui* food web, where the herbivore exploits much less than 10% of its resource, whereas an average from 25 to 50% (and occasionally up to 100%) of its larval populations are parasitized (Section II.B.3; Fig. 6).

vi. At a local scale food webs on Cardueae hosts show different degrees of stability (Sections II.C.1 and II.C.2). Some food webs on perennial host plants in relatively undisturbed habitats combine high durational stability with low fluctuations; most food webs on biennial Cardueae are characterized by medium durational stability, moderate or high fluctuations, and high turnover rates (Table IV).

vii. Herbivores in Cardueae flower heads exhibit three different types of population structure (Section II.C.3): redistribution systems (e.g., *T. conura*), metapopulation systems (e.g., food webs on *C.*

vulgare), and source–sink systems (e.g., *U. cardui* in the Belfort–Sundgau area) (Figs. 7–9).

viii. Fluctuations in the second trophic level (Section II.C.4) are often locally higher than those of herbivores (Fig. 9). Compared at a larger spatial and temporal scale parasitoid populations are usually as persistent as herbivore populations.

ix. The *U. cardui* food web is unique among the investigated systems because of distinct extensions to its range (Section II.C.5; Figs. 10 and 11) in the Belfort–Sundgau area (1970–1991) and in northern Bavaria (1977–1991).

B. Control Mechanisms

The flux of energy and matter in Cardueae food webs and its allocation to the trophic levels and different species depends essentially on the larval densities of the primary and secondary consumers. Regulation of the energy flow in food webs is therefore mainly a function of mechanisms which control population densities of larvae at the level of microhabitats (Section II.A.1). These, in turn depend on the densities and the foraging behavior of adults (Fig. 20) in the macrohabitats (Sections III,A and III.5). It has been shown (Section II.B.1) that on the level of the host plant populations the phytophagous guilds of the investigated food webs use on the average only a relatively small fraction of the net production of Cardueae flower heads. This underexploitation of the first trophic level, i.e., of plant production, is a well-known phenomenon in ecology (Hairston *et al.*, 1960).

Which processes keep the population densities of Cardueae herbivores below the carrying capacity of their resources? As discussed in Section III, single parasitoids or predators, which are often assumed to be a key mechanism of density control in plant–animal systems (Strong *et al.*, 1984) are not dominating control factors in the Cardueae food webs. A regulating, density-dependent mortality, caused by single antagonists, was not found in any of the analyzed Cardueae host–parasitoid systems (Sections III.C.1 and III.C.2). Rather than being driving forces in controlling host densities parasitoid guilds in Cardueae food webs constitute satellite systems with a tendency to inverse density dependence. Our data correspond with the conclusions of Stiling (1987, 1988) that only a small proportion of parasitoid–host systems show positive density dependence.

However, the combined impact of parasitoids, predators, and competitors has a latent potential to keep herbivore densities within certain limits. This is demonstrated by the fact that Cardueae herbivores introduced to North America (Julien, 1987; Harris, 1991) without their natural enemies and competitors (Section III.C.3, Fig. 17) can reach 5 to 15 times higher population densities than those in the complex European

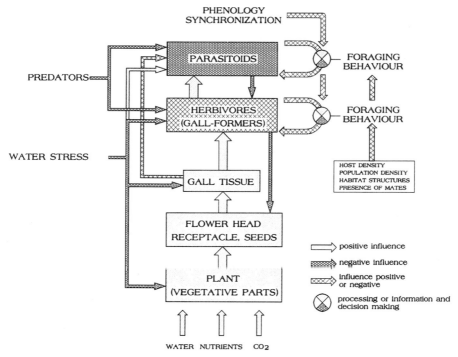

Figure 20　A summary of interactions and control mechanisms in a Cardueae food web with a gall former (*Urophora*). Foraging for oviposition sites is a major behavioural element of the multiple loop system which determines larval densities of herbivores and parasitoids. Foraging behavior depends on resource variables (phenology, concentration, distribution pattern), environmental variables, and the internal status of adult insects.

habitat are a result of relatively simple larval behavior patterns (e.g., feeding activity, pupation, induction of gall formation), processes in the macrohabitat result from complex adult behavior patterns. Individuals searching for resources (mates, host plants, or host insects) have to evaluate information and to make decisions (Roitberg, 1985; Mangel, 1989). Examples of such decisions are the determination of which patch to visit, how much time to spend in a patch, whether and how intensively to exploit an encountered resource, and whether and how intensively to fight against competitors. The available time for host searching and locating appropriate resources can be an important factor for resource utilization by adult herbivores (Straw, 1991). Our field studies of the *T. conura* food web and the *U. cardui* food web have shown that adult dispersal behavior is a particularly important, and as yet not fully predictable, factor.

Our study leads to the conclusion that when the dynamics of ecological systems with animal populations are studied, a detailed knowledge of population ecology and intra- as well as interspecific interactions is required. Such studies should not focus only on the role of predator–prey relationships and limitations by the carrying capacity. It is also important to examine to what extent population structures, types of resource exploitation, and behavior patterns related to foraging and dispersal influence the dynamics and stability of the system.

Acknowledgments

I thank Dr. Jürgen Angermann, Dr. Hanna Arnold-Rinehart, Dr. Habil, Gerhard Bauer, Thomas Baumann, Dr. Roland Brandl, Dr. Stefan Bringezu, Sabine Eber, Gunther Freese, Jürgen Herbst, Dr. Manfred Komma, Dr. Harald Michaelis, Dr. Horst Möller, Dr. Maria Romstöck-Völkl, Dr. Helmut Schlumprecht, Dorothee Schlegel, Christine Schmelzer, Professor Dr. Alfred Seitz (Universität Mainz), Dr. Bernhard Stadler, and Dr. Wolfgang Völkl, who contributed to the results summarized in this chapter. I am grateful to Gerlinde Göttfert, Mechthild Gredler, Stefan Hering, Gabi Lutschinger, Marion Preiß, Annik Servant-Miosga, Andrea Volkmann, and many summer students, who all provided valuable technical help. I am obliged to Dr. Gerhard Bauer and Dr. Trevor Petney (Heidelberg) who kindly read and corrected the manuscript. I thank my wife Uta for her assistance during 30 years of enjoyable field work, which provided the starting point of this study.

References

Angermann, J. (1984). Populationsökologische Untersuchungen an Cirsium arvense-Insekten: Ressourcennutzung und Synchronisation. Diplomarbeit, University of Bayreuth.

Angermann, J. (1986). Ökologische Differenzierung der Bohrfliegen *Xyphosia miliaria* und *Orellia ruficauda* (Diptera:Tephritidae) in den Blütenköpfen der Ackerdistel (*Cirsium arvense*). *Entomol. Gen.* **11,** 249–261.

Angermann, J. (1987). Die Blütenkopf-Insekten-Systeme der Ackerdistel (*Cirsium arvense*) und der Speerdistel (*Cirsium vulgare*). Gildenstruktur, Ressourcennutzung und Zeitstabilität. Doctoral Dissertation, University of Bayreuth.

Arnold, J. (1985). Ökologische Untersuchungen über den Insektenkomplex der *Cirsium palustre*. Diplomarbeit, University of Bayreuth.

Arnold-Rinehart, J. (1989). Histologie und Morphogense von *Urophora*- und *Myopites* Blütenkopfgallen (Diptera:Tephritidae) in Asteraceae. Doctoral Dissertation, University of Bayreuth.

Askew, R. R., and Shaw, M. R. (1986). Parasitoid communities: Their size, structure and development. *In* "Insect Parasitoids" (J. Waage and D. Greathead, eds.), pp. 225–264. Academic Press, London.

Bringezu, S. (1987). Populationsökologische Untersuchungen an Phytophagen-Entomophagen Systemen der Brennessel. Doctoral Dissertation, University of Bayreuth.

Capek, M., and Zwölfer, H. (1990). Braconids associated with insects inhabiting thistles. *Acta Entomol. Bohemoslov.* **87,** 262–277.

Cohen, J. E. (1978). "Food Webs and Niche Space," Monogr. Popul. Biol. 11. Princeton Univ. Press, Princeton, NJ.

Den Boer, P. J. (1968). Spreading of risk and stabilization of animal numbers. *Acta Biotheor.* **18,** 165–194.

Eber, S. (1993). Ökologische und populationsgenetische Muster von *Urophora cardui* als Hinweis auf Populationsstruktur und biogeographische Prozesse. Doctoral Dissertation, University of Bayreuth, in preparation.

Eber, S., and Brandl, R. (1993). Ecological and genetic spatial patterns of *Urophora cardui* (Diptera:Tephritidae) as evidence for population structure and biogeographical processes. *J. Anim. Ecol.,* in press.

Ehler, L. E. (1992). Guild analysis in biological control. *Environ. Entomol.* **21,** 26–40.

Ellenberg, H., Mayer, R., and Schauermann, J. (1986). "Ökosystem-Forschung, Ergebnisse des Solling-Projekts." Ulmer, Stuttgart.

Eschenbacher, H. (1982). Untersuchungen über den Insektenkomplex in den Blütenköpfen der Kohldistel, *Cirsium oleraceum* L. (Compositae). Diplomarbeit, University of Bayreuth.

Freese, G. (1991). Struktur- und Funktionsanalyse der Stengelboher-Gilde in Disteln (Compositae, Cardueae). Ressourcennutzung und Ressourcen aufteilung. Diplomarbeit, University of Bayreuth.

Freese, G. (1992). Biology, ecology and parasitoids of *Agapanthia villosoviridescens* DeCeer (Coleoptera, Cerambycidae) an important stem-borer of thistles. *Zool. Anz.* (in press).

Freese, G. (1993a). The insect complexes associated with the stems of 7 thistle species with notes on potential biocontrol agents. *Z. Angew. Entomol.* (in press).

Frenzel, M., Dettner, K., Boland, W., Erbes, P. (1990). Identification and biological significance of 4-methyl-3Z,5-hexadienoic acid produced by males in the gallforming Tephritids *Urophora cardui* L. and *Urophora stylata* Fab. (Diptera:Tephritidae). *Experientia* **46,** 542–547.

Gilpin, M., and Hanski, I., eds. (1991). "Metapopulation Dynamics: Empirical and Theoretical Investigations." Academic Press, London.

Goeden, R. D., and Ricker, D. W. (1985). Seasonal asynchrony of Italian Thistle, *Carduus pycnocephalus,* and the weevil *Rhinocyllus conicus* (Coleoptera:Curculionidae), introduced for biological control in southern California. *Environ. Entomol.* **14,** 433–436.

Hairston, N. G., Smith, F., and Slobodkin, L. (1960). Community structure, population control and competition. *Am. Nat.* **94,** 421–425.

Hanski, I., and Gilpin, M. (1991). Metapopulation dynamics: Brief history and conceptual domain. *Biol. J. Linn. Soc.* **42,** 3–16.

Harris, P. (1980). Effects of *Urophora affinis* Frfld. and *U. quadrifasciata* (Meig.) (Diptera: Tephritidae) on *Centaurea diffusa* Lam. and *C. maculosa* Lam. (Compositae). *Z. Angew. Entomol.* **90,** 190–201.

Harris, P. (1989). The use of Tephritidae for the biological control of weeds. *Biocontrol. News Inf.* **10,** 7–16.

Harris, P. (1990). The Canadian biocontrol of weeds program. *In* "Range Weeds Revisited" (B. R. Roche, and C. T. Roche, eds.), pp. 61–68. Washington State University, Pullman.

Harris, P. (1991). Invitation paper (C. P. Alexander fund): Classical biocontrol of weeds: Its definition, selection of effective agents, and administrative-political problems. *Can. Entomol.* **123,** 827–849.

Harrison, S. (1991). Local extinction in a metapopulation context: an empirical evaluation. *Biol. J. Linn. Soc.* **42,** 73–88.

Hassell, M. P. (1979). Non-random search in predator-prey models. *Fortschr. Zool.* **25,** 311–330.

Hassell, M. P. (1986). Detecting density dependence. *Trends Ecol. Evol.* **1,** 90–93.

Hawthorn, W. R., and Hayne, P. D. (1978). Seed production and predispersal seed predation in the biennial Composite species *Arctium minus* (Hill) Bernh. and *A. lappa* L. *Oecologia* **34**, 283–295.

Jansson, A. (1992). Distribution and dispersal of *Urophora cardui* (Diptera, Tephritidae) in Finland in 1985–1991. *Entomol. Fenn.* **2**, 211–216.

Julien, M. H. (1987). "Biological Control of Weeds: A World Catalogue of Agents and Their Target Weeds," 2nd ed. CAB Int. Inst. Biol. Control, Imperial College, Ascot, Berks, UK.

Kok, L. T., and Pienkowski, R. L. (1985). Biological control of musk thistle by *Rhinocyllus conicus* (Coleoptera:Curculionidae) in Virginia from 1969 to 1980. *Proc. Int. Symp. Biol. Control Weeds, 6th, 1984,* Vancouver, Canada, pp. 433–438.

Levins, R. (1968). "Evolution in Changing Environments." Princeton Univ. Press, Princeton, NJ.

Levins, R. (1969). Some demographic and genetic consequences of environmental heterogeneity for biological control. *Bull. Entomol. Soc. Am.* **15**, 237–240.

Lindeman, R. L. (1942). The trophic-dynamic aspect of ecology. *Ecology* **23**, 399–418.

Mangel, M. (1989). An evolutionary interpretation of the "motivation to oviposit." *J. Evol. Biol.* **2**, 157–172.

Merz, B. (1991). Die Fruchtflilegen der Stadt Zürich (Diptera:Tephritidae). *Vierteljahresschr. Naturforsch. Ges. Zürich* **136**, 105–111.

Michaelis, H. (1984). Struktur- und Funktionsuntersuchungen zum Nahrungsnetz in den Blütenköpfen von *Cirsium vulgare.* Doctoral Dissertation, University of Bayreuth.

Möller-Joop, H. (1989). Biosystematisch-ökologische Untersuchungen an *Urophora solstitialis* L. (Tephritidae): Wirtskreis, Biotypen und Eignung zur biologischen Bekämpfung von *Carduus acanthiodes* L. (Compositae) in Kanada. Doctoral Dissertation, University of Bayreuth.

Morrison, G., and Strong, D. R. (1980). Spatial variations in host density and the intensity of parasitism: Some experimental examples. *Environ. Entomol.* **9**, 149–152.

Müller, H. (1984). Die Strukturanalyse der Wurzelphytophagen-komplexe von *Centaurea maculosa* Lam. und *C. diffusa* Lam. (Compositae) in Europa und Interaktionen zwischen wichtigen Phytophagenarten und ihren Wirtspflanzen. Doctoral Dissertation, University of Bern.

Odum, H. T. (1983). "Systems Ecology." Wiley (Interscience), New York.

Peschken, D. P., and Harris, P. (1975). Host specificity and biology of *Urophora cardui* (Diptera:Tephritidae). A biocontrol agent for Canada thistle (*Cirsium arvense*). *Can. Entomol.* **107**, 1101–1110.

Peschken, D. P., Finnamore, D. B., and Watson, A. K. (1982). Biocontrol of the weed Canada thistle (*Cirsium arvense*): Release and development of the gall fly *Urophora cardui* (Diptera:Tephritidae) in Canada. *Can. Entomol.* **114**, 349–357.

Petney, T. N. (1988). Influence of insect attack on reproductive potential of thistle species in Jordan. *Entomol Gen.* **14**, 25–35.

Petney, T. N., and Zwölfer, H. (1985). Phytophagous insects associated with Cynareae hosts (Asteraceae) in Jordan. *Isr. J. Entomol.* **29**, 147–159.

Phillipson, J. (1966). "Ecological Energetics. Studies in Biology No. 1." Arnold, London.

Pimm, S. L. (1982). "Food Webs." Chapman & Hall, London.

Pimm, S. L. (1984). The complexity and stability of ecosystems. *Nature (London)* **307**, 321–326.

Pimm, S. L. (1988). 13. Energy flow and trophic structure. *Ecol. Stud.* **67**, 263–278.

Pimm, S. L., Lawton, J. H., and Cohen, J. E. (1991). Food web patterns and their consequences. *Nature (London)* **350**, 669–674.

Price, P. W. (1988). Inversely density-dependent parasitism: The role of plant refuges for hosts. *J. Anim. Ecol.* **57**, 89–96.

Price, P. W., and Clancy, K. M. (1986). Interactions among three trophic levels: Gall size and parasitoid attack. *Ecology* **67**, 1593–1600.

Price, P. W., Bouton, C. E., Gross, P., McPheron, B. A., Thompson, J. N., and Weis, A. E. (1980). Interactions among three trophic levels: Influence of plants on interactions between insect herbivores and natural enemies. *Annu. Rev. Ecol. Syst.* **11**, 41–65.

Redfern, M. (1968). The natural history of spear thistle-heads. *Field Stud.* **2**, 669–717.

Redfern, M., and Cameron, R. A. D. (1989). Density and survival of introduced populations of *Urophora stylata* (Diptera:Tephritidae) in *Cirsium vulgare* (Compositae) in Canada, compared with native populations. *Proc. Int. Symp. Biol. Control Weeds, 7th, 1988*, Rome, Italy, pp. 203–210.

Rees, N. E. (1978). Interaction of *Rhinocyllus conicus* and thistles in the Galatin Valley. *In* "Biological Control of Thistles in the Genus *Carduus* in the United States" (K. E. Frick, ed.), pp. 31–38. Science and Education Administration, U.S. Dep. Agric., Washington, DC.

Remund, U., and Zwölfer, H. (1993). The dispersal capacity of *Urophora cardui* (Diptera: Tephritidae): flightmill tests and field observations. In preparation.

Roitberg, B. D. (1985). Search dynamics in fruit-parasitic insects. *J. Insect Physiol.* **31**, 865–872.

Romstöck, M. (1982). Untersuchungen über den Insektenkomplex in den Blütenköpfen von *Cirsium heterophyllum* (Cardueae). Diplomarbeit, University of Bayreuth.

Romstöck, M. (1987). *Tephritis conura* Loew (Diptera: Tephritidae) und *Cirsium heterophyllum* L. (Hill) (Cardueae): Struktur- und Funktionsanalyse eines ökologischen Kleinsystems. Doctoral Dissertation, University of Bayreuth.

Romstöck-Völkl, M. (1990a). Population dynamics of *Tephritis conura* Loew (Diptera: Tephritidae): Determinants of density from three trophic levels. *J. Anim. Ecol.* **59**, 251–268.

Romstöck-Völkl, M. (1990b). Host refuges and spatial patterns of parasitism in an endophytic host-parasitoid system. *Ecol. Entomol.* **15**, 321–331.

Romstöck-Völkl, M., and Wissel, C. (1989). Spatial and seasonal patterns in the egg distribution of *Tephritis conura* Lw (Diptera:Tephritidae). *Oikos* **55**, 165–174.

Scheidler, M. (1989). Niche partitioning and density distribution in two species of Theridion (Theridiidae, Araneae) on thistles. *Zool. Anz.* **223**, 49–56.

Schlumprecht, H. (1989). Dispersal in the thistle gallfly *Urophora cardui* (Diptera:Tephritidae) and its endoparasitoid *Eurytoma serratulae* (Hymenoptera:Eurytomidae). *Ecol. Entomol.* **14**, 341–348.

Schlumprecht, H. (1990). Untersuchungen zur Populationsökologie des Phytophagen-Parasitoid-Systems von *Urophora cardui* L. (Diptera:Tephritidae). Doctoral Dissertation, University of Bayreuth.

Scriber, J. M., and Slansky, F. (1981). The nutritional ecology of immature insects. *Annu. Rev. Entomol.* **26**, 183–211.

Shorthouse, J. D., and Lalonde, R. C. (1984). Structural damage by *Rhinocyllus conicus* (Coleoptera:Curculionidae) within the flowerheads of nodding thistle. *Can. Entomol.* **116**, 1335–1343.

Sobhian, R., Zwölfer, H. (1985). Phytophagous insect species associated with flower heads of yellow star thistle (*Centaurea solstitialis* L.). *Z. Angew. Entomol.* **99**, 301–321.

Sokal, R. R., and Rohlf, F. J. (1981). "Biometry." Freeman, San Francisco.

Southwood, T. R. E. (1976). Bionomic strategies and population parameters. *In* "Theoretical Ecology" (R. M. May, ed.), pp. 26–48. Blackwell, Oxford.

Southwood, T. R. E. (1978). "Ecological Methods," 2nd ed. Chapman & Hall, London.

Stadler, B. (1989). Untersuchungen zur Populationsökologie von *Uroleucon jaceae* in Oberfranken. Diplomarbeit, University of Bayreuth.

Stadler, B. (1990). Relationships between host plant quality and reproductive investment in *Uroleucon jaceae*. *Acta. Phytopathol. Entomol. Hung.* **25**, 177–183.

Stiling, P. (1987). The frequency of density dependence in insect host-parasitoid systems. *Ecology* **68**, 844–856.

Stiling, P. (1988). Density-dependent processes and key factors in insect populations. *J. Anim. Ecol.* **57**, 581–593.

Stinner, B. R., and Abrahamson, W. G. (1979). Energetics of the *Solidago canadensis* stem gall insect-parasitoid guild interaction. *Ecology* **60**, 918–926.

Straw, N. A. (1991). Resource limitation of tephritid flies on lesser burdock, *Arctium minus* (Hill) Bernh. (Compositae). *Oecologia* **86**, 492–502.

Strong, D. R., Lawton, J. H., and Southwood, R. (1984). "Insects on Plants: Community Patterns and Mechanism." Blackwell, London.

Sturm, P. (1988). Vergleichende ökologische Untersuchungen an den Insektenkomplexen von *Arctium tomentosum* Mill. and *Arctium lappa* L. Diplomarbeit, University of Bayreuth.

Tischler, W. (1949). "Grundzüge der terrestrischen Tierökologie." Vieweg, Braunschweig.

Tscharntke, T. (1992). Coexistence, tritrophic interactions and density dependence in a species-rich parasitoid community. *J. Anim. Ecol.* **61**, 59–67.

Varley, G. C., Gradwell, G. R., and Hassell, M. P. (1973). "Insect Population Ecology: An Analytical Approach." Blackwell, Oxford.

Völkl, W. (1989). Resource partitioning in a guild of aphid species associated with creeping thistle *Crisium arvense*. *Entomol. Exp. Appl.* **51**, 41–47.

Völkl, W. (1990). Fortpflanzungsstrategien bei Blattlaus-Parasitoiden (Hymenoptera, Aphidiidae): Konsequenzen ihrer Interaktionen mit Wirten und Ameisen. Doctoral Dissertation, University of Bayreuth.

Völkl, W. (1991). Species-specific larval instar preferences and aphid defense reactions in three parasitoids of *Aphis fabae*. *In* "Ecology of Aphidophaga, Gödöllö" (L. Polgar, R. Chambers, A. F. G. Dixon, and I. Hodek, eds.), pp. 73–78.

Völkl, W. (1992). Aphids or their parasitoids: Who actually benefits from ant-attendance? *J. Anim. Ecol.* **61**, 273–281.

Völkl, W., Zwölfer, H., Romstöck-Völkl, M., and Schmelzer, C. (1993). Habitat management in calcareous grasslands: Effects on the insect community dwelling endophytically in flower heads of thistles. J. Applied Ecology, in press.

Walde, A. E., and Murdoch, W. W. (1988). Spatial density dependence in parasitoids. *Annu. Rev. Entomol.* **33**, 441–466.

Weis, A. E., Walton, R., and Crego, C. L. (1988). Reactive plant tissue sites and the population biology of gall makers. *Annu. Rev. Entomol.* **33**, 467–486.

Weisser, W. (1991). Das Eiablageverhalten von Blattlausparasitoiden (Hym., Aphidiidae): Welchen Einfluss haben Habitatfaktoren und das Parasitoidenalter? Diplomarbeit, University of Bayreuth.

Zwölfer, H. (1965). Preliminary list of phytophagous insects attacking wild Cynareae (Compositae) in Europe. *Commonw. Inst. Biol. Control, Tech. Bull.* **6**, 81–154.

Zwölfer, H. (1972). "Investigations on *Urophora stylata* Fabr., a Possible Agent for the Biological Control of *Cirsium vulgare* in Canada," Prog. Rep. IXXX. Commonw. Inst. Biol. Control, European Station (Delemont CH).

Zwölfer, H. (1978). An analysis of the insect complexes associated with the heads of European *Centaurea maculosa* populations. *Proc. Knapweed Symp., 1977*, Kamloops, BC, pp. 139–163.

Zwölfer, H. (1979). Strategies and counterstrategies in insect population systems competing for space and food in flower heads and plant galls. *Fortschr. Zool.* **25**, 331–353.

Zwölfer, H. (1980). Distelblütenköpfe als ökologische Kleinsysteme: Konkurrenz und Koexistenz in Phytophagenkomplexen. *Mitt. Dtsch. Ges. Allg. Angew. Entomol.* **2**, 21–37.

Zwölfer, H. (1985a). Energiefluss steuerung durch informationelle Prozesse—ein vernach-lässigtes Gebiet der ökosystemforschung. *Verh. Ges. Oekol.*, Bremen, *1983* **13**, 285–294.

Zwölfer, H. (1985b). Insects and thistle heads: Resource utilization and guild structure. *Proc. Int. Symp. Biol. Control Weeds, 6th, 1984*, Vancouver, Canada, pp. 407–416.

Zwölfer, H. (1987). Species richness, species packing, and evolution in insect-plant systems. *Ecol. Stud.* **61**, 301–319.

Zwölfer, H. (1988). Evolutionary and ecological relationships of the insect fauna of thistles. *Annu. Rev. Entomol.* **33**, 103–122.

Zwölfer, H. (1990). Disteln und ihre Insektenfauna: Makroevolution in einem Phyto-phagen-Pflanzen-System. *In* "Evolutionsprozesse im Tierreich" (B. Streit, ed.), pp. 255–278. Birkhäuser-Verlag, Basel.

Zwölfer, H., and Arnold-Rinehart, J. (1993). The evolution of interactions and diversity in plant-insect systems: The *Urophora–Eurytoma* Food Web in galls on Palearctic Cardueae. *Ecol. Stud.* **99**, 211–233.

Zwölfer, H., and Harris, P. (1966). *Ceutorhynchus litura* (F.) (Col.:Curculionidae), a potential insect for the biocontrol of thistle, *Cirsium arvense* (L.) Scop. in Canada. *Can. J. Zool.* **44**, 23–38.

Zwölfer, H., and Harris, P. (1984). Biology and host specificity of *Rhinocyllus conicus* Froel. (Col.:Curculionidae), a successful agent for biocontrol of the thistle *Carduus nutans* L. *Z. Angew. Entomol.* **97**, 36–62.

Zwölfer, H., and Peschken, D. (1993). Male-female communication and resource utilization in *Urophora cardui* (Diptera:Tephritidae). In preparation.

Zwölfer, H., and Romstöck-Völkl, M. (1991). Biotypes and the evolution of niches in phytophagous insects on Cardueae hosts. *In* "Plant-Animal Interactions, Evolutionary Ecology in Tropical and Temperate Regions" (P. W. Price, T. M. Lewinsohn, G. Wilson Fernandes, and W. W. Benson, eds.), pp. 487–507. Wiley (Interscience), New York.

12

Fluxes in Ecosystems

E.-D. Schulze and H. Zwölfer

I. Introduction

Previous chapters of this book analyzed mechanisms by which fluxes are regulated in cells, organisms, and populations, and a number of similarities of mechanisms emerged by which fluxes are regulated at different levels of organization. In the following we explore the extent to which such principles are also applicable at the ecosystem level. However, the questions asked are different at the physiological (deterministic) level of single organisms than at the population (stochastic) level of single species and ecosystems. At the primary producer level, which plant life forms and structures establish themselves in a given resource environment and thus affect other resources of the habitat are important. This may in turn lead to the decline and invasion of other species. At the consumer level the degree to which resources are used determines how predictable and stable food web structures are. At the system level numerous interactions exist not only within each trophic level but also between plants and their environment, between plants and herbivores, between plants and herbivores and their parasitoids, and between all trophic levels and the decomposer chain.

It is quite clear that an ecosystem has numerous constraints including nutrient availability, climate migration, history, and land use (Fig. 1). Within this framework of constraints the living organisms affect fluxes within and through the system by (i) their species composition, (ii) their structure, (iii) their physiological capacities to carry out certain ecosystem functions, and (iv) their population dynamics. Species composition has strong effects on structure and function of the ecosystem. Also, there are obvious effects of structure on species composition and function. In

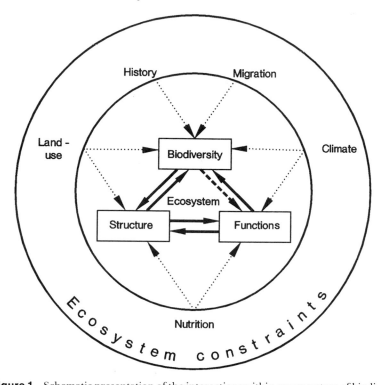

Figure 1   Schematic presentation of the interactions within an ecosystem of biodiversity, community structure, and function and effects of external ecosystem constraints on these parameters. Solid-lined arrows indicate strong feedbacks, while broken-lined arrows between biodiversity and function indicates a weak feedback. Dotted lines indicate major effects of external factors on ecosystem internal processes (after Schulze and Mooney, 1993).

contrast, present knowledge suggests that the effects of species diversity on function are weak and need further exploration. In the following we try to explain some of these interactions. The discussion is in part based on the treatment of "Biodiversity and Ecosystem Function" by Schulze and Mooney (1993), as well as on the books by Begon *et al.* (1990), Harper (1977), and Crawley (1986).

II. The Ecosystem Concept

We refer to terrestrial ecosystems as the combined communities of plant and animals and their physical environment. The boundaries are the atmosphere and the soil through which ecosystems exchange water, gases, and ions. Also species may be exchanged with surrounding ecosystems. Fluxes of particles are initiated by photosynthesis of green plants or by the weathering process in soils. These fluxes through the system are mediated by water which generally enters by wet deposition from the atmosphere. Water sustains life and drains eventually into groundwater. The time constant for processes and fluxes are different for soils, plants, animals, and microorganisms. One property of ecosystems is that a quasi-balance of fluxes exists according to demand and supply of resources. Otherwise, resources would be lost (e.g., nitrate to groundwater) or accumulated (e.g., raw humus in forests). In both cases an imbalance of demand and supply initiates a change in the community as soon as local disturbance allows for such change, and succession of species will adjust fluxes to a new level of resource demand and supply.

The relation between a system and its components may be explained by using the analogy of a car (Schulze, 1989b) which contains components which are absolutely necessary and others which only improve functioning or have a function in case of emergency. There are also components which have no function but make the car more attractive. Even if all components are present and intact the car may not "function" if not "tuned"—the concert of parts and functions is not equilibrated. The analogy of the car illustrates the variable importance of parts in a system and the role of backups (redundancy, e.g., brakes in the case of the car); however, as Schulze and Mooney (1993) pointed out, the analogy is not fully applicable. An ecosystem is not a machine constructed to accomplish a given function. Components of ecosystems operate collectively to fully utilize the available resources (e.g., water, light, and nutrients in the case of animals), but species in ecosystems have additional properties such as the ability to compensate. Fluctuations of population density and succession of species is one of the main differences between the machine analogy and an ecosystem. The functional role of a component may change depending on the activity of the neighbors. Density-dependent compensation will occur only in ecosystems (Pimm, 1984; McNaughton, 1983). Therefore, it seems to be important to understand the rules and mechanisms which lead to species organization in communities, which determine the pathway and the magnitude of fluxes, and which are the bases for stability and resilience against environmental change (Steffen *et al.*, 1992).

III. Factors Involved in Structure and Organization of Ecosystems

A. Availability of Inorganic Resources Determines Plant Cover

The temporal and spacial variations in resources especially water and nutrients seem to be the major factors that determine the distribution of plant life forms along environmental gradients (Schulze, 1982). The effect of temporal and spatial resource availability is enforced by a factor which is under plant control, namely, the ability to compete for light (Schulze and Chapin, 1987; see also Stitt and Schulze, this volume, Chapter 4). Permanent poor resource supply will favor evergreen species which are also very competitive for light. Thus, evergreen woody vegetation is typical for tropical rain forests, boreal forests, and evergreen shrublands on laterites while deciduous forests indicate increasing seasonality and supply of resources. Deciduous vegetation encourages perennial herbaceous vegetation if the competitive ability for light is weakened because of external factors such as reduced rainfall. When there is a predictable short seasonal resource supply perennial herbaceous vegetation will be dominant (e.g., in alpine meadows or savannahs), but with unpredictable shortened pulses of resources annual vegetation becomes more successful (e.g., in deserts and steppes). In many cases the natural vegetation cover reflects an evolutionary history of plant migrations. The invasion of European weeds to other continents and of *Pinus radiata* to South Africa does indicate the "relict" stage of many "natural" vegetation types (Mooney and Drake, 1984).

Avaliability of nutrients not only determines plant formations but also the coexistence and displacement of species along resource gradients (Fig. 2) where the minimum relative resource ratios which individual species are capable of utilizing will determine species separation (Tilman, 1986). At constant spatial heterogeneity many species could coexist in resource-poor but not resource-rich habitats. Such resource equilibrium could be disturbed by invaders, which exhibit different capabilities for nutrient use.

While Tilman (1986, 1993) explains species interaction mainly at the root level, the actual competition among species is often aboveground competition for light (Schulze and Chapin, 1987). For instance, the succession of woody species may be explained by their capability of gaining space for expanding leaf area per dry matter investment (Schulze et al., 1986). *Prunus spinosa,* an early successional species, gains 0.04 m^3 spatial volume kg^{-1} dry matter investment. This increases with succession to 0.08 m^3 kg^{-1} in *Crataegus macrocarpa* and 0.13 m^3 kg^{-1} in *Acer campestre* and reaches 0.55 m^3 kg^{-1} in the late successional *Fagus sylvatica.* At the same time shade tolerance increases but photosynthetic capacity decreases from 9 to 12 μmol m^{-2} s^{-1} in *Prunus* to 3 to 4 μmol m^{-2} s^{-1} in *Fagus.*

It was the combination of below- and aboveground competition which caused the change in grassland vegetation from high species diversity to dominance by a single or few species in the oldest long-term fertilization study that has been performed in ecological history at Rothamsted (Fig. 3; Tilman, 1986). In the year 1856 a meadow was divided into 20 plots

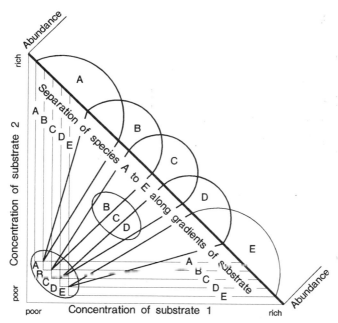

Figure 2 Isolines of the minimum concentration to which different species (A to E) can draw resources 1 and 2. Each species could potentially exist in the area described by the species-specific isolines of minimum resource use. Along an environmental gradient (oblique axis) different species are separated by competition for resources (half circles along oblique axis representing abundance of each species). Overlapping ranges exist in which different species reach maximum abundance depending on their capacity to extract resources in a range of resource conditions (ranges indicated by thick black lines). If similar-sized areas are expected (shaded areas), a higher number of species may coexist in poor resource environments than in rich resource sites (after Tilman, 1986).

each receiving a different combination of fertilizer. There was no change in species composition in unfertilized plots, while diversity fell dramatically with addition of nitrogen. Fertilized plots were eventually dominated by single species, namely *Holcus lanatus* at low pH, while *Alopecurus pratensis* dominated at higher pH. *Alopecurus* and *Arrhenaterum* coexisted at the highest pH and dominated all other species.

B. The Exploitation of Primary Production Determines the Link of Animals to Vegetation

Despite the fact that the vegetation of terrestrial ecosystems seems to be the primary food resource for animals, it is used to a surprisingly low extent and only by relatively few of higher animal taxa (e.g., the orders of the land snails, Lepidoptera, Orthoptera, ungulates, rodents, and groups within other insect orders). The exploitation of plant production as ani-

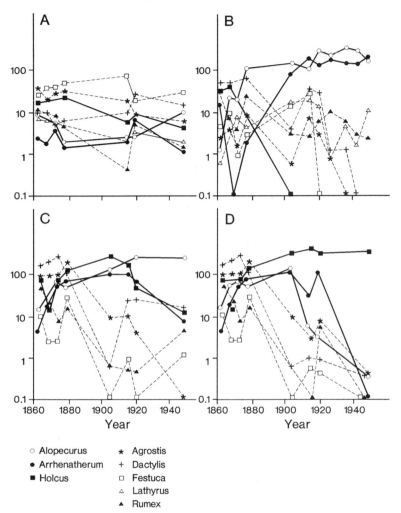

Figure 3   The dynamic change in four plots of the Rothamstad grassland experiments. (A) control plots with no change in species composition. (B) Plot limed and fertilized with nitrate (pH 6) is dominated by *Arrhenaterum* and *Alopecurus*. (C) The limed plot fertilized with ammonium (pH 4.2) is dominated by *Alopecurus* only. (D) The unlimed plot fertilized with ammonium (pH 3.7) is dominated by *Holcus* (after Tilman, 1986).

mal food presented a formidable evolutionary barrier, as has been shown by Southwood (1973) for insects. Major obstacles for herbivory were problems of nutrition caused by the relatively low protein–nitrogen levels in plant tissues and problems of physical and chemical defenses of plants. However, in animal groups which succeeded in overcoming this hurdle (e.g., by incorporating microorganisms as endosymbionts) phytophagy offered unique opportunities for evolutionary radiations. Herbivorous animal taxa and the associated predacious, parasitic, and saprophagous animals represent a tremendous biological and ecological diversity of species which outnumber all other known animal and plant species (Zwölfer, 1978; Strong *et al.*, 1984).

Two different basic strategies of exploiting the vegetation are those of (i) grazers and browsers and (ii) plant parasites. Foraging by grazers and browsers, e.g., species such as red and roe deer, voles, slugs, and grasshoppers, but also caterpillars of certain lepidopterous species, is dependent on free movements among the vegetation. Many grazing and browsing species are well adapted to cope with unpredictable resource situations, as they can concentrate on places with sufficient and qualitatively good food supply and they can abandon overexploited sites. Usually grazers and browsers do not exhibit a high degree of host specificity but they may be very selective with respect to food quality.

Plant parasitic animals (Price, 1977; Zwölfer and Herbst, 1988) differ from grazers in that they are much more tightly linked to single plant species and particular plant organs. The choice of the host plant is made by the ovipositing adults, i.e., plant parasitic animals forage for oviposition sites. The immature stages particularly in endophytic species are usually not able to abandon their resource units. The life cycle of plant parasitic animals is closely integrated into the life cycle of the host plant, and a precise synchronization with the phenology of the resource is of great importance. Because of their relative immobility many plant parasitic insects are an easy target for parasitoids and predators and the risk of intraspecific competition for food and space is greater than that in grazing or browsing insects. On the other hand, the intimate association with a given host plant allows for a high degree of niche partitioning and much more subtle ways of resource exploitation. An example is the induction of plant galls, i.e., the manipulation of the growth pattern, structures, and fluxes in an organ of the host plant (see also Zwölfer, this volume, Chapter 11).

In the context of the ecosystem a major function of herbivores is their role in making the energy and organic matter produced and stored in green plants accessible to animal species at higher trophic levels. In food chains and food webs of terrestrial communities many herbivores occupy a key position because they affect the competitive interactions at the producer level. The great diversity of such communities depends on herbivore diversity as much as on the diversity of the vegetation.

Many groups of herbivores have evolved mutualistic relationships with plants. Pollination by insects and other animals is a process which moves pollen with great accuracy from source to destination. Compared to wind pollination a smaller output of pollen is needed and insect pollination may also result in a higher diversity of the vegetation, as flowers of isolated plant individuals can be reliably fertilized. Animals such as certain species of ants, wood mice, or birds are important vectors of plant seeds. The diversity of shrub species in hedgerows or on forest borders in Europe is largely due to the fact that their seeds are packed into berries or nuts which are harvested and transported by birds and small mammals (Zwölfer and Stechmann, 1989). Pollination and transport of seeds are processes which increase the diversity of both plants and animals of a community, as they provide means of dispersal and additional food resources.

The exploitation of plants by herbivores in most cases does not significantly affect the vegetation (Hairstone *et al.*, 1960). It can, however, develop enough pressure to influence growth, production, and survival of single plant individuals as well as the species composition of the vegetation of an ecosystem. Examples are bark beetles and other forest insects with the potential to develop outbreaks (Barbosa and Schultz, 1987) and the cases of the successful control of noxious weeds by introduced phytophagous insect species (Huffaker and Messenger, 1976). Drastic but often overlooked examples of the impact of herbivores on the composition of vegetation are the considerable problems created by roe deer and red deer in many European forests which do not contain natural predators of these animals.

C. Microorganisms Are Cosmopolitans

Microorganisms are ubiquitous and cosmopolitan (Meyer, 1993). There are no effective barriers to distribution. They can easily be transported by wind, water, and animals. Transport has been demonstrated in dust particles, snowflakes, and precipitation. As a consequence of this ease of distribution, metabolic flexibility and the ability to resist environmental extremes microbial species from almost any taxonomic unit may be found "in a single gram of normal garden soil, irrespective of the location where it was collected" (Meyer, 1993). Thus, the problem in terms of functioning of the ecosystem is different for microbes and for higher organisms. It is not the problem of presence and succession, but rather the environmental conditions determining microbial activity which has strong feedback on the functioning of the whole system (see below).

D. Tight Functional Links between Species

Communities are not assemblies of species at random, not even in aquatic systems. Functional links exist for host–parasite relations, mutualistic relations of roots and mycorrhizae, plants and vectors of pollen and seeds, plant–herbivore interactions, and predator–prey relationships. If the transfer of resource is established from one organism to another additional species may follow to use a secure food source.

The effect of tight functional links is manifold. The loss of the host will result in the loss of a whole food chain, and this may have secondary compensating effects in food webs. Tight links also occur in species classified into "functional groups" and "guilds" (Root, 1967). Individual species within such groups share resources with other guild members but may be individually tightly dependent on the existence of species in quite another functional group (see also Fig. 7).

Food webs of producers, consumers, and decomposers contain nodes (Pimm, 1984) which have an effect on the functioning of the system and on species diversity which can be more than proportional if these species are lost or if a new species invades the system. Species which exert such a nonlinear effect by changing the properties of the whole system are defined as "keystone species" (Bond, 1993). Although the description of a keystone species is quite clear in theory, their identification in ecosystems is difficult and requires detailed knowledge of the system especially since the role of keystones may be system dependent. Keystone species that have particularly large impacts appear to fall into several groups such as pathogens (e.g., Dutch elm disease in combination with the elm barkbeetle), herbivores (e.g., migratory locusts, the North American spruce budworm, rodents, and ungulates), animal species that change physical properties of their ecosystem (e.g., the beaver), species that limit regeneration of a community dominant (e.g., grass vegetation which inhibits regeneration of trees under conditions of nitrogen deposition), species that affect disturbance regimes or resource stability (e.g., flammable grasses or nitrogen fixers), and species which are cultivated by man (e.g., Eucalyptus in California, spruce forest in Europe).

Tight links in food chains are most obvious in microbial decomposition (Fig. 4A). Litter fall contains organic polymers, which cannot be broken down directly or digested by exoenzymes outside of microorganisms (Meyer, 1993). Those substances which cannot be depolymerized enter into the humus as a major carbon accumulation compartment of ecosystems (see also Zech, this volume, Chapter 9). Depolymerization results in monomers which are digested in ordinary metabolic pathways and which return minerals to the primary producers of the ecosystem. The turnover of monomers is shared by a large variety of species each of which executes only a single specific step in the sequence of mineralization. For instance, in the nitrogen cycle (Fig. 4B) several separate organisms perform the individual steps of denitrification starting from nitrate and leading to N_2. The activity of the organisms in this sequence is not regulated by the presence of other organisms and their supply of metabolic products. Their activity is regulated by the chemical environment.

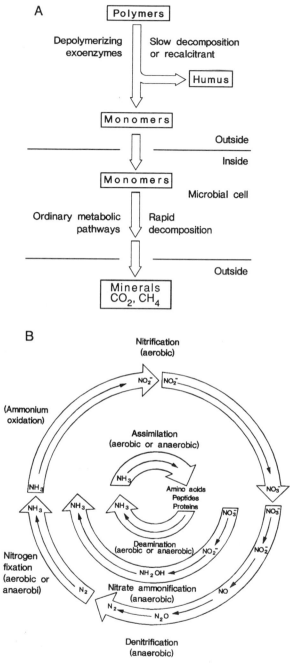

Figure 4 (A) Major steps in the microbial decomposition of natural polymers. (B) Functional groups of microorganisms in the biogeochemical cycle of nitrogen. Open arrows refer to sequences occuring within single organisms, small arrows refer to intermediates. (A and B after Meyer, 1993.)

C

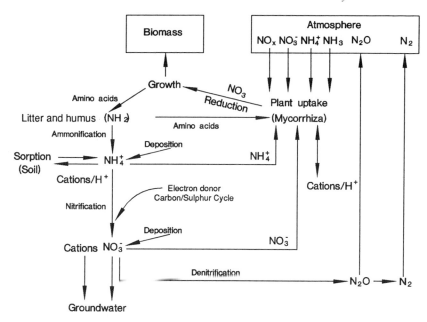

Figure 4 *(continued)* (C) Processes of nitrogen transformation in soils.

Microbial decomposition may even be analyzed in analogy to enzyme activations of metabolic pathways in cells (see also Stitt, this volume, Chapter 2). In decomposition an initial substrate is converted into an end product via a series of organisms which are linked together by metabolic intermediates. The "pathway" proceeds under given conditions at a defined rate in which the intermediate concentrations are used by different organisms in balance with the overall flux. If one organism increased its activity, there would be a net increase of a product which in turn may activate the activity of the neighbor of this sequence until there was a balance in flux between both organisms. If the supply of substrate falls, all intermediates will decrease and the system will "run down" in sources and supplies and in organism activities. Organism activity is adjusted also by the chemical conditions of the environment which may be a result of the chemical nature in the supplied polymers (e.g., needles or leaves) and of the soil-buffering range (Ulrich, 1987). Thus, the environment determines whether the linear decomposer chain branches into different end products. More carbon may enter into the "waste" compartment of humus at lower temperature and higher lignin or phenol content of the foliage; nitrogen may be lost as N_2O rather than as N_2 at lower soil pH; or nitrate may be washed out into ground waters (Fig. 4C). Each of these options has different and dramatic feedbacks on the system. Raw humus may decompose at increasing temperature and moisture after forest clear felling and this may lead to a different soil fauna and vegetation while cation and nitrate loss will accelerate soil acidification.

Tight links of matter transfer and branching into different products exist also in producer and consumer chains. Succession, e.g., from grasslands to forest, represents such a "switch" which may alter the whole sequence of element cycling at a later stage. Such switches in the resource flow may result from disturbances such as changes in land use by man. Also, processes in the early successional stage such as accumulations of products or resource losses may initiate a change in resource flow at a later point of development. Thus, not only the primary minerals in soils determine the chemical environment (Ulrich, 1987) but also the vegetation cover feeds back on the chemical and physical environment of the ecosystem and thus determines future species composition. Most obvious is the man-made change in forest cover from deciduous to coniferous forest of Europe which changed among other factors the water balance, the input and chemical quality of litter, and soil temperature. In a vegetation dominated by $Urtica$ $30 \text{ g} \cdot \text{m}^{-2} \text{year}^{-1}$ of nitrogen is added to the decomposer chain (Schulze and Chapin, 1987). This contrasts to about $1 \text{ g} \cdot \text{m}^2 \text{year}^{-1}$ of nitrogen which is transferred to the decomposers in a spruce forest. The difference is related to a much larger leaf turnover in $Urtica$ (see also Stitt and Schulze, this volume, Chapter 4). The high productivity of alluvial soils is among other factors related to this higher turnover of resources. If alluvial forests are replanted by spruce which in fact happened in large scale in Germany an ecosystem which maintained function by fast turnover is slowed down to a very slow circulation type. This results in losses of large quantities of resources which are released by decomposition until the whole system operates at a lower level of turnover. Thus, plants have a double role in ecosystem functioning. As part of the functional group of "primary producers" they initiate and maintain a flux of matter through the system, but the rate of flux depends not on the classification scheme of functional groups, but on the activity of the individual species which results from competitive interactions with other species.

This "dual role" of species becomes most obvious from species invasions (e.g., $Opuntia$ spp. in Australia and South Africa, $Hypericum$ $perforatum$ in the western US, $Carduus$ $nutans$ in central Canada). New monocultures of aggressive alien weeds may accumulate and lock up organic products, i.e., primary net production which cannot enter a grazer food chain. Biocontrol of weeds, i.e., the deliberate introduction of selected consumers (e.g., $Cactoblastis$ $cactorum$ and $Dactylopius$ spp. against $Opuntia$ spp., $Chrysolina$ $hyperici$ against $H.$ $perforatum$ (Huffaker and Messenger, 1976), or $Rhinocyllus$ $conicus$ against $C.$ $nutans$ (see also Zwölfer, this volume, Chapter 11)) demonstrated that under certain conditions single herbivore species can reverse this community structure and bring it back to the former degree of diversity and overall ecosystem flux (Harris, 1991).

IV. System Regulation of Fluxes and Stability

The word "regulation" refers to a finite process, namely, that something is being governed or directed according to a rule (Encyclopedia Brittanica). It is quite clear that this type of regulation does not exist in an ecosystem with its assembly of species because there is no preset rule (see above). However, the fluxes in ecosystems are subject to constraints. They are determined by the composition, activity, and abundance of species and individuals which exert some sort of control on the flux. However, in order to describe this observation we lack a term which is not as loaded as the word regulation. What happens at the system level is a "quasi-regulation" or a "system regulation" which refers to the fact that certain regulatory features exist such as feedback, but that this system regulation is stochastic rather than deterministic, as it takes place among species and not within individuals (Schulze and Mooney, 1992). Despite many similarities, there are important features which make system regulation different from the regulation within an organism.

A. Stochastic vs Deterministic Processes

Processes within cells (see Scheibe and Beck, this volume, Chapter 1) or organisms (see Stitt and Schulze, this volume, Chapter 4) are strongly deterministic. As living beings they are dependent on the maintenance of physiological equilibria within fairly narrow limits. Environmental influences with their often stochastic impact can temporarily disturb system variables of the organism, but the resilience of the latter is usually sufficiently effective to restore the former equilibrium (see also Stitt and Schulze, this volume, Chapter 4). These processes and their interactions and equilibria are under the control of Darwinian selection, as they are the basis of the fitness of the individual. Only a tight co-adaptation of the different physiological mechanisms and precise feedback control warrant survival and reproductive success at the system level.

While species are optimized by natural selection, ecosystems are not under the direct control of selection. They consist of a multitude of species, each of which is individually selected for maximal fitness. Some of these species are adapted for mutual cooperation (e.g., forest trees and their mycorrhiza), others are evolutionarily adjusted "modular units" which fit into food webs and food chains (see Zwölfer, this volume, Chapter 11), and still others (e.g., the elm barkbeetle or diseases of forest trees) can be rather destructive and even change the structure of the ecosystem. The interplay of all these species is strongly governed by their coincidence in time and space (Thalenhorst, 1951), their population densities, internal state variables, and environmental variables, i.e., by stochastic processes. In analogy with the central limit theorem in statistics, this multitude of possible interactions usually results in patterns which are predictable on a statistical basis only.

B. Effect of Canopy Height and Rooting Depth

The ecosystem was defined by its boundaries, namely, the atmosphere and the soil through which an exchange of gases and ions takes place (see above). These boundaries are strongly influenced by the structure and physiology of the vegetation cover.

The aboveground vegetation cover (and not the plant individual) determines the coupling to the atmosphere and thus the partitioning between latent and sensible heat during dissipation of the incoming energy from the sun. The coupling of the vegetation to the atmosphere is mainly a function of the roughness of the vegetation cover, which in turn depends on plant height. Jarvis and McNaughton (1986) defined a decoupling factor which decreases while aerodynamic conductance increases with canopy height. This means that an increasing fraction of energy is dissipated as latent heat (Fig. 5). This has consequences for all other gaseous fluxes such as CO_2, gaseous pollutants, and aeosols. However, the coupling is under physiological control by the plant cover. A *Nothofagus* forest responded like a meadow in the morning and like a coniferous forest in the afternoon because of stomatal closure (Kelliher *et al.*, 1992; see Schulze, this volume, Chapter 7). Also, the "living space" for organisms and the within canopy climate is a function of vegetation height. On a clear summer day the temperature difference in 2 m height aboveground in a grassland and a forest cover is equivalent to an elevational gradient of about 1000 m (Schulze, 1982).

While aboveground structure sets the boundaries of the ecosystem to the atmosphere, belowground structure is just as important for water and nutrient acquisition (Kelliher *et al.*, 1993). In a feedback canopy height is related to rooting depth. It is obvious that rooting depth of trees vs grasses determines the zonation of savannas, shrublands, and grasslands in subtropical climates (Walter, 1964), and influences by man through grazing and fire will alter this balance.

In addition to its effects on boundary processes vegetation structure determines several other general cause/effect relationships between form and function. For instance, the tie between woodiness and size predicts not only patterns of energy and water exchange but also interrelated ties because of unavoidable trade-offs among alternative patterns of alloca-

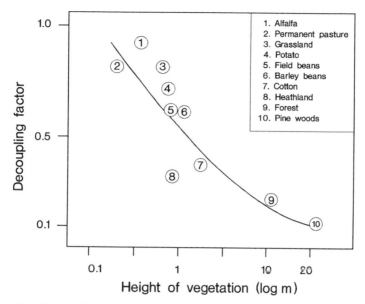

Figure 5 Change of the decoupling factor with canopy height (data from Jarvis and McNaughton, 1986).

tion, such as growth vs defense or reproduction, roots vs shoots, or competitive ability vs colonizing potential (Hobbie *et al.*, 1993). These are statistical relations which are based, however, on a physiological background and a regulation within the organism (see also Stitt, this volume, Chapter 2, and Stitt and Schulze, this volume, Chapter 4).

C. Redundancy of Species

Effects of species diversity and species packing on ecosystem fluxes are probably the least-understood area of system regulations. To our present knowledge, the direct effect of species diversity on ecosystem function is weak (Schulze and Mooney, 1993, see also Fig. 1). The few studies in which diversity of natural or agricultural ecosystems has been manipulated suggest that increases in plant species from 1 to 10 alters ecosystem function but there is very little effect on nutrient cycling of increasing plant species numbers beyond this point (Vitousek and Hooper, 1993). This observation seems to be supported by comparisons between the species-poor deciduous forests of Europe and the species-rich deciduous forests of North America. There are no obvious differences in processes or in the maintenance of processes at different intensities, and we may conclude that tree species are redundant in species-rich systems. This is supported by the observation that if a single tree species disappears, such as *Ulmus* by Dutch elm disease, the functioning of the system is resumed by compensatory expansion of other species. However, the example of the elm disease shows an important function of redundancy, namely, to supply security in events of emergency. The effect was described by Ehrlich and Ehrlich (1981) as the "rivet popper" hypothesis which suggests that if one "rivet" (in analogy to a species) after the other is taken from a well-made airplane, at a given point the machine will fail. Thus, redundancy is thought to provide stability if one component fails.

Woodward (1993) asked the reverse question, namely, how few species are necessary to make a functional ecosystem and he found that at least six species coexist in a lichen-dominated community in the Antarctica. However, these are not six species of lichens but species which exert different ecosystem function, such as photosynthesis and decomposition. In addition, each other group of lichens may consist of a different set of species. Thus, it is not the number of species, but their quality in the concert of the functioning of the whole community that is important. Schulze and Mooney (1993) described this parameter as "system-diversity" which quantifies the decreasing returns to the system of additional species of the same functional group and the gain to the system when adding an additional species of a different functional group.

Obviously, according to our present knowledge of ecosystems numerous species appear to be redundant in ecosystems if similar trophic levels are inspected and if ecosystem functions are defined only as fluxes of matter within and through the system (Lawton and Brown, 1993). Redundancy occurs despite the fact that species packing may not have reached an upper limit of niche exploitation in many systems (Zwölfer and Arnold-Rinehart, 1993). Apparently some loss of species may not affect ecosystem function, as measured, e.g., by nutrient cycling, but the decrease in biodiversity may affect the system in quite different ways. Only a few species reach large enough numbers of individuals to actually impact ecosystem processes. Most species are quite rare without a large effect on the overall flux. However, the rare species are the resource of an ecosystem which may take over if the present dominant group of species deteriorates under conditions of land use or climatic change. Thus, the presence of diversity and redundancy is of paramount importance during environmental shifts. If there are several species in a functional group, each of which responds individually to its environment, it is likely that some species in each functional group will survive environmental extremes. The greater the "redundancy" within a functional group the more effective this insurance which is in essence the message of the rivet hypothesis.

While redundancy obviously exists with regard to the ecosystem function of matter transfer, there is no "genetic redundancy" in a community, as every species is genetically unique. Species diversity also always means genetic diversity and consequently diversity of evolutionary potential. Ecological systems poor in species have fewer evolutionary options than species-rich systems. (A good demonstration is the rich and ecologically highly diversified Palearctic fauna of *Cirsium* and the depauperate Nearctic *Cirsium* fauna which originate from very few immigrant species which invaded North America via Beringia in Pliocene and Pleistocene; Zwölfer, 1988.) Species diversity may therefore also be seen as a genetic safeguard which allows an ecological system to respond evolutionarily to environmental changes.

Ecosystems are able to recover from all sorts of perturbations if they are formed by many species each of which has a unique set of characteristics. The role played by species diversity may thus be a dual role (Solbrig, 1993). On the one hand, species provide the units through which matter and energy flows which are observed as functional properties. On the other hand, they provide the system with resilience to respond to unpredictable perturbations. Thus, the stochastic nature of species interaction In competition and the large fluctuations of presence and activity of individuals may lay the basis for the overall stability of the whole functional group in matter transformation.

D. Feedforward Effects at the Ecosystem Level

Feedforward at the ecosystem level is generally connected with tight links between species and with the establishment or disappearance of a constant food source. If a keystone species of a community disappears, for instance, by disease, the food chain which is linked to this species also disappears (see above). Parasites and diseases may have such a dramatic effect, especially if they are introduced into new areas, e.g., *Phytophtora* to Australia (Burdon, 1993).

Besides this ultimate effect by extinction or invasion, there are numerous feedforward regulations through visual and chemical signals. The induction of plant galls by salivary enzymes of insect larvae; the induction of mating and oviposition, larval growth; pupation, and the emergence of adults in holometabolic insects; and determination and ending of diapause are examples of ecological and ecophysiological processes in which feed-forward control dominates (see also Zwölfer, this volume, Chapter 11).

E. Feedback Responses at the Ecosystem Level

Interactions at the system level are generally described by feedback. The abundance of species in a system is determined by processes that impinge on the population, whether the population reaches the limits of its environmental resources or suffers from effects of overcrowding by density-dependent forces (Begon *et al.*, 1990).

In plants the trade-offs between resource acquisition and density-related factors are most clearly apparent from Yoda's law of self-thinning (Yoda *et al.*, 1963; Westoby, 1984). An empirical equation was developed which described the relation between size and density which results in an equation with a constant slope (0.5 log N, where N is the number of individuals per area) for very different types of species. An example for the interaction between density-dependent factors and resource supply is obvious during the development of *Urtica* stands (Teckelmann, 1987; see also Stitt and Schulze, this volume, Chapter 4). There is a linear relation between maximum leaf area index and stem number for stands growing under different regimes of light and nutrition which represents the resource supply function (Fig. 6A). During stand development, stem numbers decrease until they reach the crossover point to this resource line. The change in stem numbers during this process follows the self-thinning rule (Fig. 6B). In early spring the stand starts with a high number of shoots, which very soon build a closed canopy. At this point stem numbers decrease continually while total biomass increases with a slope of about 0.5 log N. Depending on light availability each of the stands develops at a different level of biomass, but follows thereafter the self-thinning rule with its constant slope.

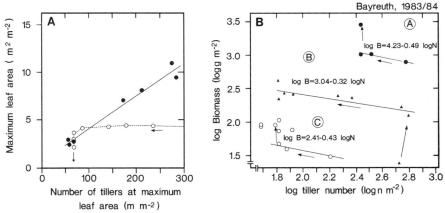

Figure 6 (A) Maximum leaf area index of closed *Urtica dioica* canopies as related to shoot numbers in stands of different light and nutrient availability. The dashed line indicates the seasonal development of a single stand in reaching the maximum leaf area index. (B) Biomass as related to stem number of three closed stands of *Urtica* representing different levels of light climate. The lines are regressions of different harvests and indicate the self-thinning process according to the self-thinning rules of Westoby (1984). (data from Teckelmann, 1987.)

In plant–animal or predator–prey systems feed-forward control by reproduction (with the physiological potential for exponential growth) is coupled with a complex system of negative feedback processes. The most simple form of these, overexploitation of the resources and starvation, is relatively rare in phytophagous insects (see also Zwölfer, this volume, Chapter 11). Classical feedback control in the form of the Lotka-Volterra equations for predator–prey or host–parasitoid systems seems to occur much less frequently than formerly assumed. Territoriality and density-dependent dispersal (i.e., stabilizing behavior patterns at the consumer level), diffuse predator and parasitoid pressure, and plant reactions have been found to be possible negative feedback mechanisms in Cardueae insects. It must, however, be underlined that the effectiveness of these control mechanisms depends on the population structures (metapopulations and source–sink systems) and on the interplay between processes in the microhabitat (e.g., Cardueae flower heads) and the macrohabitat (ecosystem with host plant populations).

The balance between resource and density limitations which exist for each species is the basis for feedbacks between species at the ecosystem level. Depending on a change in supply of resources, which may be a change in nutrients, water, light, and energy, different components of the system increase or decrease in abundance (i.e., they exert feedback on the whole system). Based on these rules we are able to draw larger pictures of feedbacks in ecosystems and predict what may happen if CO_2 and temperature increase further. As far as we can predict, elevated CO_2 will have positive feedback on photosynthesis which will interact with the nutrient cycle and negative effects on leaf conductance which will interact with the water cycle. The nutrient cycle is affected through feedbacks on litter quality which will eventually decrease nutrient losses because of an increased nutrient demand. In contrast, leaf conductance will affect transpiration and thus soil water, and this will affect root growth and in turn nutrition. It is expected that the interaction of the nutrient and water cycle will feed back on shoot growth and, in the long-term, on species composition (Hobbie *et al.*, 1993).

Although these interactions are based on physiological knowledge, it is not yet possible to develop a model, because in contrast to laws in physics the parameterization of these physiological responses is based on statistical probabilities and regression analysis and will thus depend on species and on the starting conditions of the system.

F. Element Cycling

Ulrich (1987) hypothesized that ecosystems function by maintaining a matter balance, i.e., under steady-state conditions resource supply and use are equilibrated, or the system is undergoing change. Thus the input–output analysis of the mineral budget contains information about the dynamics of the system. However, the magnitude of signal transformations between inputs and outputs (storage of substances and generation of new products like nitrate) cannot be explained without including "information" about the species composition of the observed system, but the signal transformations themselves may contain information about future development of the system irrespective of whether change is activated by geological substrate (e.g., cation supply), species action (e.g., N fixation), or climate.

G. System Response

Following various mechanisms of system regulation, we expect systems to respond if environmental conditions change. Responses may become apparent at the single-plant level, at the species population level, or at the whole-ecosystem level.

If we consider responses at the plant level, it is mainly resource allocation (root/shoot/fruits) and leaf area which are controlled by plants in response to factors acting on the plant at the system level. An example of such response may be the phenomenon of defoliation during forest decline (Schulze, 1989a). Acid rain and nitrogen inputs caused a decrease in the supply of cations while demand for growth was increasing. This imbalance of resources resulted in a decrease in foliage biomass. Unless decreased acid inputs allow weathering of primary silicates to restore the cation balance the loss of biomass will increase soil leaching because of decreased canopy transpiration and decreased nutrient demand by sparser vegetation. This spiral of decline may result eventually in a loss of the forest canopy and a replacement by grasses in the herb layer.

On the consumer level chain reactions which eventually lead to a decline in the numbers and densities of beneficial arthropods in agricultural ecosystems are well known as a consequence of the application of pesticides (Krieg and Franz, 1989). Other examples are the extinction of species and the drastic loss of faunal diversity which is caused by the steadily increasing eutrophication of grasslands or stream water and by the acidification of lakes (Blab, 1986; Plachter, 1991).

At the whole-system level, e.g., forest ecosystems, accumulation and loss of resources initiates succession. Leaching of nitrate will cause a depletion of calcium, which decreases the resilience (a measure of how fast a disturbed system returns to its former equilibrium value) and buffer capacity against further acidification processes. Also accumulation of biomass will cause an increasing risk if rapid desintegration takes place with loss of stability of the existing vegetation. Ulrich (1987) hypothesized that high stability and high resiliance of temperate forests is reached only in the carbonate buffer range. If this forest is harvested, it will go through a period of humus disintegration with high levels of cation loss. At this point a branching pattern may exist. If carbonate is present the system will return to its original point after a period of aggregation. In contrast, if the carbonate buffer is lost, the mineral soil will continue to lose cations with subsequent harvests and eventually reach the aluminium buffer range. Initially, all sensitive organisms of the decomposer chain accumulate in the humus layer which remains temporarily at high pH. The system may accumulate cations again from silicate weathering, but with subsequent events of cation loss it will further acidify and eventually introduce a change in species composition. Only over long periods at low acid inputs will the system regenerate again.

According to the hypothesis of Ulrich (1987) soil processes feed back on the organisms. We hypothesize that similar changes may occur and may be driven also by the plant cover. Land use by man determines the flux rates through the system, which in combination with the uptake capacity of the vegetation will determine nutrient losses or accumulation of resources. The change from forests to grassland and to field as well as the changes in forests from deciduous to evergreen species imposed changes of magnitude similar to those presently caused by acid rain.

If we try to combine the information about tight links between species, compensatory responses of individual species, and ecosystem fluxes a picture emerges in which stability of fluxes at the system level is based on or due to stochastic instability at the organism level which compensates for environmental perturbations within the range of genetic resources and species adaptations to extremes. As an illustration (Fig. 7), species of different functional groups (primary producers, herbivores, consumers) maintain a flux of energy and matter because tight functional links exist between individual species of different functional levels. In this case plant A is linked to herbivore A1, which is linked to predator A2, and so on for B and C. We are dealing only with linear interactions, while recognizing that network interactions exist in reality. If plant species A is affected by environmental conditions or even goes extinct (e.g., Dutch elm disease), the whole food chain is lost which may result in a compensatory response of plant B. This may even result in an increased competitive affect on species C and its associated food chain. In this situation, the overall flux could be constant (driven by the structure of the whole canopy as in the case of transpiration), and the classification into functional groups would regard species A, B, and C as being redundant. However, this neglects the fact that the competitive instability at the species level with its ability for compensation is the basis for stability at the system level.

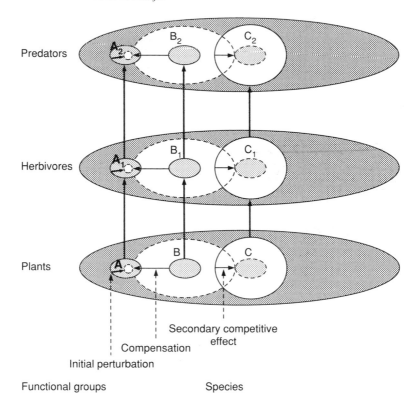

Figure 7 Schematic diagram of fluxes between species and functional groups of species in ecosystems. In this example, plant A, B, and C are tightly linked to herbivore A1, B1, and C1 respectively, and these are tightly linked to specific predators A2, B2, and C2, respectively. If environmental conditions cause a decrease of the activity in A, species B may resume activity by competitive, density-related compensation (dashed lines). This may directly or indirectly via responses through higher trophic levels affect the activity of species C (dashed line). Viewing species as part of functional groups, plant A, B, and C, and their herbivorous and predatorial food chain appear to be redundant. However, because of diversity at each level, compensatory density-related fluctuations exist, which compensate for change in the activity of individual components.

VI. Conclusions

In this chapter we gave an overview of the way that the availability of resources determines plant cover and how animals are linked with plants. We analyzed major functional links between species and species groups of the ecosystem. In contrast to the deterministic processes of physiological systems, processes at the population level of the ecosystem are stochastic. Important aspects of the boundary processes in ecosystems are the effects of canopy height and rooting depth of the vegetation. Another character of most ecosystems is the diversity of species, many of which appear to be redundant with regard to ecological functions but not with respect to system stability. In case of disturbances this redundancy provides alternative pathways for the flow of energy and matter and, thus, provides a safeguard against the hazard of environmental changes. Moreover, species diversity increases the evolutionary potential of ecosystems. At the ecosystem level system regulation is achieved by combinations of feedforward and feedback control in a multiple loop system. Anthropogenic changes of environmental conditions can lead to chain reactions which become manifest in the soil system, the vegetation, and the fauna. Species thus exhibit a dual role in ecosystem fluxes. As part of functional groups they initiate and maintain a quasi-steady state of fluxes within and through ecosystems. As individual organisms and species they maintain a highly fluctuating competitive interaction with neighbors, and it is the compensatory response which maintains the flux density on a ground area basis. Thus, instability at the organism scale contributes to the overall stability at the system scale.

References

Barbosa, P., and Schultz, J. C., eds. (1987). "Insect Outbreaks." Academic Press, San Diego.

Begon, M., Harper, J. L., and Townsend, C. R. (1990). "Ecology, Individuals, Populations and Communities." Blackwell, Oxford.

Blab, J. (1986). "Grundlagen des Biotopschutzes für Tiere." Kilda-Verlag, Bonn.

Bond, W. J. (1993). Keystone species. *Ecol. Stud.* **99**, 237–254.

Burdon, J. J. (1993). The role of parasites in plant populations and communities. *Ecol. Stud.* **99**, 165–180.

Crawley, M. J. (1986). "Plant Ecology." Blackwell, Oxford.

Ehrlich, P. R., and Ehrlich, H. A. (1981). "The Causes and Consequences of the Disappearance of Species." Random House, New York.

Hairstone, N. G., Smith, E., and Slobodkin, I. (1960). Community structure, population control and competition. *Am. Nat.* **94**, 421–425.

Harper, J. L. (1977). "The Population Biology of Plants." Academic Press, London.

Harris, P. (1991). Classical biocontrol of weeds: Its definition, selection of effective agents, and administrative-political problems. *Can. Entomol.* **123**, 827–849.

Hobbie, S. E., Jensen, D. B., and Chapin, S. F., III (1993). Resource supply and disturbance as control over past and future plant diversity. *Ecol. Stud.* **99**, 385–408.

Huffaker, C. B., and Messenger, P. S., eds. (1976). "Theory and Practice of Biological Control." Academic Press, New York.

Jarvis, P. G., and McNaughton, K. G. (1986). Stomatal control of transpiration: Scaling up from leaf to region. *Adv. Ecol. Res.* **15**, 1–49.

Kelliher, F. M., Köstner, B. M. M., Hollinger, D. Y., Beyers, J. N., Hunt, J. E., McSeveny, T. M., Meserth, R., Weir, P. L., and Schulze, E.-D. (1992). Evaporation, xylem sap flow, and tree transpiration in a New Zealand broad-leaved forest. *Agric. Forst. Met.* **62**, 53–73.

Kelliher, F. M., Leuning, R., and Schulze, E.-D. (1993). Evaporation and canopy characteristics of coniferous forest and grassland. *Oecologia* (in press).

Krieg, A., and Franz, J. M. (1989). "Lehrbuch der biologischen Schädlingsbekämpfung." Parey, Hamburg.

Lawton, J. H., and Brown, V. K. (1993). Redundancy in ecosystems. *Ecol. Stud.* **99**, 255–270.

McNaughton, S. J. (1983). Serengeti grassland ecology: The role of composite environmental factors and contingency in community organization. *Ecol. Monogr.* **53**, 291–320.

Meyer, O. (1993). Functional groups of microorganisms. *Ecol. Stud.* **99**, 67–96.

Mooney, H. A., and Drake, J. A. (1984). "Ecology of Biological Invasions of North America and Hawaii," Ecol. Stud. No. 58. Springer-Verlag, New York.

Pimm, S. L. (1984). The complexity and stability of ecosystems. *Nature (London)* **307**, 321–326.

Plachter, H. (1991). "Naturschutz UTB Taschenbücher." Fischer, Stuttgart.

Price, P. W. (1977). General concepts on the evolutionary biology of parasites. *Evolution (Lawrence, Kans.)* **31**, 405-420.

Root, R. (1967). The niche exploitation pattern of the blue-grey gnatcatcher. *Ecol. Monogr.* **37**, 317–350.

Schulze, E.-D. (1982). Plant life forms and their carbon, water and nutrient relations. *Encycl. Plant Physiol., New Ser.* **12B**, 615–676.

Schulze, E.-D. (1989a). Air pollution and forest decline in a spruce (Picea abies) forest. *Science* **244**, 776–783.

Schulze, E.-D. (1989b) Ökosystemforschung—Die Entwicklung einer jungen Wissenschaft. *In* "Wie die Zukunft Wurzeln schlug" (R. Gerwin, ed.), pp. 55–64. Springer-Verlag, Berlin.

Schulze, E.-D., and Chapin, F. S., III (1987). Plant specialization to environments of different resource availability. *Ecol. Stud.* **61**, 120–148.

Schulze, E.-D., and Mooney, H. A. (1993). Ecosystem function of Biodiversity: A summary. *Ecol. Stud.* **99**, 597–510.

Schulze, E.-D., Küppers, M., and Matyssek, R. (1986). The role of carbon balance and branching pattern in the growth of woody species. *In* "On the Economy of Plant Form and Function" (T. J. Givnish, ed.), pp. 585–602. Cambridge Univ. Press, Cambridge.

Solbrig, O. T. (1993). Plant traits and adaptive strategies: Their role in ecosystem function. *Ecol. Stud.* **99**, 97–116.

Southwood, T. R. E. (1973). The insect/plant relationship—an evolutionary perspective. *Symp. R. Entomol. Soc. London* **6**, 3–30.

Steffen, W. L., Walker, B. H., Ingram, J. S., and Koch, G. W. (1992). "Global Change and Terrestrial Ecosystems: The Operational Plan." IGBP-Secretariat, The Royal Swedish Academy of Sciences.

Strong, D. R., Lawton, J. H., and Southwood, T. R. E. (1984). "Insects on Plants: Community Patterns and Mechanisms." Blackwell, Oxford.

Teckelmann, M. (1987). Kohlenstoff-, Wasser- und Stickstoffhaushalt von *Urtica dioica* L an natürlichen Standorten. Doctoral Thesis, University of Bayreuth.

Thalenhorst, W. (1951). Die Koinzidenz als gradologisches Problem. *Z. Angew. Entomol.* **32**, 1–48.

Tilman, D. (1986). Resources, competition and the dynamics of plant communities. *In* "Plant Ecology" (M. J. Crawley, ed.), pp. 51–76. Blackwell, Oxford.

Tilman, D. (1993). Community diversity and succession: The role of competition, dispersal, and habitat modification. *Ecol. Stud.* **99**, 327–348.

Ulrich, B. (1987). Stability, elasticity, and resilience of terrestrial ecosystems with respect to matter balance. *Ecol. Stud.* **61**, 11–49.

Vitousek, P. M., and Hooper, D. U. (1993). Biological diversity and terrestrial ecosystem biogeochemistry. *Ecol. Stud.* **99,** 3–14.

Walter, H. (1964). "Die Vegetation der Erde in öko-physiologischer Betrachtung Bd I: Die tropischen und subtropischen Zonen." Fischer, Stuttgart.

Westoby, M. (1984). The self-thinning rule. *Adv. Ecol. Res.* **14,** 167–225.

Woodward, F. I. (1993). How many species are required for a functional ecosystem? *Ecol. Stud.* **99,** 271–292.

Yoda, K., Kira, T., Ogawa, H., and Hozumi, H. (1963). Self-thinning in overcrowded pure stands under cultivated and natural conditions. *J. Inst. Polytech., Osaka City Univ., Ser. D* **14,** 107–129.

Zwölfer, H. (1978). Mechanismen und Ergebnisse der Co-Evolution von phytophagen und entomophagen Insekten und höheren Pflanzen. *Sonderb. Naturwiss. ver. Hamburg* **2,** 7–50.

Zwölfer, H. (1988). Evolutionary and ecological relationships of the insect fauna of thistles. *Annu. Rev. Entomol.* **33,** 103–22.

Zwölfer, H., and Arnold-Rinehart, J. (1993). The evolution of interactions and diversity in plant-insect systems: The *Urophora-Eurytoma* food web in galls on palearctic Cardueae. *Ecol. Stud.* **99,** 211–236.

Zwölfer, H., and Herbst, J. (1988). Präadaptation, Wirtskreiserweiterung und Parallel Cladogenese in der Evolution von phytophagen Insekten. *Z. Zool. Syst. Evol.-Forsch.* **26,** 320–340.

Zwölfer, H., and Stechmann, D. H. (1989). Struktur und Funktion von Feldhecken in tierökologischer Sicht. *Verh. Ges. Oekol.* **17,** 643–656.

13

Adjustment of Gene Flow at the Population, Species, and Ecosystem Level: Thistles and Their Herbivores

U. Jensen and H. Zwölfer

I. Introduction

In the previous chapters the principles of the flow of matter and energy plus competition strategies in ecological systems have been analyzed and discussed leaving aside the genetic constitution and variability of the taxa included. Therefore, besides the flux and flux control aspects in physiological and synecological systems, in this chapter the specific characters of the genetic system are analyzed as far as they are constituents of the flux control mechanisms in ecosystems. They regulate or change the phenotypic properties and control the physiological and ecological potential of the organisms whether of a population, a species, or an ecosystem.

The genetic system governs the phenotypic appearance of the taxa. Its information determines the ontogenetic and morphogenetic manifestations of structures and processes in the organisms. Of course, organisms in one taxon which is the subject of a physiological or ecological research project will phenotypically differ in only minor components. The principles in ecosystems therefore concern an average phenotype. However, when extrapolating in time the variability and the permanent fluctuation of the genetic components are of essential importance as evolutionary aspects come into play.

During the period of a research project the genetic system appears to be stable. In detail, however, this is not true. The genetic system can be regarded as a steady-state system which is continuously fluctuating. There is a considerable flux of genetic information on the level of the population and the ecosystem (being the integral of populations from different taxa, such as topo- and ecodemes which colonize a defined biotope). Each pollination event in plants and all sexual reproduction in animals recombine the genetic material of the parents into new genetic sets. Therefore, in the case of outcrossing within one population no plant or animal is identical to another. The phenotypes, however, are much less visibly variable and change only slightly during longer periods, as most phenotypic characters are determined by a multitude of genes. For evolutionary time scales the continuous existence of a well-circumscribed phylogenetic taxon is warranted. All members of such a taxon exhibit a consistent organization (structural and functional) scheme ("Bauplan"). Such a "Bauplan" is significant for each taxonomic level.

A species includes individuals with many important—at least superficially—identical characters, biological structures, or processes. The genetic differences are ordinarily expressed in different frequencies of alleles. Although these alleles at least in the case of outcrossing give rise to the huge variability of genotypes, the specific character expressions are more or less alike and concern mostly quantitative phenomena. It is a usual assumption of a physiological and ecological research project that individuals of a species are identical. Usually, this simplification can be accepted for practical purposes, and it is especially appropriate if many individuals and consequently many genotypes of one taxon are included in the sampling procedure.

On the higher taxonomic levels of genera, families, orders, etc. a characteristic structural organization is also known, but the number of marker characters as well as the basic genetic correspondances decreases with increasing taxonomic levels. In the same direction the phenotypic and genotypic range of the entire physiological and ecological properties becomes enlarged. This in consequence leads to the challenge of a broader sampling strategy.

The phenomena of phenotypic plasticity are mentioned here because they do not exceed the limits of the basic structural organization. However, within a genetically determined range different phenotypes may be caused by differences in light, temperature, nutrition, and many other components. Under extreme field conditions some extraordinary phenotypes can dominate. The organismic response to such modifying conditions is normally a quantitative one. Each modification of the Bauplan characters is connected with a modified expression of genes and the production of gene products. As a rule this is governed by a complicated system of interdependent gene effects.

Also individual gene products, i.e., enzymes, are regulated by environmental factors. In the case of gene families a shifting of paralogous proteins within one protein family can be observed. The formation of such gene and protein families in evolution is understandable as the formation of an adequate reaction range to increase the individual fitness facing the different environmental influences.

High levels of nitrogen supply, for example, result in a preferential increase of prolamines in both barley and wheat (Abrol *et al.,* 1971; Koie *et al.,* 1976). It is interesting to research what kinds of molecular changes might be tolerated by the developing seed. For many seed storage proteins, glycosylation and post-translational modifications are not really essential. This shift can also be the product of a shift between the gene products of "related" genes. Sulfur deficiency in soil results in a reduction of storage proteins that are more rich in sulfuric amino acids (i.e., legumins) and an increase of low sulfur proteins (i.e., vicilins; Blagrove *et al.,* 1976; Spencer, 1984). A similar influence has been reported for phosphorus and potassium (Randall *et al.,* 1979).

If longer periods of time are considered, the evolutionary process gains importance. Such processes are regulated by selection factors. Their progress can be very different in speed and extent and depends on the selective power of the environmental factors involved as well as on other evolutionary events. The response of the genetic system may be an altered gene frequency or the acquisition of additional alleles, gene loci, altered chromosomes, or multiplied genomes (polyploidy).

II. Adjustment of Gene Flow

A. No Selection: The Molecular Clock

The molecular clock describes the phenomenon of equal rates of genetic change during geological time periods. In a geological time frame the molecular clock proceeds uniformly, i.e., gene changes occur at a constant rate. This is possible when a gene is exposed to constant or no selection pressure. In this context the theory of Kimura (1968, 1987) states that, especially at the molecular level, a great deal of the evolutionary changes are not mediated by Darwinian selection. By comparing sequence data of such genes and proteins which have been correlated with the diversification within fossil phylogenies, the speed of such a molecular clock has been determined for several proteins and genes. For plants Martin *et al.* (1989) have recently calculated such constant molecular changes for the GAPDH gene. Since these data only approximately fit the fossil knowledge, they have been repeatedly critisized. Probably a uniformly running molecular clock is an approach valid only for a limited time and for a broad geological scale.

Also for isozymes a molecular clock has been proposed as Nei distance indices between two taxa used for the time estimation of their phylogenetic diversification. A Nei value of 1.0 has been proposed to correspond to values between ca. 14 million years (Larson *et al.*, 1981) and 18 million years of divergent evolution (Thorpe, 1982). If applied to the data of Fig. 5 the separation of the Arctotideae and Cardueae (respectively, their *Larinus* phytophages) might have occurred ca. 28–36 million years ago.

B. Selection Regulates and Limits the Gene Flow

In contrast to regular genetic changes, there exist molecular processes which produce a permanent genetic fluctuation at a different rate and intensity. Mutation, recombination, and crossing over are the most frequent events to produce large variability in new genotypes. These are subjected to selection processes which constrain the variability of phenotypes and genotypes and put limits on genetic variation. Such regulating processes provide an optimal adaptation of the entire taxon to the prevailing environmental conditions. Selection may be directed toward stabilization and exclude extreme or newly generated variation patterns. Directional selection gradually leads to an evolutionary transformation of the taxa and may result in development of new species. These changes are especially rapid for apomictic populations, founder populations, or those populations which are believed to depend strongly on coevolutionary interactions.

1. Selection on Host Plants by Phytophages: The Hypothesis of Coevolution

Our special interest is directed to those selection forces which govern the interdependence of a host plant and its phytophages in "thistleheads" which represent mini-ecosystems. We examined the hypothesis that these plant–insect systems are the result of coevolutionary processes. The term coevolution has been coined by Ehrlich and Raven (1964) for reciprocal evolutionary feedback processes which adapt interacting species to each other. This term has been used in diverse ways. In an extreme case the word "coevolution" was used for mutual selection processes at the molecular level of interacting strains of two species (Mode, 1961). Also, it has been said that "all evolution is coevolution" (von Wahlert, 1978). Mode (1961) applies the term coevolution in a narrow sense for processes where species A evolves specific adaptations to or defenses against species B and vice versa. Examples are the evolution of species-specific pollination systems in *Ficus* and fig wasps (Wiebes, 1979) or the evolution of resistance genes in plant cultivars and the corresponding evolution of anti-resistance genes in pathogens or biotypes of phytophagous insects (Mode, 1961). Some authors (Ehrlich and Raven, 1964) use coevolution in a broader sense to describe evolutionary interactions between groups of species, i.e., the mutual adaptations of grazing animals and grasses. Janzen (1980) called this type of coevolution "diffuse coevolution," as it involves many species and interactions in a stochastic rather than a deterministic way. Diffuse coevolution often tends to be asymmetric with regard to the driving forces, i.e., one group of partners exerts a stronger selection pressure than the other. The extreme case is the sequential evolution sensu Jermy (1976). A sequential evolution is one-sided with regard to selection pressures. One group of organisms (e.g., phytophagous insects) adapts itself continuously to the evolutionary changes of another group (e.g., host plants) without influencing the evolution of this group. Figure 1 summarizes the mechanisms involved in "strict coevolution," "diffuse coevolution," and "sequential evolution" for plant–herbivore systems.

Coevolutionary processes between a host plant and its herbivores have been repeatedly demonstrated. Well-known examples are the evolution of a flower form, shape, or color adapted to the pollinating organism and the production of new repellents with increasing toxicity (e.g., cumaroid substances in the Apiaceae; Berenbaum, 1982).

Our research has concentrated on the question of whether the evolution of pseudanthium characters of thistle-like Cichorioideae host plants is controlled by phytophagous herbivores acting as predators upon the flower heads. From the concept of coevolution presented by Ehrlich and Raven (1964) it is expected that the attack on flower heads by phytophages may force the host plant to react rapidly in developing defensive, morpho-

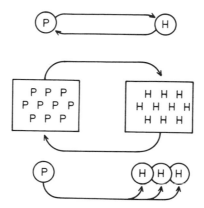

Figure 1 Strict coevolution (sensu Mode, 1961), diffuse coevolution (sensu Janzen, 1980), and sequential evolution (sensu Jermy, 1976). The arrows indicate the direction of selection pressures. P = plants, H = herbivores.

logical, or chemical properties. On the other hand, these protective adaptations could have forced the phytophagous partners to evolve countermeasures. Evolutionary feedback or diffuse coevolution can be shown for Cardueae–herbivore systems. An example being the spines on the bracts of *Centaurea solstitialis* flower heads which as in many other Cardueae are a protective device against browsing mammals. It can be experimentally demonstrated that these spines elicit oviposition responses by *Urophora sirunaseva*, an important phytophagous enemy of *C. solstitialis* (Zwölfer, 1969). This plant species evolved a defensive structure against vertebrate herbivores, and specialized phytophagous insects evolved an orientation pattern which uses this structure as one of the decisive host recognition signals. Other cases which suggest the effect of a herbivore-mediated selection on Cardueae species are the existence of extrafloral nectaries on the bracts of certain *Centaurea* spp. or the sticky exudates on the bracts of the North American *Cirsium discolor* and *Cirsium flodmani* (Willson *et al.*, 1983).

All these cases cannot be interpreted as coevolution in the strict sense. We therefore investigated the problem of a coevolution sensu strictu of Cardueae host plants and their flower head inhabitants. In order to obtain a more inclusive insight into the occurrence of coevolutionary processes of this plant–insect system, the population structure of the host plants and their phytophages have been investigated and correlated. If any coevolutionary processes exist, genetically defined biotypes divergent within and congruent between both plants and animals should be detected. Isozymes and their allelic structures have been used as markers for the detection of such genetic diversity.

A prerequisite for the detection of evolutionary events on the infraspecific level is the presence of sufficient genetic variability. In our case it would have been preferable to study the variability of genetic factors governing phenotypic characters of the host plants which are under selection pressure of the phytophages (i.e., the morphological defense): (1) transformations of the shape of bracts which impede the host recognition process by tephrids (for which bracts are important token stimuli; Zwölfer, 1987), (2) an elongated inclusion of the young flower heads into bud leaves, (3) production of epidermal wax layer, and others. Chemical defense can involve production of new olfactory components within the superficial etherical oil glands or a changed spectrum of substances within the feeding tissue.

Since direct access to this genetic basis was difficult, as a substitute the variability of those genes was used, which can be easily and exactly analyzed via their proteins (allozymes). Although the evolutionary transformation of properties primarily is concentrated on those unknown parts of the genome exposed to selection, the less selection-dependent enzymatic genes will show changes which should be relative to the level of evolution. Based on this assumption, enzyme analysis is effectively used in evolutionary biology (Nevo *et al.*, 1981; Gottlieb, 1981, 1984; Crawford, 1983; Soltis and Soltis, 1987).

Experimental analyses have concentrated on the infraspecific genetic differentiation within both host plant and phytophages. The fruit fly *Tephritis conura* Loew and the common hosts *Cirsium palustre*, *Cirsium helenioides*, and *Cirsium oleraceum* have been investigated in detail. The species *T. conura* consists of morphologically (different wing length) and genetically (dominance of either the A or the B hexokinase allele) distinct races (Seitz and Komma, 1984; Komma, 1990)—one attacking *C. helenioides* and *C. palustre*, the other *C. oleraceum*, and some related species. This fact has been interpreted as an adaptation of *T. conura* toward a different time window for oviposition given by the two host plant groups (earlier on *Cirsium heterophyllum* and *C. palustre*, later on *C. oleraceum*). This adaptation is part of an evolutionary process which has been called subsequent evolution (Brandl and Steinert, 1993) or sequential evolution sensu Jermy (1976).

Another factor deserving more attention is that *T. conura* frequently occurs throughout large areas of Europe where *C. helenioides*, *C. palustre*, and *C. oleraceum* are sympatric or parapatric, but *C. palustre* is only attacked in northern Great Britain. Here a *T. conura* biotype exists which exploits both *C. helenioides* and *C. palustre* (Romstöck and Arnold, 1987). For the host plant *C. palustre* the observed difference in the mode of *Tephritis* attack should cause a specific evolutionary response in the populations of northern Great Britain (i.e., genetic differences), if the concept of coevolution is valid. Indeed, the genetic analysis of the *C. palustre* populations from different parts of the range of distribution unequivocally demonstrates the separation of the northern Great Britain populations (and a northern Swedish population) from the main European ones (Steinert, 1992; Fig. 2).

Steinert (1992) has shown that the separation of the northern Great Britain populations is restricted to a decline of the allozyme spectrum. This has been demonstrated by the presence of only 12.7% polymorphic loci in northern Great Britain against 22.17% in Central Europe. No additional (marker) alleles have been found. Steinert (1992) explains this observation by the absence of gynodioeceous flowers which otherwise are common within the continental populations. This leads to the assumption of a higher rate of inbreeding in the populations of North Britain than in those of Central Europe. Thus the genetic differentiation of the northern British populations cannot be interpreted in terms of a progressive evolution of the genetic system due to the invasion of *T. conura*. It was probably during the late glacial recolonization from the

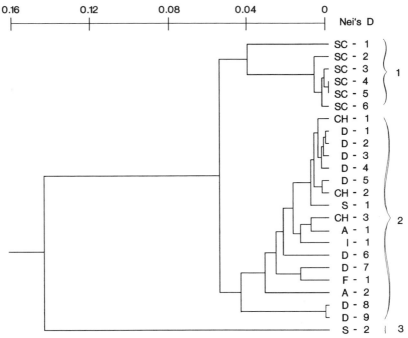

Figure 2   Nei distances for isozyme data (16 loci) from 24 *Cirsium palustre* populations from Scotland (Sc), Switzerland (Ch), Germany (D), Sweden (S), Austria (A), Italy (I), and France (F) indicating the genetic separation of the Scottish (1) and northern Swedish populations (3) from the other ones (2) (from Steinert, 1992).

southern European relic areas that the northern British populations as well as the northern Scandinavian population lost a part of their original allele spectrum.

These findings do not demonstrate any microevolutionary variation of the genetic potential of the host plant in consequence of a phytophagous selection pressure. Data from other *Cirsium* species lead to the same conclusion (Brandl and Steinert, 1993). The phytophagous insects associated with Cardueae hosts evolved physiological and behavioral adaptations to use *Cirsium* species for their larval development, but they did not visibly affect the genetic constitution of their host plants. Although no arguments for coevolutionary events on the plant population level were found in *Cirsium*, in their phylogenies parallel cladistic processes at the plant and the insect level seem possible.

2. Selection by Pollinators   For many flowers, the evolutionary interdependence of pollinators has been demonstrated (e.g., Raven *et al.*, 1982). Many floral structures are only understandable in view of strong mutual adaptation processes. In some cases such evolution was relatively rapid, e.g., in the spur-flowered genus *Aquilegia* vs the unspurred *Semiaquilegia* (Jensen and Penner, 1980).

The evolution of ray flowers in Asteraceae has been interpreted in terms of increased attraction for pollinating insects. Whether this is true for the genetically identical outcrossing species *Centaurea nigra* and *Centaurea jacea* is a matter for investigation. The flower head of *C. jacea* is characterized by typical ray flowers, which are missing in *C. nigra*. Although *C. nigra* shows a western distribution in Europe and *C. jacea* an eastern distribution, sympatric populations are found along the French–German border. For such mixed populations containing the typical rayless *C. nigra* flower-head plants, the ray-flowered *C. jacea*, and their hybrid swarms, the frequency of bees visiting the flower heads was registered (Sommer, 1990). No differences were observed in the visiting frequency of the bees in relation to flower-head types. Since these observations were statistically significant, no selection advantage can be assumed for *C. jacea* vs *C. nigra* in respect to this trait at least when bees are the main pollinators. If there is any selection advantage for the *C. jacea* pseudanthium, it has not been verified.

3. Selection by Environmental Factors   Since allozymes have been used for the detection of genetic variability, the selection effects on these molecular plant characters were investigated. Whether allozymes are selectively neutral has been the subject of controversy among researchers (Nevo *et al.*, 1981; Kimura, 1968, 1987). We were interested in the Cardueae taxa, because the distribution of their allozymes has been used for the evaluation of evolutionary events. A prerequisite for this evaluation is selective neutrality. In the experiment the allele frequency of subpopulations of a *Centaurea pseudophrygia* population was investigated. From the topography and floristic composition two different microhabitats within the population were presumed to exist. The allele frequences of PGI and PGM for the subpopulations A, B, M, and N were determined (observe the overlapping of the subpopulations according to Fig. 3). The resulting pairwise Nei distances were $D = 0.0002$ for A/B, and $D = 0.0754$ for M/N. Also the distribution of the alleles was almost identical in A/B; however, significant differences in M/N occur, using the homogenity G test (Sachs, 1984). It has been concluded that the gene flow is not limited among the subpopulations. Thus, the habitat should have a selective influence on the composition of the genotypes. This conclusion must be restricted to the investigated system (*C. pseudophrygia*, alleles PGI and PGM, microhabitat). The data have demonstrated that allozyme data must be used cautiously for drawing evolutionary conclusions.

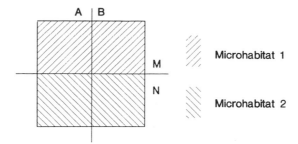

Figure 3 Subpopulations A, B, resp., M, N, within one *Centaurea pseudophrygia* population. For explanation see text.

C. Fusion and Interaction of Gene Flows

Whereas at an ecological time scale the gene pools of populations and entire taxa may appear to be independent units, their fluxes may interact at an evolutionary time scale. Formerly independent gene pools may fuse or they may undergo parallel evolution.

1. Fusion of Two Evolutionary Lines   Evolutionary lines can fuse, if two previously separated populations (or related plant species) become sympatric again and a gene flow among them was retained during separation. New recombinants are expected and thus new morphotypes, especially the ecologically interesting heterozygotes. If the two populations did belong to two separate species, hybridization events would be possible in plants. Such a hybridization can lead to new plant taxa, especially if polyploidization is involved. At the molecular level a summation of homologous gene products (proteins) has been demonstrated in electrophoresis experiments. In animals speciation by hybridization or by polyploidization is a very rare event.

In *Centaurea* section *Jacea* our experiments (Sommer, 1990) have indicated such a fusion. On the diploid level populations of *C. jacea* and *C. nigra* were morphologically distinct, and in isozyme analysis they showed genetic differentiation. In contrast, in tetraploid populations morphological intermediates between *C. jacea* and *C. nigra* occurred and no genetic differentiation among populations could be observed. This can be explained by the quarternary history of the taxa. Originally *C. jacea* and *C. nigra* were mostly diploids which had geographically disjunct diluvial refugial areas. After glaciation new habitats occurred favoring tetraploids and a broad dispersal of both species. In sympatric areas hybridization and gene flow on the tetraploid level occurred resulting in a common gene pool.

2. Parallel Evolutionary Pathways of Two Taxa   Even without a reciprocal coevolutionary interdependence a parallel cladogenesis is possible in cases where one taxon (predators, parasites, or parasitoids) exploits another taxon (prey, host animals, or host plants). An excellent example of a parallel cladogenesis (sensu Regenfuss, 1978) has recently been demonstrated by Mitter *et al.* (1991) involving the chrysomelid genus *Phyllobrotica* and the plant genus *Scutellaria.* The taxonomic relationships of the aphid genus group *Anuraphis–Macrosiphum–Acyrtosiphum* are congruent to the cladogenesis of the Rosaceae host plants (Hille Ris Lambers, 1979). Eastop (1979) found parallels between the phylogenies of aphids and of the "Zweite Hauptgruppe" of the angiosperms (sensu Huber, 1991). Roskam (1985) detected at least partial parallel cladogenesis between gall midges and several angiosperm host taxa. Pschorn-Walcher (1969) described a parallel cladogenesis between sawflies and some of their parasitoids.

We investigated the phylogeny of the thistle-like tribes Cichorioideae (Fam.: Asteraceae) and the *Larinus–Rhinocyllus–Bangasternus* complex of the tribe Lixinae (Curculionidae) (Herbst, 1993). Two sets of cladograms were obtained which were examined for parallelisms in the evolution of both host plants and phytophages (Fig. 4). It has been suggested by Zwölfer and Herbst (1988) that the Cardueae and related tribes were used as platforms for the evolutionary radiation of the genera *Rhinocyllus*, *Bangasternus,* and *Larinus* (Fig. 5). If a parallel cladogenesis between weevils and Cardueae occurred at all, it should be found in these three closely related insect genera.

The phylogenetic origin of the tribe Lixini (Curculionidae) is relatively well known (Zwölfer and Harris, 1984; Sobhian and Zwölfer, 1985; Herbst, 1993). For the host plant taxa (i.e., the tribe Cardueae) as well as for the other thistle-like Asteraceae tribes (Cichorioideae) the phylogenetic relationships were dubious. Therefore, Fischer (1990; see also Fischer and Jensen, 1990, 1992) investigated and compared the similarity of molecular legumin characters within this taxonomic group. Legumin is the major storage protein in angiosperms which has been proved to be a significant marker molecule for evolutionary relationships (Fairbrothers, 1983; Cristofolini, 1980; Jensen, 1968). From the legumin results, further evidence has been deduced for the separation of the subtribes Carduinae and Centaureinae as sister groups, both included in the tribe Cardueae. This finding confirms the evidence for a relationship based upon flower morphology (Dittrich, 1977).

The fact that the status of the subtribes Carduinae and Centaureinae as sister taxa could be corroborated by our analysis of storage legumins (Fischer, 1990; Fischer and Jensen, 1992) is of interest with regard to

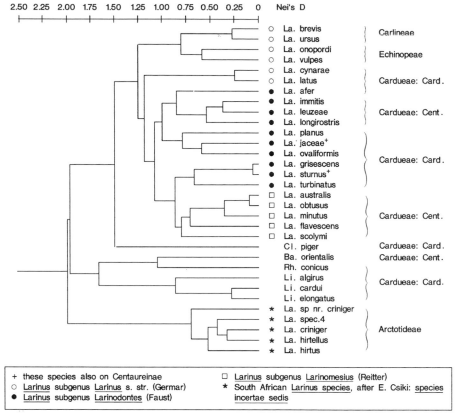

Figure 4 Phenogram of genetic distances (UPGMA cluster based on Nei's *D*, modified after Hillis) among phytophagous species belonging to the genera *Larinus* (La.), *Cleonus* (Cl.), *Bangasternus* (Ba.), *Rhinocyllus* (Rh.), and *Lixus* (Li.) based on isozyme data (data from Herbst, 1993). Their host plant taxa (Asteraceae tribes) are indicated marginally; for the Cardueae tribe the subtribes Carduinae and Centaureinae are given.

the weevil genera *Rhinocyllus* and *Bangasternus*. Both genera share striking morphological and bionomic synapomorphies which show that they are sister taxa which must have developed from a common ancestor on Cardueae hosts with an aberrant mode of oviposition (eggs are deposited on and not into plant tissue and are covered with a protective plate). Herbst (1993) who analyzed enzyme patterns of *Bangasternus* and *Rhinocyllus* came to the same conclusion. The host range of *Rhinocyllus* is restricted to the subtribe Carduinae where the genera *Carduus, Cirsium, Silybum, Notobasis,* and *Onopordum* are exploited (Klein, 1991), and the range of *Bangasternus* is restricted to the subtribe Centaureinae (host records from *Centaurea* and *Carthamus*).

The separation of the subtribes Carduinae and Centaureinae from a common Cardueae ancestor has been placed into the late Oligocene or Miocene by Small (1919). Gel-electrophoretic data by Herbst (1993) indicated a Nei distance between the genera *Rhinocyllus* and *Bangasternus* and the subgenera of *Larinus* which suggests a separation of the genera 28–36 million years ago. This means that the splitting of the common ancestor and the origin of *Rhinocyllus* and *Bangasternus* should have taken place during the second half of the Oligocene. Thus, not only the taxonomic relationships and the host patterns, but also the presumable evolutionary age of Carduinae and Centaureinae and the genera *Rhinocyllus* and *Bangasternus* are consistent with the hypothesis of a parallel cladogenesis.

According to the serological data of Fischer (1990; see also Fischer and Jensen, 1992) all thistle-like Asteraceae tribes are closely related including the lactiferous tribe Lactuceae (Fig. 6) which has been separated as a subfamily or even family of its own in the past (e.g., Frohne and Jensen, 1985). Recent investigations of other molecular characters (rbcL sequences, M. Chase, personal communication; RFLP of the cpDNA, Jansen *et al.*, 1990) support this interpretation.

In comparing the genetic Nei distances for the *Larinus* subgenera calculated from the allele frequencies (Herbst, 1993) to the relationships of the thistle-like Asteraceae, an astonishingly tight correlation is found. Using the known taxonomic data from both the Asteraceae and the *Larinus–Rhinocyllus–Bangasternus* group, the following concept of a parallel cladogenesis of Cichorioideae and the *Larinus–Rhinocyllus–Bangasternus* group becomes conceivable: The most ancient tribe of the Asteraceae is the South American tribe Mutisieae because of the missing cpDNA inversion in some part of the taxon (Jansen and Palmer, 1988). If we determine the origin of the Cichorioideae to be South America, the development of the Mutisieae and Vernonieae on the South American continent and the Arctotideae in South Africa can be anticipated. Owing to the effective wind dispersal adaption (pappus!) the Cichorioideae should have reached the westasiatic-mediterranean area during the early Tertiary where the open areas of the prairie grasslands favored the radiation of Cichorioideae. In the Eocene (54–37 million years ago) grazing mammals particularly ungulates and their ancestors evolved and established considerable selection pressure on the ancient Cichorioideae populations (Zwölfer and Herbst, 1988; Zwölfer, 1990). The plants' response was the evolution of two defense strategies: (1) the lactiferous system in Lactuceae and (2) the thistle-like habit in Cardueae, Echinopeae, and Carlineae, and also in the Arctotideae genus *Berkheya*. In contrast, the Asteroideae subfamily evolving on the American continents accumulated natural compounds such as sesquiterpene lactones, poly-

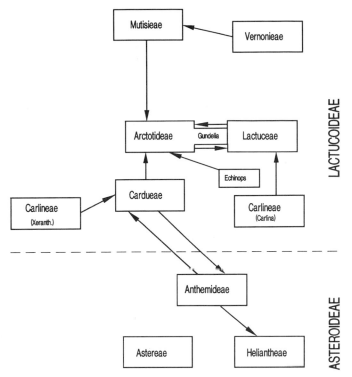

Figure 6 Serological similarities for the legumin proteins among tribes of the Asteraceae-Cichorioideae including three Asteroideae tribes as outgroups. Arrows indicate the serologically detected "closest related neighbors" (from Fischer and Jensen, 1992).

ines, and others used as a "chemical club" for defense purposes (Frohne and Jensen, 1992).

If a Nei distance of one unit is taken for a time span of 14–18 million years (Thorpe, 1982), the Nei calculations of Herbst (1993) date the first separation of the genus *Larinus* back into the oligocene period. This is consistent with the fact that there exist several fossil records of members of the genus *Larinus* from the lower Oligocene and the middle Oligocene (i.e., 30 and 35 million years ago) (Scudder, 1891). If the South African *Larinus* species are a monophyletic group and they really were the first *Larinus* taxa which separated from the ancestor as the *Larinus* line (Herbst, 1993), then all other existing *Larinus* subgenera would form a sister taxon to the South African *Larinus* sp. on *Berkheya*. A more recent diversification (Fig. 4) occurred in the *Larinus* subgenera *Larinomesius* (on *Centaurea* hosts) and *Larinodontes* (mainly associated with Carduinae hosts). The fact that within the *Larinodontes* species *Larinus jaceae* and *Larinus sturnus*

Figure 5 *Larinus cynarae* (top left; photo, Jörg Herbst), a phytophagous insect feeding on Cardueae, *Cirsium spinosissimum* in the Central Alps (top right; photo, Ingrid Steinert), *Cirsium eriophorum* in the Central Alps (bottom left; photo, Ingrid Steinert), and *Centaurea Nigra* in Western France (bottom right; photo, Sylvia Sommer).

biotypes exist on Carduinae as well as on Centaureinae hosts (Zwölfer and Romstöck-Völkl, 1991) indicates that at least for some *Larinus* species a host transfer from one Cardueae subtribe to another is possible even in recent times. A colonization of the Lactuceae by members of the genus *Larinus* never occurred. It is interesting that the evolution of the *Larinus* subgenera as well as that of the genera *Rhinocyllus* and *Bangasternus* lags behind the evolution of the host plant taxa. This delay in the evolution of specialized phytophagous genera with regard to the evolution of the host taxa is a common observation in the evolution of herbivore–host associations (Mitter and Farrell, 1991) and has also been found in many other genera of the Cardueae fauna (Zwölfer, 1988). In parasitology this phenomenon is known as "Fahrenholz' rule."

Since at the population level no indications have been detected for the occurrence of a coevolutionary process, we interpret the apparent correlations in the cladograms from phytophages and their host plants in terms of a parallel cladogenesis, where evolutionary diversification of the phytophagous insects follows that of the host plant taxa with a time lag (sequential evolution sensu Jermy, 1976). The Cardueae subtribes Centaureinae and Carduinae provided major radiation platforms for the specialized phytophagous fauna of thistles. This is also true for many insect parasitoids exploiting host insects in Cardueae flower heads. It is often overlooked that coevolution as used by Ehrlich and Raven (1964) is not restricted to plants and herbivores. It also includes predators and parasites, i.c., the third trophic level of ecosystems. In the Cardueae–insect system investigated by us, Capek and Zwölfer (1991) showed that many braconid parasitoid species exploiting insect hosts in Cardueae flower heads are "niche specific", i.e., for their host ranges the microhabitat "flower head" is more important than the occurrence of a particular host insect species.

Although we failed to demonstrate coevolutionary defense and adaptation mechanisms of Cardueae host plants against their phytophagous insects and a strict reciprocal evolution of the plant–insect system, the concept of coevolution (Ehrlich and Raven, 1964) as a general "escape and radiation" model (Mitter and Farrell, 1991) is not invalidated. At least for the "thistle/phytophage system" no strict reciprocal evolution of the ecologically intimately connected organisms has been detected. For the Cardueae host plants evolution is governed by the sum of many influences which include competition with the other plants of the ecosystem, exploitation of site resources, maintenance of mating systems, and an average reaction of the chemical defenses against the numerous bacteriae, fungi, arthropods, and mammals. The evolutionary response is rather undirected and long-termed. Genetic variation within the geographic distribution of a Cardueae species is evidently low; generally the genetic differences between populations are in the same range as within populations. In some cases (e.g., *C. jacea;* Jensen *et al.*, 1987) clines were detected showing a decrease in the number of alleles toward northern samples, i.e., in the direction of the migration after the diluvial regression. Historical reasons can also be discussed in other cases; The genetic separation of the circumalpine and disjunctly distributed *Cirsium erisithales* populations in the western alps from the other populations was probably caused by a long-term geographic separation of a relic distribution, an assumption which is supported by the genetic differences between the *T. conura* biotypes associated with the western and the eastern *C. erisithales* populations (Komma, 1990). Also the genetic differentiation of the most northern population (northwest Finland) of *C. helenioides* can be explained by its having a different origin, namely from southeastern glacial relic populations (Steinert, 1992). Thus, our investigations show that the thistles evolved mainly by selection components other than specific phytophage pressure.

III. Conclusions

The goal of our investigations was to analyze factors which control the gene flow in Cardueae–insect systems. This was done on the population level by an attempt to detect genetic variations which could be interpreted as mutual evolutionary dependences and at a level of species and genera, where we compared the phylogenies of hosts and herbivores.

At the population level it was not possible to demonstrate a specific and direct selective impact of phytophagous insects or pollinators on their Cardueae host plants. Where genetical variation was found in plant populations it could be attributed to other factors such as the microhabitat or the post-Pleistocene history of the host taxon. On the other hand, an investigation of the numerous biotypes of phytophagous insects associated with Cardueae host plants showed that host plants at the population level were a major evolutionary factor for insects. Zwölfer and Romstöck-Völkl (1991) discussed such microevolutionary processes as consequences of sympatric or parapatric host shifts in Cardueae insects. Examples are (1) "resource tracking" of the tephritid *T. conura* or the weevil *L. sturnus* at the latitudinal or altitudinal distribution border of *Cirsium* host plants, (2) regional differences in host plant abundance which explain host shifts in *Urophora* and *Rhinocyllus,* (3) differences in phenologies of host plants of *T. conura,* and (4) adaptations to structural pecularities of host plants (e.g., the inflorescences of *Cirsium spinosissimum*).

The host patterns of the investigated Cardueae fauna result from different processes which are summarized in Fig. 7. Type A in Fig. 7 represents a parallel cladogenesis. Such a pattern would explain the high correspondence between the phylogenies of the *Larinus–Rhinocyllus–Bangasternus* group (insects) and the Cichorioideae (plants). In this group of Cardueae insects the evolutionary pathways were largely determined by genetic constraints and an opportunistic exploitation of different Cardueae subtribes as radiation platforms (Zwölfer, 1988). This parallel cladogenesis at higher taxonomic levels can be interpreted as a particular type of sequential evolution sensu Jermy (1976). Type B in Fig. 7 shows the diversification and splitting of an euryphagous herbivorous ancestor due to specialization processes. An example is the aphid subspecies *Aphis fabae cirsiiacanthoidis* Scop. on *Cirsium* and *Carduus* (Völkl, 1989) with the broadly polyphagous *Aphis fabae* as ancestor species. Other examples can be found in the tephritid genus *Urophora*, where an evolutionary trend toward the formation of complex plant galls restricted an originally relatively broad host range on the genus *Centaurea* (Arnold-Rinehart, 1989). Type C in Fig. 7 shows the extension of an originally narrow host association. A possible example is the Scottish biotype of *T. conura*, which regionally exploits both *C. palustre* and *C. helenioides* (Romstöck and Arnold, 1987). Type D represents a host transfer to another plant taxon followed by the development of a new evolutionary line of the phytophage. Examples of such shifts to other Cardueae genera or subtribes are found among members of the weevil genus *Larinus* (Herbst, 1993) or the tephritid genus *Urophora* (Pönisch and Brandl, 1992).

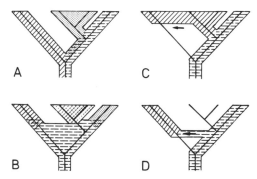

Figure 7 Four basic types of the evolution of host patterns in phytophagous taxa. The lines represent the cladogenesis of a host plant taxon, the overlaid signatures refer to different phytophagous species which in all four graphs originate from a common ancestor associated with the ancestor of the plant taxon. (A) Parallel cladogenesis; (B) restriction of an originally broad host range. The arrow in (C) symbolizes an extension of the host range; the arrow in (D) a host transfer with subsequent speciation.

Acknowledgments

We acknowledge the efforts of our collaborators Hanna Arnold, Roland Brandl, Hilde Fischer, Jörg Herbst, Manfred Komma, Maria Romstöck-Völkl, Alfred Seitz, Sylvia Sommer, and Ingrid Steinert, who were engaged in the population and coevolutionary research. Major technical assistance was provided by Ms. Servant-Miosga, Erika Schill, and Josef Ring. The English manuscript was been corrected by Professor D. E. Fairbrothers.

References

Abrol, Y. P., Uprety, D. C., Ahuja, V. P., and Naik, M. S. (1971). Soil fertility levels and protein quality in wheat grains. *Aust. J. Res.* **22,** 197–202.

Arnold-Rinehart, J. (1989). Histologie und Morphogenese von *Urophora* und *Myopites*-Blütenkopfgallen (Diptera:Tephritidae) in Asteraceae. Doctoral Dissertation, University of Bayreuth, Germany.

Berenbaum, M. (1982). Patterns of furanocoumarin distribution and insect herbivory in the Umbelliferae: Plant chemistry and community structure. *Ecology* **62,** 1254–1266.

Blagrove, R. J., Gillespie, J. M., and Randall, P. J. (1976). Effect of sulfur supply on the seed globulin composition of *Lupinus angustifolius. Aust. J. Plant Physiol.* **3,** 173–184.

Brandl, R., and Steinert, I. (1993). The plant genus *Cirsium:* a case study in plant insect evolution. In preparation.

Capek, M., and Zwölfer, H. (1991). Braconids associated with insects inhabiting thistles (Asteraceae, Cynaroideae). *Acta Entomol. Bohemoslev.* **87,** 262–277.

Crawford, D. J. (1983). Phylogenetic and systematic inferences from electrophoretic studies. *In* "Isozymes in Plant Genetics and Breeding" (S. D. Tanksley and J. Orton, eds.), Part A, pp. 257–289. Elsevier, Amsterdam.

Cristofolini, G. (1980). Interpretation and analysis of serological data. *In* "Chemosystematics: Principles and Practice" (F. A. Bisby, J. G. Vaughan, and C. A. Wright, eds.), pp. 269–288. Academic Press, New York.

Dittrich, M. (1977). Cynareae—systematic review. *In* "The Biology and Chemistry of the Compositae" (V. H. Heywood, J. B. Harborne, and B. L. Turner, eds.), Academic Press, New York.

Eastop, V. F. (1979). Deductions from the present day host plants of aphids and related insects. *Symp. R. Entomol. Soc. London* **6,** 157–178.

Ehrlich, P. R., and Raven, P. H. (1964). Butterflies and plants: A study in coevolution. *Evolution (Lawrence, Kans.)* **18,** 586–608.

Fairbrothers, D. E. (1983). Evidence from nucleic acid and protein chemistry, in particular serology, in angiosperm classification. *Nord. J. Bot.* **3,** 35–41.

Fischer, H. (1990). Vergleichende serologische Untersuchungen am Samen-Reserveprotein Legumin in der Familie Asteraceae, Unterfamilie Cichorioideae (sensu Bremer). Doctoral Dissertation, University of Bayreuth.

Fischer, H., and Jensen, U. (1990). Phytoserological investigation of the tribe Cardueae s.l. (Compositae). *Plant Syst. Evol., Suppl.* **4,** 99–111.

Fischer, H., and Jensen, U. (1992). Utilization of proteins to estimate relationships in plants: Serology; A discussion based on the Asteraceae-Cichorioideae. *Belg. Journ. Bot.* **125,** 243–255.

Frohne, D., and Jensen, U. (1985). "Systematik des Pflanzenreichs," 3rd ed. G. Fischer, Stuttgart.

Frohne, D., and Jensen, U. (1992). "Systematik des Pflanzenreichs," 4th ed. G. Fischer, Stuttgart.

Gottlieb, L. D. (1981). Electrophoretic evidence and plant populations. *Prog. Phytochem.* **7**, 1–46.

Gottlieb, L. D. (1984). Genetics and morphological evolution in plants. *Am. Nat.* **123**, 681–709.

Herbst, J. (1993). Biosystematik und Evolutionsgeschichte der Wirtspflanzenassoziation in der Rüsselkäfergattung *Larinus* (Col. Curcul.). Doctoral Dissertation, University of Bayreuth.

Hille Ris Lambers, D. G. (1979). Aphids as botanists? *Symb. Bot. Ups.* **22**, 114–119.

Huber, H. (1991). "Angiospermen. Leitfaden durch die Ordnungen und Familien der Bedecktsamer." G. Fischer, Stuttgart and New York.

Jansen, R. I., and Palmer, J. D. (1988). Phylogenetic implications of chloroplast DNA restriction site variation in the Mutisieae (Asteraceae). *Am. J. Bot.* **75**(5), 753–766.

Jansen, R. K., Holsinger, K. E., Michaels, H. J., and Palmer, J. D. (1990). Phylogenetic analysis of chloroplast DNA restriction site data at higher taxonomic levels: An example from the Asteraceae. *Evolution (Lawrence, Kans.)* **44**, 2089–2105.

Janzen, D. H. (1980). When is it coevolution? *Evolution (Lawrence, Kans.)* **34**, 611–612.

Jensen, U. (1968). Serologische Beiträge zur Systematik der Ranunculaceae. *Bot. Jahrb. Syst. Pflanzengesch. Pflanzengeogr.* **88**, 204–268.

Jensen, U., and Penner, R. (1980). Investigation of serological determinants from single storage plant proteins. *Biochem. Syst. Ecol.* **8**, 161–170.

Jensen, U., Sommer, S., ad Steinert, I. (1987). Isozymanalysen von *Centaurea* sect. Jacea-Populationen in Zentraleuropa. *Bot. Jahrb. Syst. Pflanzengesch. Pflanzengeogr.* **108**, 239–250.

Jermy, T. (1976). Evolution of insect/host-plant relationship. *Am. Nat.* **124**, 609–630.

Kimura, M. (1968). Evolutionary rate at the molecular level. *Nature (London)* **217**, 624–626.

Kimura, M. (1987). "Die Neutralitätstheorie der molekularen Evolution." Parey, Berlin.

Klein, M. (1991). "Populationsbiologische Untersuchungen an *Rhinocyllus conicus* Fröl. (Col. Curculionidae). "Wissenschafts-Verlag Dr Wigbert Maraun, Frankfurt.

Koie, B., Ingversen, J. M., Anderson, A. J., Doll, H., and Eggum, B. O. (1976). Compositional and nutritional quality of barley proteins. *In* "Evaluation of Seed Protein Alterations by Mutation Breeding," pp. 55–61. Int. At. Energy Agency, Vienna.

Komma, M. (1990). Der Pflanzenparasit *Tephritis conura* und die Wirtsgattung *Cirsium*. Doctoral Dissertation, University of Bayreuth.

Larson, A., Wake, D. B., Maxson, L. R., and Highton, R. (1981). A molecular phylogenetic perspective on the origins of morphological novelties in the salamanders of the tribe Plethodontini (Amphibia, Plethodontidae). *Evolution (Lawrence, Kans.)* **35**, 405–422.

Martin, W., Gierl, A., and Saedler, H. (1989). Molecular evidence for pre-Cretaceous angiosperm origins. *Nature (London)* **339**, 46–48.

Mitter, C., and Farrell, B. (1991). Macroevolutionary aspects of insect/plant interactions. *In* "Insect-Plant Interactions" (E. Bernay, ed.), Vol. 3, pp. 35–78. CRC Press, Boca Raton, FL.

Mitter, C., Farrell, B., and Futuyma, D. J. (1991). Phylogenetic studies of insect-plant interactions: Insight into the genesis of diversity. *Trees* **6**, 290–294.

Mode, C. J. (1961). A generalized model of a host-pathogen system. *Biometrics* **17**, 386–404.

Nevo, E., Brown, A. H. D., Zohary, D., Storch, N., and Beiles, A. (1981). Microgeographic edaphic differentiation in allozyme polymorphisms of wild barley (*Hordeum spontaneum,* Poaceae). *Plant Syst. Evol.* **138**, 287–292.

Pönisch, S., and Brandl, R. (1992). Cytogenetics and diversification of the phytophagous fly genus *Urophora* (Tephritidae). *Zool. Anz.* **228**, 12–25.

Pschorn-Walcher, H. (1969). Die Wirtsspezifität der parasitischen Hymenopteren in ökologisch-phylogenetischer Betrachtung. *Ber. Wandervers. Dtsch. Entomol.* **10,** 55–63.

Randall, P. J., Thomson, J. A., and Schroeder, H. E. (1979). Cotyledonary storage proteins in *Pisum sativum.* IV. Effects of sulphur, phosphorus, potassium and magnesium deficiencies. *Aust. J. Plant Physiol.* **6,** 11–24.

Raven, P. H., Evert, R. F., and Curtis, H. (1982). "Biology of Plants." Worth, New York.

Regenfuss, H. (1978). Ursachen und Konsequenzen einer parallelen phylogenetischen Aufspaltung von Parasiten und Wirten. *Sonderb. Naturwiss. Ver. Hamburg* **2,** 83–99.

Romstöck, M., and Arnold, H. (1987). Populationsökologie und Wirtswahl bei *Tephritis conura* Loew–Biotypen (Dipt. Tephritidae). *Zool. Anz.* **219,** 83–102.

Roskam, J. C. (1985). Evolutionary patterns in gall midge-host plant associations (Diptera, Cecidomyiidae). *Tijdschr. Entomol.* **128,** 193–213.

Sachs, L. (1984). "Angewandte Statistik." Springer-Verlag, Berlin.

Scudder, S. H. (1891). Index of the known fossil insects of the world. *Geol. Surv. Bull. (U.S.)* **71,** 189–696.

Seitz, A., and Komma, M. (1984). Genetic polymorphism and its ecological background in tephrid populations. *In* "Population Biology and Evolution" (K. Wöhrmann and V. Loeschke, eds.), pp. 143–158. Springer-Verlag, Heidelberg.

Small, J. (1919). "The Origin and Development of the Compositae." W. Wesley, London.

Sobhian, R., and Zwölfer, H. (1985). Phytophagous insect species associated with flower heads of Yellow Star Thistle (*Centaurea solstitialis* L.). *Z. Angew. Entomol.* **99,** 301–321.

Soltis, P. S., and Soltis, D. E. (1987). Population structure and estimates of gene flow in the homosporous fern *Polystichum munitum. Evolution (Lawrence, Kans.)* **41,** 620–629.

Sommer, S. (1990). Isozymanalyse zur Ermittlung genetischer Variabilität und mikroevolutiver Prozesse bei *Centaurea* sect. Jacea (Asteraceae). Doctoral Dissertation, University of Bayreuth, Germany.

Spencer, D. (1984). The physiological role of storage proteins in seeds. *Philos. Trans. R. Soc. London, Ser. B* **304,** 275–285.

Steinert, I. (1992). Populationsbiologische Untersuchungen an Arten der Gattung *Cirsium* Miller. Doctoral Dissertation, University of Bayreuth, Germany.

Thorpe, J. P. (1982). The molecular clock hypothesis: Biochemical evaluation, genetic differentiation and systematics. *Annu. Rev. Ecol. Syst.* **13,** 139–168.

Völkl, W. (1989). Resource partitioning in a guild of aphid species associated with creeping thistle *Cirsium arvense. Entomol. Exp. Appl.* **51,** 41–47.

von Wahlert, G. (1978). Co-Evolution herrscht überall. 20. Phylogen. Symposium (Hamburg, 1975). *Sonderb. Naturwiss. Ver. Hamburg* **2,** 101–125.

Wiebes, J. T. (1979). Co-evolution on figs and their insect pollinators. *Annu. Rev. Ecol. Syst.* **10,** 1–12.

Willson, M. F., Anderson, P. K., and Thomas, P. A. (1983). Bracteal exudates in two *Cirsium* species as possible deterrents to insect consumers of seeds. *Am. Midl. Nat.* **110,** 212–214.

Zwölfer, H. (1969). *Urophora sirunaseva* (Hg) (Dipt.: Trypetidae), a potential insect for the biological control of *Centaurea solstitialis* L. in California. *Commonw. Inst. Biol. Control, Tech. Bull.* **11,** 105–155.

Zwölfer, H. (1987). Species richness, species packing and evolution in insect-plant systems. *Ecol. Stud.* **61,** 301–319.

Zwölfer, H. (1988). Evolutionary and ecological relationships of the insect fauna of thistles. *Annu. Rev. Entomol.* **33,** 103–122.

Zwölfer, H. (1990). Disteln und ihre Insektenfauna. *In* "Evolutionsprozesse im Tierreich (B. Streit ed.), pp. 255–278. Birkhäuser Verlag, Basel.

Zwölfer, H., and Harris, P. (1984). Biology and host specificity of *Rhinocyllus conicus* Froel.

(Col. Curculionidae), a successful agent for biocontrol of the thistle *Carduus nutans* L. *Z. Angew. Entomol.* **97,** 36–62.

Zwölfer, H., and Herbst, J. (1988). Präadaptation, Wirtskreiserweiterung und Parallel-Cladogenese in der Evolution phytophager Insekten. *Z. Zool. Syst. Evol. Forsch.* **26,** 320–340.

Zwölfer, H., and Romstöck-Völkl, M. (1991). Biotypes and evolution of niches in phytophagous insects on Cardueae hosts. *In* "Plant-Animal Interactions: Evolutionary Ecology in Tropical and Temperate Regions" (P. W. Price, M. Lewinsohn, and G. W. Fernandes, eds.), pp. 487–507. Wiley, New York.

V

Flux Control in Biological Systems: A Comparative View

14

Flux Control in Biological Systems: A Comparative View

E.-D. Schulze, E. Beck, E. Steudle, M. Stitt, and H. Zwölfer

I. Introduction

Biological systems are characterized by their ability to maintain specific functions and structures across a broad range of conditions encountered in their environment. The individual components of the system themselves have a wide variety of functions, catalyze many different processes, operate in a complex network of compartments and pathways, and may exhibit a larger variation over time than the integrated system. The maintenance of system function is critically dependent on its ability to adjust, which (i) affects the capacity of subsystems, (ii) modifies the exchange of matter and energy between compartments, (iii) alters the pools of intermediates within the system, and (iv) alters the flow of resources into and out of the system. Even within a single organism, interactions between different processes will generate a complex overall pattern of regulation despite the fact that all cells have the same genome. In populations or ecosystems, variations in the gene pool of species make this adjustment even more complicated.

In addition to purely thermodynamic (passive) processes adjustments may take place as control, which is the transduction and translation of information (signals) and its transformation into an action. The term "regulation" contains elements of "control," but it is used if processes are "direct or governed according to a rule, principle or system" (Webster's New World Dictionary). This definition introduces the notion of comparing the status of a process with a set point and that this process is steered with the aid of a regulating device such that certain desired parameters may be maintained within certain limits (steady state).

In the following we use the term regulation in biological systems up to the level of organisms, because they are deterministic systems which contain intrinsic goals, such as surviving, occupying or retaining niches, or reproducing under conditions of competition and alterations in the surrounding world. Also we can invoke the concept of natural selection, up to and including the level of an organism, to explain why mechanisms could have evolved in a biological system which are analogous to those which we would expect if an engineer had sat down and logically planned a "functioning" machine.

In contrast, the use of the term regulation in more complex biological systems which contain different species (populations, ecosystems) is ambiguous, because these systems have not been exactly constructed and there is no set point at which they operate. They contain interactions between individual organisms which are not deterministic but rather stochastic, but which nevertheless may affect the performance of other organisms in the same system. In ecosystems the functional role of a component may change depending on the activity of neighbors, and subsystems (species) may even replace each other while "ecosystem functioning" is unchanged. We consider ecosystem functioning as interactions between organisms resulting in a stabilization of a system with respect to fluxes within or through the system on a ground area basis. Compensatory responses of populations are part of this. We suspect that similar phenomena are possible also in cells at the enzyme level. However, in addition, animals may even have a choice, being able to make alternative decisions in response to signals from the system. In this case the use of the term regulation becomes even more problematic. In human life, we talk about "traffic control" of cars in a city which channels the flux of cars by traffic lights to avoid traffic congestion, although each driver can also make a decision to use a different road. Although in this case traffic control has a finite deterministic "aim," we still use the term control rather than regulation when dealing with systems above the level of organisms. A system may contain elements analogous to regulation (system regulation as defined by Schulze and Mooney, 1992) when it exhibits some flexibility in its response to changes in the external conditions toward maintaining performances such as nutrient cycling. Thus in contrast to using concepts like control and regulation of internal processes, we may consider the relation between the system and its environment to involve adjustment. Organisms can react to such changes in their environment by regulation within organisms, control by species interaction, or changes at the ecosystem level.

The following analysis is based on a comparative view of process regulation or control in organelles, cells, whole organisms, populations, and ecosystems, which should help us to discover if there are analogous patterns of regulation or control and how they change both in simple systems as well as in complex ones. We focus on the mechanisms which allow self-regulation in response to a variable environment. We do not deal with the problem of regulation at the genome level which is extremely important at the organism level. Also, we do not consider processes which determine species diversity and which were discussed elsewhere (Schulze and Mooney, 1992). Rather, we aim to understand basic principles which describe the control of flux in metabolic or organismic pathways, in individual organisms, in species of populations, and in ecosystems. These control mechanisms contain feedforward and feedback as common control features, but additionally there are the phenomena of shared control, futile cycles, and switches, as well as features of synchronization and coordination, which facilitate and stabilize functioning and interactions at the micro (cellular)- and macro (population)-scale.

This is a controversial subject, and we intend to invoke discussion and stimulate interest in the problem of how biological systems are regulated or controlled.

II. Hierarchy of Resource Limitations for Individual Processes or Organisms

Ecosystems can be described as a hierarchy of levels (soil decomposers, primary producers, herbivores, and predators) which have switch points, at which the flux of energy and matter is channelled either to a higher level in the hierarchy or switched back to the lower level of the decomposers. In such a system, switches at a lower rank (primary producers) reduce the options at levels of higher rank (predators). For example soil abiotic factors have a great effect on vegetation, this in turn has great impact on herbivores, and these in turn impact the remaining faunal groups. However, there is less evidence for feedback from the top layers to the lower layers (see also Zwölfer, this volume, Chapter 11), i.e., if a certain vegetation type disappears (forest changed into grassland) most other species disappear, especially those which were tightly linked to the original vegetation type. In contrast if forest herbivores disappear in forests there may be a small effect on fundamental processes such as nutrient cycling, although there is also evidence that megaherbivors, such as the elephant, may have shaped the vegetation of savannahs, while the New Zealand flora is assumed to have evolved without vertebrate herbivory. In contrast to this "vertical" structure of biological systems, within each layer (plants, herbivores and others) "horizontal" effects exist between individuals of the same species or between different species by density-related parameters such as crowding (Begon *et al.*, 1990).

From the structure of the ecosystem we learn that the performance of a single organism is "constrained" by vertical interactions, such as the availability of resources (food, substrate) or the pressure of enemies, and by horizontal interactions of density-dependent parameters which may in turn lead to secondary limitations by resources or diseases or by behavioral limitations (Schulze and Zwölfer, this volume, Chapter 12). The latter could in fact become more effective than primary resource limitations (Zwölfer, this volume, Chapter 11). The relative position of the performance of the organism within these limits determines its success and, in particular, its fecundity and mortality. It may cause populations to exhibit vast fluctuations or greater stability (Begon *et al.*, 1990).

Responses to supply and demand or to resource limitation, and the balancing effects such as increased mortality or reduced size of individuals because of overcrowding in populations, are examples of self-regulation. The latter is based on (i) the selfishness of individuals to obtain resources, (ii) tight links and their irreversible steps in pathways or networks of interactions, and (iii) antagonisms with other components of larger systems. Tight links of food webs and metabolic pathways are especially important. At the cellular level, all catalyzed reactions are linked in pathways of enzymatic processes which may branch and join, but which always channel the flow of substances in distinct directions due to the thermodynamics of the process and to the specificity of the enzyme. Analogously, at the ecosystem level, many organisms are specialists. They are bound into distinct patterns and positions in the flow of energy or substances in a food chain (decomposers; Meyer, 1992) or food web (Zwölfer, this volume, Chapter 11); there are fewer "generalists" which can consume or make use of "any" resource (like man, crows, raccoons, starlings, rats, cockroaches).

III. Feedforward and Feedback

Although demand and supply will affect the level of operation (the actual activity), we need to understand the mechanisms which control the flow rate. Positive or negative feedback (product regulation) and feedforward control (substrate regulation, e.g., by allosteric effects on enzymes) have been described repeatedly as a key mechanism of process control at the molecular, cellular, and the population level (e.g., Ziegler, 1991; Begon *et al.*, 1990; Schulze and Mooney, 1992). The interesting feature which emerges from this book is that very often feedforward and feedback act together on the same control node. For example, Stitt (this volume,

Chapter 2) and Stitt and Schulze (this volume, Chapter 4) demonstrate that some of the enzymes which lead to the formation of sucrose in the cytosol, namely, cytosolic fructose-1,6-bisphosphatase and sucrose-phosphate synthase are regulated in their activity by specific effectors (external regulators, in this case fructose-2,6-bisphosphate), by the product (feedback), and also by the substrate (feedforward). In a very similar way nitrate reductase activity is feedforward regulated by substrate (nitrate availability) and feedback regulated by the carbohydrate availability (Stitt and Schulze, this volume, Chapter 4). Such regulatory sequences which operate through an external effector are called open cycles (Crabtree and Newsholme, 1987). Also, in whole-cell complexes (stomata; this volume Chapter 7) or organs (growth; this volume, Chapter 8) the joint action of feedforward and feedback can be demonstrated. An analogous situation exists in food webs. The influence of plant activity and of predators on the performance of herbivores acts as a cooperation of feedforward and feedback control mechanisms. The interaction of feedforward and feedback mechanisms will contribute to stability under some conditions; however, it could also lead to fluctuations, depending on how the regulatory loops are integrated.

One prerequisite for feedforward and feedback regulation is the existence of a signal which induces change and a receptor which is sensitive to such signal. An example for a signal is the salivary gland enzyme of gall-forming insects which increases the sink strength for assimilates of the tissue of the receptacle in the Cardueae flower head and thus induces gall formation which is critical for the development of the insect (Zwölfer, this volume, Chapter 11). At the enzyme level products, substrates, or unrelated metabolites may influence allosteric enzymes and are an example of such a signal (Scheibe and Beck, this volume, Chapter 1; Stitt, this volume, Chapter 2). At the level of cells, turgor pressure is a signal which drives growth (Steudle, this volume, Chapter 8). At the whole-plant level, the production and flux of phytohormones have the function of signal transfer and regulation between different places of action (Beck, this volume, Chapter 5; Schulze, this volume, Chapter 7), and at the population level numerous behavioral traits of species may respond to structural signals (Zwölfer, this volume, Chapter 11).

In order to exert control signal transduction requires specific receptors and mechanisms of signal conversion. For this purpose structure becomes an essential component of regulation. An impressive example of the importance of structure for regulation is described by Scheibe and Beck (this volume, Chapter 1): In microorganisms the enzyme malate dehydrogenase has a very high catalytic capacity that cannot be regulated. With evolution of higher plants, the corresponding enzyme in the chloroplast obtained additional peptides at both termini, which turned out to be the basis for allosteric effects which in turn regulate the activation and inactivation of the enzyme. In multispecies systems a comparable feature may be seen in structures which result in a greater variety of organisms at a certain trophic level.

The cost of regulation at the enzyme level is a loss in maximal velocity. In metabolic pathways (see also Stitt, this volume, Chapter 2), regulation results in a loss of performance of the individual component while the flux through the metabolic chain is increased. Regulation results in coordination and thus in an increased flux through the whole system. Similarly, traffic control may reduce the speed of individual cars, but increase the total flux of traffic. It is not clear whether this occurs also in more complex systems, but it is likely that in a food chain the total flux through the chain of consumers is highest when the individual species operates below its maximal consumption capacity.

IV. Shared Control

Although the classical view of control is centered around the concept of key enzymes and species, we think that it is even more important to recognize that control is generally shared by a number of enzymes in a pathway or by a sequence of species in a food chain which are involved in the transformation of substances in an ecosystem.

Resource limitation is the simplest form of a shared control. Crabtree and Newsholme (1987) gave a clear description of shared control by resource limitation in a metabolic chain. An initial substrate is converted into an end product via a series of reactions linked together by metabolic intermediates. In this case the intermediate concentrations balance each individual reaction to the overall flux. If the first reaction were to proceed faster than the second reaction there would be a net increase of intermediate product. If the second reaction can respond to this increase in substrate, it will increase its rate until both reactions obtain equal rates. At this point the pool of intermediate product will remain constant. The process of balancing adjacent reactions will enable the flux to be transmitted through the entire chain and the concentrations of the intermediates to be determined completely by the flux. Crabtree and Newsholme (1987) therefore defined these intermediates as "internal effectors." The process of balancing a chain reaction in fact requires that the single reactions are not saturated with their pathway substrates. However, it also is clear that this description of the pathway cannot explain the initial pool of the substrate. In fact, it would require that the first reaction be unresponsive to changes in substrate, i.e., the first reaction may approach saturation with its substrate. Also, if this initial substrate declines, the steady state of the metabolic chain will decline, similar to running down a battery (Crabtree and Newsholme, 1987). Also, the flux must be protected from effects of the end product.

In order to describe the contribution of individual steps in a chain of reactions, Kacser and Porteous (1987) introduced the idea of a shared control. We suggest that this concept can also be applied in systems which involve interactions of different species. We learn from microbial processes in soils (Meyer, 1992) that, in analogy to metabolic pathways, chains of microorganisms operate in concert when mineralizing nitrogen or sulfur in soils.

At the enzyme level (Stitt, this volume, Chapter 2) shared control means that (i) more than one enzyme has regulatory function in the same pathway, (ii) several effectors exist which can affect the activity of these enzymes, (iii) amplifications occur due to activation or inhibition of neighboring enzymes, and (iv) balance and coordination occur in a metabolic chain. As a result, all enzymes do not operate at maximum activity but at a balanced rate of substrate turnover. Finally (v) the extent to which each enzyme exerts control over the whole metabolic chain can be variable and depend, for example, on the environmental conditions. Shared control allows self-regulation as an adjustment to variable external factors and the maintenance of a functional state of stable operation. We think that the situation is quite similar in multispecies systems, where "enzyme" could be replaced by "species" in the above definitions to elucidate the analogous situation (Schulze and Zwölfer, this volume, Chapter 12). In metabolic pathways, reactions having a high negative free energy change, i.e., irreversible reactions, are frequently coupled with others where even in a steady state the change of free energy is small. By contrast, energy flow in ecological systems among species takes place always via irreversible steps only. However, the analogy of flux control in both types of systems is provided at least with respect to features of pool and population sizes. Also, even at the ecosystem level cycling (see below) will allow such thermodynamically irreversible steps to be reversed.

At the cellular level shared control may be observed during extension growth which depends on water supply, special ions, and the mechanical extensibility of cell walls. Depending on the conditions, each of these parameters may limit extension growth in a different way. While water shortage may result in a reduced growth rate which could be overcome by solute accumulation (osmoregulation), the metabolically regulated cell wall extensibility may limit growth at water saturation (Steudle, this volume, Chapter 8).

An important property of shared control is redundancy which is found at the biochemical level (for exmple, plants have multiple and alternative pathways for carbohydrate breakdown; Stitt, 1990), at the genetic level (isoenzymes; Jensen and Zwölfer, this volume, Chapter 13), as well as at the population level (multispecies complex of herbivores; Zwölfer, this volume, Chapter 12). As explained by Schulze and Mooney (1992) redundancy provides the security of ecosystems to maintain their functions even under extreme conditions. At the population level, the balanced action of herbivores is lost, if redundancy is lost. For example, in biological control a single preditor may control the population of a host (pest) if it does not have to compete with other enemies (Zwölfer, this volume, Chapter 11) and this effect may even be more important for the overall consumption by herbivores than the effect of predators. Thus species play several roles in ecosystems: (i) they act in concert when maintaining a flux per ground area. This "function" leads to the notion that species may be redundant. However, in addition to this (ii) density-related processes will compensate for loss of activity or presence of species if conditions change. Therefore, the fluctuations and the apparent instability at the species (organismic) level contributes to the stability and quasi-steady state of fluxes per ground area at the system level. In fact, (iii) species composition will determine the magnitude of the flux (see also Schulze and Zwölfer, this volume, Chapter 12).

V. Futile Cycles

In addition to control of linear processes in a food chain or in a metabolic pathway, it appears that futile cycles represent a general and very sensitive mechanism for the onset of control in a complex arrangement of catalytic processes in organelles, cells, and whole systems (Scheibe and Beck, this volume, Chapter 1; Komor, this volume, Chapter 6; Schulze, this volume, Chapter 7). In metabolism futile cycles are such reactions where the net flux is zero, but a product is shuttled between two stages, i.e., the reactions starting from S_0 in one direction (V^{+1}) and resulting in S_1 and starting at S_1 and producing S_0 (V^{-1}) occur simultaneously and at comparable rates. In such a system a small change in V^{+1} or V^{-1} can result in a big change of S_0, which may be the active form of an enzyme or a product and/or in a big change of the net flux ($V = V^{+1} - V^{-1}$). The cycle allows rapid regulation (enzyme modulation) or rapid availability of substances (e.g., mobilization of sucrose) because it relies on a large latent pool whose potential capacity can suddenly be exploited for a particular process. In this sense the term "futile" is misleading since the cycling reaction has a regulatory function.

At the metabolic level, futile cycling of compounds is driven by the concurrent action of enzymes or by simultaneous activity of pathways that operate in opposite directions. The way in which futile cycles provide a means for flux control is illustrated by the chloroplast NADP-malate dehydrogenase. This enzyme is considered the pacemaker for the export of excess photosynthetically produced reduction equivalents from the chloroplast and thus contributes to the maintenance of a steady state of photosynthetic electron flow. To avoid consumption and subsequent efflux of reduction equivalents which are required in the chloroplast, its activity must be exactly and rapidly adjusted to the situation in the chloroplast. Continuous inactivation of the MDH by oxygen and immediate activation by photosynthetic reduction provides the basic machinery for regulation. The component that controls the extent of reduction (= activation) is the concentration of NADP which inhibits activation. In this case the futile cycle is the basis for a rapid availability or disappearance of a catalytic capacity under conditions of environmental change, such as changes of light. Among other processes, it may be also the basis for avoiding damage (Schäfer, this volume, Chapter 3). Other futile cycles in carbon metabolism are important at the level of flux regulation, e.g., a cycle between sucrose synthesis and sucrose degradation allows the net rate of sucrose synthesis or degradation to be very rapidly altered (Geigenberger and Stitt, 1991).

A very interesting futile cycle at the whole-plant level is mediated by the long-distance transport system of xylem and phloem, which circulate specific nutrients and phytohormones (Komor, this volume, Chapter 6; Schulze, this volume, Chapter 7). It is the basis for a constant supply of substances to sinks. Transfer of these compounds between phloem and xylem and vice versa occurs via the free space of the apoplast and the uptake activity of the surrounding cells will determine the fate of the respective solute, namely, uptake and utilization (metabolism or deposition), or recirculation (Komor, this volume, Chapter 6). In this model the apoplast is a free space for exchange of substances.

Ultimately, cycling elements in an ecosystem may also be interpreted as an analogy to a futile cycle, at least in the sense that greater amounts of elements may flow through the system than would be required for the function of the individual component (e.g., cations or nitrate which is not reduced but stored in the vacuole and shed with leaf fall). It was demonstrated by Schulze and Chapin (1987) that in *Urtica* vegetation nitrogen may circulate three to five times during a single growing season through the plant and the decomposer chain. It passes repeatedly through the "free space" of the soil solution or it may be lost to the groundwater. In this sense the free space in the apoplast is analogous to the soil solution at the ecosystem level, where physicochemical processes of the soil interact with the microbial processes of decomposition, with uptake by roots and recycling to the atmosphere via transpiration, and with losses to the groundwater. Furthermore, several components of the soil solution are coupled to an analogue of futile cycles, namely of litter fall, decomposition, and uptake.

VI. Compartmentation and Accumulation

Structural components are important because they enable flow of material to be directed along specific routes and they allow regulation. Related to this is compartmentation, which separates pathways and which plays an important role in osmotic processes of plants such as in coupled flows of water and solutes during growth, phloem transport, and nutrient uptake by roots. The exchange of solutes between apoplasmic and symplasmic compartments of tissues causes an osmotic disequilibrium and changes of turgidity which drives important biological processes. The solute levels (osmotic concentrations) in the compartments are used for regulation. Changes of the apoplasmic solute concentration are important because this compartment is small and thus allows a regulation which could be more effective and less expensive than the futile cycling of solutes (Steudle, this volume, Chapter 8).

Compartmentation is also a basis for separation of species and populations of the same species at the herbivore level. Zwölfer (this volume, Chapter 11) described a variety of species which feed in the same thistle flower head, but in separate well-defined compartments: the seed, the flower base, the stem. Most compartments are separated by well-defined boundaries, such as the gall structure. Thus, compartmentation is the prerequisite for autonomy, as well as for cooperation, if compartments share the same metabolites or species in spatially separated populations. Therefore, if a special component in one compartment falls short, replenishment by import from neighboring units will assist the affected compartment to preserve its function or ecological role. In that sense, compartmentation is a means for spreading of risk and of compensatory responses.

One result of compartmentation is the ability to accumulate substances. In plants these may be used later in the life cycle (Chapin *et al.*, 1990). Examples are starch formation in the chloroplast, nitrate accumulation in the vacuole (Stitt and Schulze, this volume, Chapter 4), and sucrose accumulation in internodal cells of sugar cane (Komor, this volume, Chapter 6). These all reflect overproduction of one part of the system at one time and utilization by another system or part of a system at a different time and under different conditions.

Accumulation of substances should not only be considered as a "deficiency" or "limitation" of regulation, because in the evolutionary view accumulations turned out to be of great importance in a changing environment. Constant rates of growth depend on accumulation of starch during the day or nitrate during the night to allow sucrose export over the entire diurnal rhythm and to provide sufficient nitrate for reduction during the day. The same is true for longer life cycles when storage in general becomes significant for the transfer of products from one season into the other in order to support reproduction (Stitt and Schulze, this volume, Chapter 4).

Also at the ecosystem level accumulations are rather common, and humus may serve as an example of an accumulation in response to biotic and abiotic factors (Zech and Kögel-Knagner, this volume, Chapter 9). A chain of microorganisms is involved in decomposing organic matter in soils. The initial step is carried out by species which are able to degrade macromolecules only via exoenzymes (Meyer, 1992). Following the initial step the macromolecules are further decomposed by organisms which eventually produce CO_2, methane, or nitrate (Schulze and Zwölfer, this volume, Chapter 12). Some of the macromolecules produced by plants such as lignin are rather resistant to microbial attack. As a consequence such aromatic macromolecules are altered with respect to the chemical structure, but are only very slowly degraded. Eventually, the products form a new class of soil organic matter and humus which play an important role in the cation exchange capacity of the soil and thus has substantial impact on plant growth (Zech and Kögel-Knagner, this volume, Chapter 9).

VII. Coordination and Synchronization

In addition to the direct effects of interactions of chemical reactions or organisms, one area of control has not been mentioned explicitly so far. This is the problem of coordination at the cellular level and of synchronization at the population level.

At the metabolic level, coordination is achieved by regulatory mechanisms acting simultaneously at several points. One example would be the simultaneous light activation of several photosynthetic enzymes by thioredoxin and changes of stromal pH (Scheibe and Beck, this volume, Chapter 1). Coordination is a major task at the cellular level. To coordinate processes which operate in the cytosol with those in the chloroplast, stable operation and interaction in both compartments is necessary. This is achieved by additional control features such as the demand for phosphate in cell metabolism and a series of related regulation mechanisms (see Stitt, this volume, Chapter 2; Stitt and Schulze, this volume, Chapter 4). In this sense, coordination at the cellular level involves feedback and feedforward regulation of enzyme activity. However, coordination and synchronization are also achieved via regulation of gene expression. This is extremely important in an organism, because it determines what structural and system features are present at a given time and in a given place. This aspect is not really dealt with in this book, but we are aware of the fact that some metabolites, e.g., sucrose, regulate directly gene expression and thus achieve gene expression without phytohormonal action.

At the population level, synchronization is a qualitatively different process from coordination in cells (Zwölfer, this volume, Chapter 11). In some situations, herbivores can only attack the host plant and predators can only attack the herbivore in a small time "window" during which the plant has not yet developed structures to avoid intrusion by herbivores, or before gall-defensive structures have developed to limit access by the predator. Limitations due to low synchronization are one of the reasons why herbivores do not completely exhaust their host resources. Depending on weather, plant development may vary from season to season. Thus synchronization at the population level would require evolutionary changes to overcome the stochastic nature of the resource supply. Under conditions where the synchronization by one species is not reached, we may observe a "switch" from one species to others in the use of resources which has consequenses for the whole food chain considering the hierarchical order of system organization. Again we may draw the analogy to the malate valve (Scheibe and Beck, this volume, Chapter 1) which acts as a switch and redirects the metabolic pathway to very different products of the photochemical process.

VIII. Boundaries to Our Knowledge

A detailed understanding of systems becomes more and more complicated the closer we search for "mechanisms" governing processes. In contrast, the similarities of regulation principles allow us to question whether similar principles of self-regulation take place at all levels of organization. This would justify a procedure which describes fluxes at higher levels of organization, in a simplified nested model. For example, determining relative growth rates in relation to specific leaf weight allows us to predict biomass production without knowing all the underlying processes and interactions at the cellular scale. Knowing the energy partitioning of canopies (Kelliher *et al.*, 1993) allows us to predict transpiration without detailed information of stomatal functioning.

The similarity of regulation at different levels of organization form a basis for the management of plants and animals in forestry and agriculture. For example, yield tables have been the scientific basis for forest management for more than a century. However, we should be aware that, to date, we are only able to manage very simplified systems. We are far away from knowing how to manage complex systems such as the tropical rain forest or even nature reserves in Europe. Obviously, there is a limitation to the empirical approach of conventional land management because it is based on the invalid assumption that the resource environment does not change (or that it can be totally manipulated). However, in natural systems there is a limit to regulation also. Indeed, it is the limited ability of biological systems for regulation that allows evolutionary processes to manifest new adaptations and exploit new resources. In natural systems limited regulation generally results in accumulations (litter, humus) that are the basis for loss of resources to the groundwater or that initiate succession toward other species. Thus, the limitations in system regulation are seen in the invasion and extinction of species and in changes in species composition, and this makes populations and ecosystems different from the cellular or organismic system. By introducing new species or by loss of species, new levels of fluxes may be initiated, which may cause dramatic differences in the overall partitioning of energy and matter across the system (Bond, 1992).

There is an intrinsic limitation to the perspective that species may be managed in a simple way since, for example, herbivores and their predators are generally connected during evolution by tight links which makes it impossible to exchange one species for another without changing this part of the food web (Jensen and Zwölfer, this volume, Chapter 13). If a basic component of the system is lost, the whole satellite system of dependent species is lost as well. This is one of the reasons why it seems very difficult, at present, to predict changes in ecosystems or populations following global climate changes.

The other function which makes predictions impossible at this stage are unpredictable secondary effects of certain processes. When we reduced the activity of the CO_2-fixing enzyme in tobacco, we expected that nitrate would accumulate. However, we did not expect that the nitrate concentration in young leaves would affect extension growth of leaves and would thus change leaf area and maintain the total carbon uptake of these plants almost at a constant level (Stitt and Schulze, this volume, Chapter 4). At present, we may arrive at an idea about direct interactions and regulations in pathways of existing systems, but there is, to date, no method to describe and predict secondary effects of regulation of future systems. Perhaps, our imagination is too limited to realize all indirect effects of perturbations. For example, this makes it difficult to predict with certainty responses of genetically manipulated organisms as well as dramatic changes in the external environment as they are produced by human activity.

In addition, we must be aware, that, as we leave the level of organisms and enter the level of populations, the effect of deterministic processes decreases and the effect of stochastic events increases. This means that there may be an apparent stability at the macroscale (a forest will maintain nutrient cycling at a certain rate for decades); however, there is a large degree of variability at the patchscale (seasonal and annual changes in insect populations or pests). It is quite likely that this instability at the patchscale, which is stochastic in nature as an adjustment to external perturbations such as weather, is the basis for the stability and a quasi-steady state of fluxes at the macroscale (Schulze and Zwölfer, this volume, Chapter 12). Instability at the microscale compensates for variations in the use of resources and thus it is a prerequisite for the stability of fluxes at the macroscale. It is this instability at the patchscale which allows for invasion at the population level and for evolution at the organism level.

The main uncertainty in understanding regulation at the system level may be based on the fact that we cannot define ecosystem functioning precisely. We assume that all biological actions will be transposed one way or another into a change in ecosystem flux per ground area. However, this may not cover all ecosystem functions. In fact we are aware that there are ecosystem functions which go beyond the study of biogeochemical cycling and which would also include features such as behavior and experience at higher trophic levels.

IX. Conclusions

The study of control and regulation of fluxes at different levels of biological organization has led to the following general observations:

• Similar principles for control exist at vastly different levels of organization. The principles of control are analogous at the ecosystem, population, organism, and even of enzymatic reaction level.

• The smaller the system, the more significant is regulation in a deterministic sense. In contrast, the larger the system, the more "regulation analogous mechanisms" exist which generally are based on stochastic events.

• Microinstability (imbalance) is a requirement for regulation and it appears to be the basis for stability at the macroscale.

• Regulation and process control of biological systems is limited. This is important because it results in evolution proceeding beyond the natural plasticity of species.

• Accumulations of stored resources occur at the micro- and the macroscale. Accumulations preceed and are the basis for growth and development in organisms (e.g., vegetative to reproductive growth) and they may even result in major changes of the system (succession) unless other mechanisms eliminate accumulations from the system (e.g., diurnal or seasonal cycles, peat, or wood formation).

• Similarity of the principle components and mechanisms of regulation at the micro- and the macroscale generates the functional basis for empirical and simplified models which describe fluxes at the macroscale and neglect the involved microscale mechanisms. Although such models are suitable for predictions, they are valid only as long as the environment, species composition, and interactions at the microscale remain constant.

• Despite our knowledge of regulation we still cannot make predictions at the macroscale on the basis of mechanisms at the microscale, because of secondary and tertiary effects which accompany certain regulatory processes.

References

Begon, M., Harper, J. L., and Townsend, C. R. (1990). "Ecology, Individuals, Populations and Communities." Blackwell, Oxford.

Bond, W. J. (1992). Keystone species. *Ecol. Stud.* **99,** 237–254.

Chapin, F. S., III, Schulze, E.-D., and Mooney, H. A. (1990). The ecology and economics of storage in plants. *Annu. Rev. Ecol. Syst.* **21,** 423–447.

Crabtree, B., and Newsholme, E. A. (1987). A systematic approach to describing and analyzing metabolic control systems. *Trends Biochem. Sci.* **12,** 4–12.

Geigenberger, P., and Stitt, M. (1991). A futile cycle of sucrose synthase and degradation is involved in regulating partitioning between sucrose, starch and respiration in cotyledons of germinating *Ricinus communis* L. seedlings when phloem transport is stopped. *Planta* **185,** 81–90.

Kacser, H., and Porteous, J. W. (1987). Control of metabolism: What do we have to measure? *Trends Biochem. Sci.* **12,** 5–14.

Kelliher, F. M., Leuning, R., and Schulze, E.-D. (1993). Evaporation and canopy characteristics of coniferous forest and grassland. *Oecologia* (in press).

Meyer, O. (1992). Functional groups of microorganisms. *Ecol. Stud.* **99,** 67–96.

Schulze, E.-D., and Chapin, F. S., III (1987). Plant specialization to environments of different resource availability. *Ecol. Stud.* **61,** 120–148.

Schulze, E.-D., and Mooney, H. A., eds. (1992). "Biodiversity and Ecosystem Function," Ecol. Stud. No. 99. Springer-Verlag, Heidelberg.

Stitt, M. (1990). Fructose-2-6-bisphosphate as a regulatory molecule in plants. *Annu. Rev. Plant Physiol. Mol. Biol.* **41,** 153–185.

Ziegler, H. (1991). Physiologie. *In* "Lehrbuch der Botanik" (P. Sitte, H. Ziegler, F. Ehrendorfer, and A. Bresinsky, eds.), pp. 239–170. Fischer, Stuttgart.

Index

Abiotic condensation models, 313
Abscisic acid
  fluxes, root signal modulation,
    217–222
  stomatal conductance relationship,
    210–212, 214–217
Accumulation
  comparative aspects, 480–481
  definition, 61
Acidity balance, in nutrient uptake,
  184–186
ADP glucose pyrophosphorylase, flux
  control coefficients, 23–24
Aggregation, soil
  aeration
    soil structure effects, 347–351
    texture effects on gas transport,
     345–347
  functions, 358
  hydraulic conductivity, 341–342
  hydraulic gradient, 342–343
  processes, 336–340, 358
  thermal aspects, 343–344
  water fluxes
    in lysimeters, 352–357
    modeling, 343
  water retention curve, 340
Alkyl carbons, in forest soils, 321–322
Amino acids
  specificity of phloem loading for,
    159–160
  transport systems, kinetics and sub-
    strate specificity, 180–182
Ammonogenic carbon sink hypothesis,
  126–128
Annual plants, growth and storage,
  84–91
Anoxia, effects on root water, 286–288
Apoplasts
  phloem loading via, 161–169
  solute concentration control, 175–178

Arabidopsis thaliana, starchless mutants,
  growth and allocation, 75–78
Assimilation, CO_2 and N, relationship to
  growth, 58–61
ATP, and NADPH balance, 3–6

Biennial plants, growth and storage,
  91–99
Biopolymer degradation models, 313
Branch points, metabolic, 25–26

Canopy height, relationship to rooting
  depth, 434–435
Carbon
  distribution, cytokinin effects
    mechanisms, 137–138
    molecular biological approach,
     139–143
  sinks, ammonium-induced, morpho-
    genic effectiveness, 126–128
Carbon cycle, soil, 304–306
Carbon dioxide
  assimilation and investment, 58–61
  and water vapor-coupled fluxes, effects
    on leaf conductance, 222–224
Carbon transport
  and growth, 186–188
  long-distance solute transport
    apoplastic solute concentration con-
     trol, 175–178
    effect of water potential gradients,
     171–173
    phloem loading, 173–174
    solute circulation, 173–174
  phloem loading, 154–155
    apoplastic and symplastic routes,
     161–169
    concentration dependence, 158–161
    regulation by substrate interaction,
     169–170